COFFEE

Volume 3: Physiology

COFFEE

Volume 1: Chemistry
Volume 2: Technology
Volume 3: Physiology
Volume 4: Agronomy
Volume 5: Related Beverages
Volume 6: Commercial and Technico-Legal Aspects

COFFEE

Volume 3: PHYSIOLOGY

Edited by

R. J. CLARKE

Formerly of General Foods Ltd, Banbury, UK

and

R. MACRAE

Department of Food Science and Technology,
University of Reading, UK

ELSEVIER APPLIED SCIENCE
LONDON and NEW YORK

ELSEVIER APPLIED SCIENCE PUBLISHERS LTD
Crown House, Linton Road, Barking, Essex IG11 8JU, England

' *Sole Distributor in the USA and Canada*
ELSEVIER SCIENCE PUBLISHING CO., INC.
52 Vanderbilt Avenue, New York, NY 10017, USA

WITH 45 TABLES AND 39 ILLUSTRATIONS

British Library Cataloguing in Publication Data

Coffee
Volume 3: Physiology
1. Coffee
I. Clarke, R. J. (Ronald James *1919*)
II. Macrae, R.
641.3′373

Library of Congress Cataloging-in-Publication Data
(Revised for vol. 3)

Coffee.

Includes bibliographies and index.
Contents: 1. Chemistry—v. 2. Technology
—v. 3. Physiology—v. 5. Related beverages
1. Coffee. I. Clarke, R. J. (Ronald James)
II. Macrae, R.
TP645.C64 1985 641.3′373 85-6976
ISBN 1-85166-186-7 (v. 3)

Printed in Great Britain by Galliard (Printers) Ltd, Great Yarmouth

Preface

The present volume, Volume 3 in a series on coffee, takes a marked change of technical direction from the preceding volumes on its chemistry and technology in that it now deals with the physiological effects of coffee products at their ultimate destination, the human body. Coffee brews, from either the roasted or instant product, have been and are consumed for pleasure (an important subject covered in Chapter 6), world-wide for many years. However, very marked scientific and public interest has been taken in the details of its physiological effects on the human body. This interest has been directed particularly towards establishing whether this drink has any harmful effects in the long or short term, not least as a result of the many negative assertions that have been made—primarily in the popular media. As will be seen, the amount of scientific work that has been carried out world-wide, using all the relevant disciplines of the life sciences to study every aspect of the effect of coffee drinking on the human and animal body and to refute unscientific allegations as necessary, is truly impressive.

It is the aim of the General Editors to provide all these studies, which contain information much of which is not easily available, in a way that has not been done before, within the compass of a single volume solely devoted to the subject. They have been fortunate enough to have been able to call upon the services of many international experts, each specialising in a particular field.

The introduction gives a broad, succinct yet detailed survey across the whole area of concern, including not only the particular role of caffeine as the most physiologically active component (apart from flavour), but also those of other components present, including possible contaminants present in very small amounts. The next chapter logically deals with the metabolism of caffeine, and of other substances present. The general physiological action of coffee is presented in two chapters, Chapter 3, describing the numerous epidemiological studies, and Chapter 4, describing studies within the individual human body. A further chapter (Chapter 7) deals with the extensive experimental work on animal feeding studies of both coffee brews and of instant coffee; and yet another (Chapter 8), on studies on a very modern subject, possible mutagenic effects. As befits a volume on physiological effects, the nutritional factors in coffee (which importantly include those of any added milk or sugar) and the physiology of flavour (taste and aroma) assessment, both of which are without doubt positive factors, are described. In all the controversy on the other physiological effects of coffee, we leave it to the reader to come to his or her own conclusions about the informed observations of our distinguished contributors. Finally, coffee can carry a number of so-called exogenous or contaminant substances, in particular polycyclic hydrocarbons formed in exceedingly small quantities under roasting conditions (even less so than in other roasted foodstuffs) that pose a potential health hazard, and of mycotoxins (the latter occur only in very mouldy green coffee). Green coffee beans (especially their dust) have also been reported as causing allergies amongst green coffee handlers. The two final chapters (Chapters 9 and 10) cover these topics in extensive detail; the General Editors thought it worth including them at the expense of overall balance of subjects since, especially in respect of allergens, all this particular information has not been brought together in this way before.

Physiology as a scientific discipline necessarily uses a very large number of different terms, which may need some explanation for those not familiar with them, e.g. scientists and other readers for whom physiology or medicine is not their primary discipline. It is especially noteworthy how many of these apparently obscure terms are Greek or less frequently Latin based, so that knowledge of their origins makes them more immediately understandable. For this reason, we have included a Glossary of terms and also a short summary showing etymological derivations of interest.

R. J. CLARKE
R. MACRAE

Contents

Preface v

List of Contributors xiii

Chapter 1 *Physiologically Active Substances in Coffee* R. VIANI

1. Introduction 1
2. Coffee Constituents 2
   - 2.1. Caffeine and trace purine alkaloids 2
   - 2.2. Chlorogenic acids and related substances 3
   - 2.3. Glycosides 8
   - 2.4. Lipids 9
   - 2.5. Trigonelline and niacin 13
   - 2.6. Volatile substances 14
   - 2.7. Minerals and related substances 16
   - 2.8. Partially identified constituents 17
3. Exogenous Compounds 17
   - 3.1. Polycyclic aromatic hydrocarbons 17
   - 3.2. Pesticides 19
   - 3.3. Mycotoxins 19
4. Summary and Conclusions 24

   References 25

Chapter 2 *The Metabolism of Coffee Constituents*
M. J. ARNAUD

1. Introduction 33
2. Minerals 34
3. Carbohydrates 34
   - 3.1. Polysaccharides 34

3.2. Mono- and disaccharides 34
3.3. Phytate 35
4. Proteins and Amino Acids 35
5. Lipids 35
5.1. Fatty acids 35
5.2. Waxes 36
5.3. Sterols 36
5.4. Terpenes 37
5.5. Kahweol, cafestol and diterpene glycosides 39
6. Aliphatic Compounds 40
7. Aromatic Compounds 40
7.1. Benzo[*a*]pyrene 40
7.2. Chlorogenic acids 41
7.3. Eugenol 43
7.4. Guaiacol 44
8. Heterocyclic Compounds 45
8.1. Maltol 45
8.2. 2-Methylfuran 45
8.3. 5-Hydroxy-2-furaldehyde 45
8.4. Maillard reaction products 46
8.5. Pyridines 46
8.6. Other heterocyclic compounds 47
9. Alkaloids 47
9.1. Trigonelline 47
9.2. Caffeine 50
10. Conclusions 51
References 52

Chapter 3 *Epidemiological Studies on the Effects of Coffee Drinking*
S. HEYDEN

1. General Introduction 57
2. Study of Coffee Consumption and Mortality Rates among 10 000 Hypertensives 59
2.1. Introduction 59
2.2. Methodology 59
2.3. Results. 61
2.4. Discussion 62
3. Coffee Consumption and Fibrocystic Breast Disease 63
3.1. Introduction 63
3.2. Patients and methods 64
3.3. Results. 65
3.4. Discussion 67
3.5. Review of literature 68
3.6. Summary and acknowledgement 70
4. Habitual Coffee Consumption and Cholesterol Levels 71
4.1. Introduction 71
4.2. World-wide studies 71

4.3. Discussion . . . 76
4.4. Summary of findings . . . 77
References . . . 77

Chapter 4 *Physiological Effects of Coffee and its Components* H. MILON, R. GUIDOUX and J. A. ANTONIOLI

1. Introduction . . . 81
2. Central nervous system . . . 81
   2.1. Introduction . . . 81
   2.2. Behavioural effects . . . 82
   2.3. Biochemical effects . . . 89
   2.4. Central respiratory effects . . . 93
   2.5. Conclusions . . . 93
3. Cardiovascular System . . . 94
   3.1. Introduction . . . 94
   3.2. Studies on animals . . . 96
   3.3. Studies in man . . . 104
4. Coffee and the Gastrointestinal Tract . . . 107
   4.1. Introduction . . . 107
   4.2. Effects of coffee . . . 109
   4.3. Conclusions . . . 114
   References . . . 116

Chapter 5 *Nutritional Factors* R. MACRAE

1. Introduction . . . 125
2. Vitamins . . . 126
   2.1. Fat-soluble vitamins . . . 126
   2.2. Water-soluble vitamins . . . 127
3. Mineral Content . . . 129
4. Dietary Fibre . . . 132
5. Additives . . . 133
   5.1. Introduction . . . 133
   5.2. Coffee consumption . . . 135
   5.3. Coffee consumption patterns . . . 135
   5.4. Nutritional composition of additives . . . 137
6. Conclusion . . . 138
   References . . . 138

Chapter 6 *The Physiology of Flavour: Taste and Aroma Perception* B. HEATH

1. Sensory Perception . . . 141
   1.1. Distinctions between odour, taste and flavour . . . 141
   1.2. Sensory perceptions that influence flavour . . . 142
   1.3. Sensory stimuli associated with odour and taste . . . 143
   1.4. Nature of coffee aroma and flavour . . . 144

2. Taste: Physiological Mechanisms of the Tongue 144
2.1. Taste receptors—the taste buds 144
2.2. Taste perception 147
2.3. Theories of taste perception 148
2.4. Molecular structure and taste 149
2.5. Taste interrelationships 153
3. Odour: Physiological Mechanisms of the Nasal Cavity . . 153
3.1. Odour receptors in man 153
3.2. Odour perception 156
3.3. Molecular structure and odour 159
3.4. Coffee aroma 159
4. Coffee Flavour 161
4.1. Flavour integration 161
4.2. Basic flavour of coffee brews 162
4.3. Other variants of coffee flavour 164
5. Applications to Sensory Analysis of Coffee 165
5.1. British Standards on sensory testing 165
5.2. Environment for sensory assessment 166
5.3. Factors affecting sensory judgements 166
5.4. Panels vs. expert judgement 168
References 168

Chapter 7 *Animal Feeding Studies with Coffee*
H. P. WÜRZNER

1. Introduction 171
2. Chronic Toxicity and Carcinogenicity 172
2.1. Introduction 172
2.2. Studies on rats 173
2.3. Studies on mice 174
2.4. Caffeine, coffee and tumour promotion and inhibition . 181
2.5. Caffeine studies 182
2.6. Conclusions 183
3. Teratology and Reproduction 185
3.1. Introduction 185
3.2. Studies on rats 185
3.3. Studies on mice 185
3.4. Caffeine studies 189
3.5. Conclusions 191
4. Final Remarks and Outlook 191
References 192

Chapter 8 *Mutagenicity of Coffee* H. U. AESCHBACHER

1. Introduction 195
2. Mutagenicity Evaluation of Whole Coffee 195
2.1. Bacterial tests 196
2.2. Mammalian tests *in vitro* 199

2.3. Tests under *in vivo* conditions 202
2.4. Mammalian tests *in vivo* 204
3. Mutagenicity Evaluation of Coffee Components 205
3.1. Aliphatic carbonyls 206
3.2. Methylxanthines 207
3.3. Phenolic compounds 207
3.4. Hydrogen peroxide 208
4. Interaction 208
4.1. Interaction between coffee constituents 209
4.2. Interaction of coffee constituents with other food components. 209
5. Risk Assessment 210
6. Summary 211
References 211

Chapter 9 *Allergens and Mould Toxin Contaminants*
R. G. K. STROBEL

1. Introduction and Background 215
1.1. General 215
1.2. Importance of fungi 217
1.3. Growth of fungi 220
1.4. Fungal-based diseases and disorders 228
1.5. Immunology 243
2. Allergens 267
2.1 Respiratory system 267
2.2 Microscopic analysis of green coffee bean factory dust . 272
2.3 Chemical analysis of green bean coffee dust 274
2.4. Case and clinical studies on coffee workers handling green and roasted coffee 276
2.5. Other occupational pulmonary diseases 287
3. Mycotoxins Found in Coffee Beans 290
3.1. Mould and other microbial growth of coffee beans at elevated relative humidities 291
3.2. Ochratoxin in mould-contaminated coffee beans. . . . 292
3.3. Mycotoxin production in non-decaffeinated and decaffeinated green coffee beans 293
3.4. Sterigmatocystin in coffee beans 298
3.5. Fate of mycotoxins during roasting 301
3.6. Incidence of moulds in coffee beans (surface and subsurface populations). 302
3.7. Mould and other microbial growth potential on coffee beans from production site to roaster 304
3.8. Storage and transport of coffee beans (storage at production site) 307
3.9. Sea transport of green coffee beans 308
4. Conclusions. 310
References 313

Chapter 10 *Polycyclic Aromatic Hydrocarbon Contaminants in Coffee* R. G. K. STROBEL

1. Introduction . . . 321
   1.1. General . . . 321
   1.2. Early history of cancer . . . 322
   1.3. Discovery of 3,4-benzpyrene . . . 324
   1.4. Physical, viral and chemical carcinogens . . . 326
   1.5. Carcinogenic activity and structure of polynuclear aromatic carcinogens . . . 329
   1.6. Carcinogenic activity and metabolism . . . 332
   1.7. Two-stage theory and quantification of carcinogenesis . . . 334
   1.8. Enzymatic transformation of polynuclear aromatic hydrocarbons into carcinogens . . . 338
   1.9. Detoxification and nucleic acid repair and inhibition . . . 343
   1.10 Carcinogenic polynuclear hydrocarbons present in the human environment . . . 346
   1.11. Formation of 3,4-benzpyrene . . . 346
   1.12. Carcinogenic constituents of coal tar, industrial, automotive and residential effluents and tobacco smoke . . . 348
2. Benzo[*a*]pyrene content of coffee products and by-products . . . 351
3. Conclusions . . . 358
   References . . . 360

Appendix 1 *Glossary of Some Terms and Abbreviations Used* . . . 365

Appendix 2 *Etymology of Some Physiology/Medical Terms* 373

Index . . . 377

List of Contributors

H. U. AESCHBACHER

Nestlé Research Centre, Nestec Ltd, Vers-chez-les-Blanc, CH-1000 Lausanne 26, Switzerland

J. A. ANTONIOLI

Nestlé Research Centre, Nestec Ltd, Vers-chez-les-Blanc, CH-1000 Lausanne 26, Switzerland

M. J. ARNAUD

Nestlé Research Centre, Nestec Ltd, Vers-chez-les-Blanc, CH-1000 Lausanne 26, Switzerland

J. G. FODOR

St. John's Memorial University of Newfoundland, Canada

R. GUIDOUX

Nestlé Research Centre, Nestec Ltd, Vers-chez-les-Blanc, CH-1000 Lausanne 26, Switzerland

H. B. HEATH

35 Port Hill Gardens, Shrewsbury, Shropshire SY3 8SB, UK

S. HEYDEN

Department of Community and Family Medicine, Duke University Medical Center, Box 2914, Durham, North Carolina 27710, USA

R. MACRAE

Department of Food Science and Technology, University of Reading, Whiteknights, PO Box 226, Reading RG6 2AP, UK

J. B. MARTIN

Program in Nutrition and Dietetics, School of Allied Health Science, University of Texas, Health Sciences Center at Houston, Texas Medical Center, Houston, Texas 77225, USA.

H. MILON

Nestlé Research Centre, Nestec Ltd, Vers-chez-les-Blanc, CH-1000 Lausanne 26, Switzerland

J. RUH

University of Newfoundland, Canada

K. A. SCHNEIDER

University of Newfoundland, Canada

R. G. K. STROBEL

The Procter & Gamble Company, 6110 Center Hill Road, Cincinnati, Ohio 45224, USA

R. VIANI

Technical Division, Nestec Ltd, Avenue Nestlé 55, CH-1800 Vevey, Switzerland

H. P. WÜRZNER

Nestlé Research Centre, Nestec Ltd, Vers-chez-les-Blanc, CH-1000 Lausanne 26, Switzerland

Chapter 1

Physiologically Active Substances in Coffee

R. VIANI
Technical Division, Nestec Ltd, Vevey, Switzerland

1. INTRODUCTION

Since 1820 when Runge[1] isolated coffee's active principle, caffeine, a great deal of research has been carried out to identify and evaluate other constituents that may help to explain the fascination the brew has held over the last five centuries.

The substances present in coffee that are described as physiologically active can be subdivided as follows:

(a) Substances naturally present in the green bean, including caffeine, responsible for psychic well-being; chlorogenic acid; the glycosides and lipids; and those transformed or formed at roasting such as trigonelline and the volatile constituents, which create the unique flavour.

(b) Exogenous substances, the ubiquitous polycyclic aromatic hydrocarbons, the pesticides and the mycotoxins which may sometimes contaminate coffee.

Generally speaking, the main effects are due to the presence of the stimulant caffeine, which, when made palatable by the subtle flavour notes, explains the popularity of coffee. No other important contribution to the physiological effects of coffee has been described.

2. COFFEE CONSTITUENTS

2.1. Caffeine and the Trace Purine Alkaloids

Caffeine is the best studied constituent of coffee, and its pharmacological activity, summarised in Volume 1 of this series,[2,3] has been the subject of several recent reviews.[4–11] It is consumed by people of practically all age groups in coffee, tea, chocolate, soft drinks and common medications. The caffeine content of common beverages, foods and drugs ranges from 1 mg per cup for decaffeinated coffee up to 200 mg per tablet in some non-prescription drugs, with a content of 40–180 mg per cup* of brewed and 30–120 mg per cup† of instant regular (caffeinated) coffee. According to the FDA in the USA, 'the evidence received does not suggest that caffeine at present levels of consumption poses a hazard to public health'.[12]

Caffeine is a bitter substance with a threshold value in water of 0·2–1·8 mM litre^{-1},[13] i.e. of the same order of magnitude as its concentration in a cup of regular coffee (1–4 mM), and accounts for 10–30% of the bitterness of the brew detected by taste.[14,15] Two other main fractions responsible for more than 50% of the bitterness of a cup of coffee have been isolated but not yet identified.[15] Components which also contribute to the bitterness are chlorogenic acid and trigonelline (discussed later).

In addition to caffeine, **I** (Fig. 1), minute amounts of the related purine alkaloids, theobromine,[16] **II**, and theophylline,[17,18] **III**, have been found in the coffee bean. Recently, four more alkaloids (which were already known to be present in the leaf), paraxanthine, **IV**, theacrine, **V**, liberine, **VI**, and methylliberine, **VII**, have been identified in the seed of various *Coffea* species, and in particular in *Coffea arabica* and in *Coffea canephora* (robusta)[19,20] (see Table 1).

Considering the quantities present, the question of the contribution of these minor alkaloids to the physiological activity of a cup of coffee made from ripe beans is purely academic: theobromine, theophylline and paraxanthine have effects of the same type, but different intensity, as caffeine of which they are metabolites.[21] Among the other substances identified, only theacrine has been studied. It has been found to be an efficient solubilising agent for many carcinogenic substances, particularly polycyclic aromatic hydrocarbons, and to produce chromosomal aberrations in plant and animal cells as well as mutations in *Escherichia coli*; only a little is known about its biochemical and pharmacological action.[22]

* Based as elsewhere on a 150 ml cup and 10 g roasted and ground coffee.
† Based as elsewhere on a 150 ml cup and 2 g instant coffee.

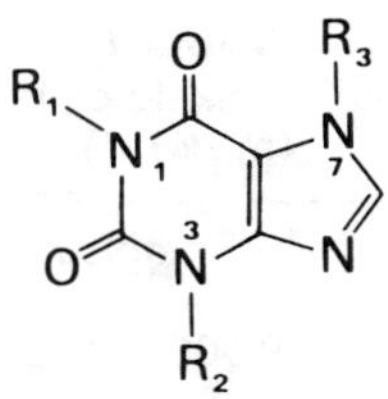

$R_1 = R_2 = R_3 = CH_3$: Caffeine, 1,3,7-trimethylxanthine, **I**
$R_1 = H$, $R_2 = R_3 = CH_3$: Theobromine, 3,7-dimethylxanthine, **II**
$R_1 = R_2 = CH_3$, $R_3 = H$: Theophylline, 1,3-dimethylxanthine, **III**
$R_1 = R_3 = CH_3$, $R_2 = H$: Paraxanthine, 1,7-dimethylxanthine, **IV**

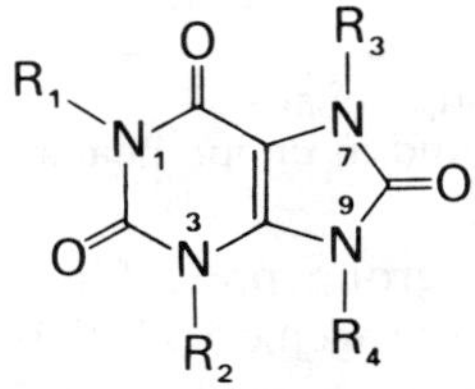

$R_1 = R_2 = R_3 = R_4 = CH_3$: Theacrine, 1,3,7,9-tetramethyluric acid, **V**

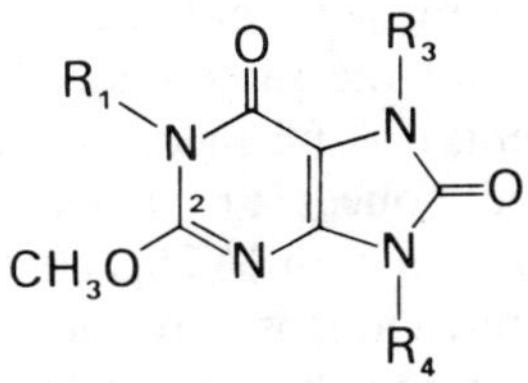

$R_1 = R_4 = CH_3$, $R_3 = H$: Liberine, O(2),1,9-trimethyluric acid, **VI**
$R_1 = R_3 = R_4 = CH_3$: Methylliberine, O(2),1,7,9-tetramethyluric acid, **VII**

Fig. 1. Structures of purine alkaloids in coffee.

2.2. Chlorogenic Acids and Related Substances

Although chlorogenic acids are common in vegetable foodstuffs, the relatively large quantities present in coffee make it an important dietary source (Table 2). From these data the content of chlorogenic acid can be estimated as varying between 15 and 325 mg per cup of brew (from 10 g of roasted and ground coffee at 85% recovery[26]) and between 55 and 240 mg per cup of instant infusion (from 2 g of powder).

Differences in taste and pharmacological activity have been recorded between the different members of this group of substances;[27] earlier pharmacological data were generally measured on mixtures enriched in the

Table 1
Content of purine alkaloids in green coffee
(mg kg⁻¹)

Alkaloid	*Arabica*	*Robusta*
Caffeine, **I**	9 000–14 000	15 000–26 000
Theobromine, **II**	36–40	23–82[a]
Theophylline, **III**	7–23	86–344[a]
Paraxanthine, **IV**	3–4	8–9
Theacrine, **V**	0	11[a]
Liberine, **VI**	5	7–110[b]
Methylliberine, **VII**	0	3[a]

[a] Only found in unripe beans.
[b] Higher figures found in unripe beans.

main component of the group present in coffee, 5-caffeoylquinic acid, referred to hereafter as chlorogenic acid, **VIII**, and on the related phenolic acids, caffeic, **IX**, and ferulic, **X**, acids (Fig. 2).[28]

Oral ingestion by man of 200 mg of chlorogenic acid was found to stimulate stomach secretion with enhancement of hydrochloric acid production. Caffeic acid had the same effect, and quinic acid, **XII**, was inactive.[29] This experiment, if confirmed, may help to explain why both regular and decaffeinated coffees are twice as potent as caffeine in stimulating gastric secretion in man.[30] This effect has also been attributed to an irritation of the stomach mucosa produced by chlorogenic acid or by its phenolic derivatives formed during roasting,[31–34] and explained by their tannin-like nature.[35] However, chlorogenic acid does not possess tanning properties, and the question has still to be settled.[36]

Choleretic effects with stimulation of biliary flow, observed in rats after i.v. and i.d. administration of chlorogenic, caffeic and ferulic acids,[37–40]

Table 2
Chlorogenic acid content of coffee

Product	*No. of samples*	*%*	*Reference*
Roasted coffee	8	0·2–3·5[a]	23
Roasted coffee	19	1·6–3·8[a]	24
Instant coffee	63	3·2–12	25
Instant coffee	16	2·7–8·9	24

[a] The lower figures were found in dark roasted samples.

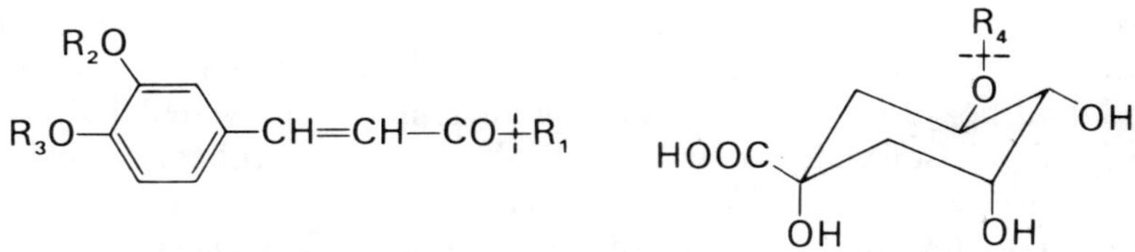

A($R_2 = R_3 = H$): Caffeoyl

B: Quinyl

$R_1 = B$, $R_2 = R_3 = H$, $R_4 = A$:	Chlorogenic acid, **VIII**
$R_1 = OH$, $R_2 = R_3 = H$:	Caffeic acid, **IX**
$R_1 = OH$, $R_2 = H$, $R_3 = CH_3$:	Ferulic acid, **X**
$R_1 = OH$, $R_2 = CH_3$, $R_3 = H$:	Isoferulic acid, **XI**
$R_4 = H$:	Quinic acid, **XII**

Fig. 2. Structures of a chlorogenic acid (taken as 5-caffeoylquinic acid) and component acids.

became less evident when chlorogenic acid was applied in the stomach or in the intestine;[41] the substance, which could not be detected in the serum and in the bile after absorption into the gastrointestinal tract, is rapidly hydrolysed into its two moieties, quinic acid and caffeic acid, which are further metabolised.[42]

Quinic acid, which is pharmacologically inactive, is transformed by the intestinal microflora and appears in the urine of both man and the rat as the aromatic substances catechol, **XV**, and hippuric acid, **XVI** (Fig. 3).[43,44] A

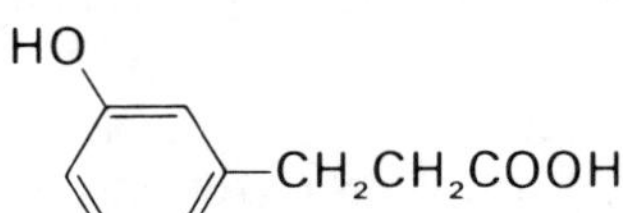

Meta-hydroxyphenylpropionic acid, **XIII**

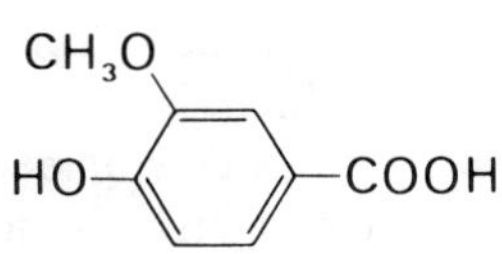

Vanillic acid, **XIV**

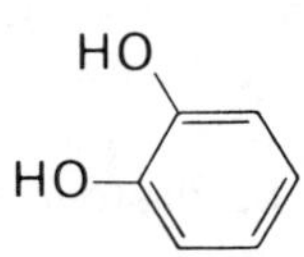

Catechol, **XV**

Hippuric acid, **XVI**

Fig. 3. Structures of some metabolic products of chlorogenic acid.

small part of the caffeic acid is excreted unchanged in the urine, most is transformed by gut bacteria into *meta*-hydroxyphenylpropionic acid, **XIII**, and decarboxylated,[45–47] and a minor part only is methylated in the liver to ferulic, **X**, isoferulic, **XI**, and vanillic, **XIV**, acids,[37] probably explaining why more significant choleretic effects are observed after i.v. or i.d. than after oral administration of chlorogenic, caffeic and ferulic acids.

The hypothesis that chlorogenic acid delays the clearance of caffeine by slowing down its resorption from the intestine[48] needs confirmation, as the complex between caffeine and chlorogenic acid does not survive at the concentrations and in the conditions of the gastric system.[49]

Caffeic and chlorogenic acids stimulate intestinal motility, whereas ferulic acid inhibits it.[50,51]

It had been found that the thiamin content of serum decreased rapidly after administration of coffee and chlorogenic or caffeic acids: the total thiamin excretion of 13 out of 15 human volunteers was lowered by almost 50% within 3 h after consumption of 1 litre of coffee with respect to the quantity measured after intake of the same quantity of water.[52] These results could not be repeated, and long-term experiments with rats and mice fed a diet containing 5% of instant coffee did not show any symptoms of vitamin deficiency, an antithiamin effect being provoked in rats only at the unphysiological dose of 3 g kg^{-1}.[53] No chemical interactions could be seen between thiamin and caffeic or chlorogenic acids.[54,55] It may be concluded that, under realistic conditions of coffee intake, there is neither a change in the urinary output,[56] nor disturbance of thiamin bioavailability.[57]

Chlorogenic acid has been described as convertogenic in yeasts, and as clastogenic with induction of chromosomal breaks in CHO cells, but as non-mutagenic in the *Salmonella* bioassay both in the presence and in the absence of the S-9 mix.[58] Most of the genotoxicity of chlorogenic acid was found in the caffeic acid moiety of the depside, which is an *ortho*-dihydroxyphenol and readily auto-oxidises in *in vitro* systems at neutral and alkaline pH, forming quinones and liberating hydrogen peroxide,[59] particularly in the presence of transition metals.[60] Some clastogenic activity might remain after treatment with the enzyme catalase, which destroys hydrogen peroxide; therefore, caffeic acid itself or some of its degradation products, formed at pH 7 or higher, might be clastogenic.[61] Also the question of whether chlorogenic acid itself is the hydrogen peroxide generating system in coffee has not yet been completely elucidated, as this substance, said to be generated on standing in an infusion of roasted coffee, is not formed in an unroasted extract, which contains even more chlorogenic acid.[62]

Caffeic and chlorogenic acids inhibit nitrosation at acidic pH. The alarming hypothesis that N-nitrosoamines can be readily formed in physiological conditions by reaction of chlorogenic acid, at the concentrations found in a cup of coffee, with secondary amines in the presence of nitrite[63] and catalysed by thiocyanate[64] from saliva, was therefore disproved, and the presence of nitrosoamines in model systems was explained by their artefactual formation during analysis.[65] Furthermore, both chlorogenic acid and instant coffee reduce the mutagenic activity of a mixture of methylurea and nitrite, probably by scavenging electrophiles and free radicals;[66] caffeic, chlorogenic and particularly the dicaffeoylquinic acids inhibit lipid peroxidation induced by ADP and ascorbic acid and by NADPH in rat liver microsomes.[67] Chlorogenic acid blocks the formation of mutagenic substances resulting from the pyrolysis of proteins,[68] and caffeic, ferulic and chlorogenic acids inhibit the mutagenic activity of B[*a*]P-7,8-diol-9,10-epoxide-2, the ultimate mutagenic and carcinogenic metabolite of benzo[*a*]pyrene.[69,70]

Medium-term rat feeding experiments with caffeic acid at 0·5% in the diet lowered the incidence of spontaneous tumours[71] and with chlorogenic acid at 0·4% showed no adverse effects on growth rate, feed consumption, health and behaviour;[72] in addition, 'populations consuming large amounts of phenolic substances ... have been shown by epidemiological studies to be at lower risk of developing gastrointestinal cancer'.[47] From these data it would appear that, in the normal conditions of the gastric system, chlorogenic, caffeic and ferulic acids have a protective effect by behaving as 'blocking agents, which prevent carcinogens from reacting with critical target sites'.[73] It is only after auto-oxidation, particularly in model systems at alkaline pH, that they may exhibit toxic effects, such as catalysis of nitrosoamine formation[63,64] or the recently described antithyrotropic and antigonadotropic activities.[74,75]

The effect of chlorogenic acid on the central nervous system is still unclear: on the one hand, an effect six times smaller than that of caffeine was measured after oral application of 3·75–15 mg kg^{-1} of the substance; the effect was found to be dose related in one study,[76] but not in another,[77] and subjects kept on a restricted phenolic diet suffered from a pronounced lethargy and mood depression;[47] on the other hand, no stimulating effect on the EEG from coffee constituents other than caffeine has been observed.[78]

Allergic reactions among workers manipulating raw coffee had been attributed to the presence of chlorogenic acid in the dust.[79] Recent work has confirmed that green coffee bean allergens are in fact proteins of a

Table 3
Content of free and bound atractyligenin (expressed in mg kg^{-1} of atractyligenin[84–89])

Product	*Atractyligenin*		*Glycoside*		
	Free[a]	*Total*[b]	***XVIII***[b]	***XIX***[b]	***XX***[b]
Raw arabica	0–10	320–1 000	140–435	5–30	0–240
Raw robusta	0	0–20	5–30	0	0–10
Roasted arabica	25–125	445–1 370[c]	30–230	30–55	60–160
Roasted robusta	0–20	0–20	ND	ND	ND
Extract arabica	25–30	415–465	300–660	40–50	50–550

ND = Not determined.
[a] Increases with roasting.
[b] Decreases with increasing roasting.
[c] Higher figures with roasted samples are explained by a better analytical technique.

heterogeneous nature with a molecular weight range of 50 000 to 500 000 daltons,[80] partially destroyed by roasting[81] and not chlorogenic acid.[82,83]

2.3. Glycosides

Atractyligenin, **XVII**, a nor-diterpenoid substance of the (−)-kaurane series, has been found in coffee free[84] and as the aglycone of three glycosides,[85,86] **XVIII–XX** (Table 3). (See Fig. 4.)

The content of atractyligenin can be estimated at 2·9–11·5 mg per cup of roasted arabica brew, at 0–0·2 mg per cup of roasted robusta brew (10 g of coffee, 66–84% recovery[89]), and at 0·8–0·9 mg per cup of instant arabica coffee (2 g of powder).

Atractyligenin, **XVII**, and two of its metabolites, dihydroatractyligenin (**XXI**) and 2β-hydroxy-15-oxoatractylan-4α-carboxylic acid (**XXII**), were identified as glucuronides in the urine of coffee drinkers.[84,90]

Atractyligenin is also the aglycone of a glycoside, atractyloside, **XXIII**, formed by decarboxylation from carboxyatractyloside, **XXIV**, found in the sweet-tasting rhizomes of *Atractylis gummifera* L. (Table 4). Ingestion of the rhizomes, which grow spontaneously in the Mediterranean basin, produces an initial hyperglycaemia, followed by hypoglycaemia and convulsions that may result in death, and the toxic effect is produced by carboxyatractyloside through inhibition of oxidative phosphorylation and energy transfers in liver mitochondria;[91] however, carboxyatractyloside is absent in coffee.

Table 4
Toxicity of substances isolated from *Atractylis gummifera* L.

Product	*Effect*
Carboxyatractyloside, **XXIV**	Non-competitive inhibition; LD_{50} (rat) i.p. 2·9–10·7 mg kg^{-1}
Atractyloside, **XXIII**	Competitive inhibition; LD_{50} (rat) i.p. 124–210 mg kg^{-1}: 10 times less toxic than **XXIV**
Atractyligenin, **XVII**	Competitive inhibition: 150 times less toxic than **XXIV**

Coffee atractyloside, **XIX**, also a competitive inhibitor, corresponds to the desulphated form of atractyloside, **XXIII**; however, it is 10 times less toxic than the latter,[92] and therefore 100 times less toxic than carboxyatractyloside.

Other glycosides have been identified in arabica coffee; they are cofaryloside,[93] **XXV**, which on roasting gives 16,17-dihydroxy-9(11)-kauren-18-oic acid,[94] **XXVI**, whose pharmacological effects have not yet been studied, cafestolone,[95] **XXVII**, related to mascaroside,[96] **XXVIII**, and the isomer of cafamarine, **XXIX**, the bitter principles of *Mascarocoffea*.[97]

2.4. Lipids

Coffee oil contains a relatively high level of unsaponifiable matter rich in two diterpenes specific to coffee, kahweol, **XXX**, and cafestol, **XXXI**, related to the aglycones of the glycosides just discussed; they are present as fatty esters, mainly of palmitic and linoleic acids (see Fig. 5).[98] The content of oil and of kahweol with cafestol of various coffee products is as shown in Table 5.[99]

The acute toxicity of coffee oil is rather low with an LD_{50} (mice) p.o. of 5 g kg^{-1}. More than 5% addition of coffee oil to the diet has, however, some toxic effects with loss of weight and increased lethality.[100]

Kahweol palmitate, **XXXII**, and to a lesser extent cafestol palmitate, **XXXIII**, inhibit tumour formation in rodents when administered prior to highly carcinogenic agents such as benzo[*a*]pyrene[101] and 7,12-dimethylbenz[*a*]anthracene.[102] The diminished incidence is explained by an enhancement in GSH S-transferase activity, and by a reduction of the reactive ultimate forms of the carcinogens with inhibition of the microsomal mono-oxygenase system. Green coffee beans at 20% in the diet,

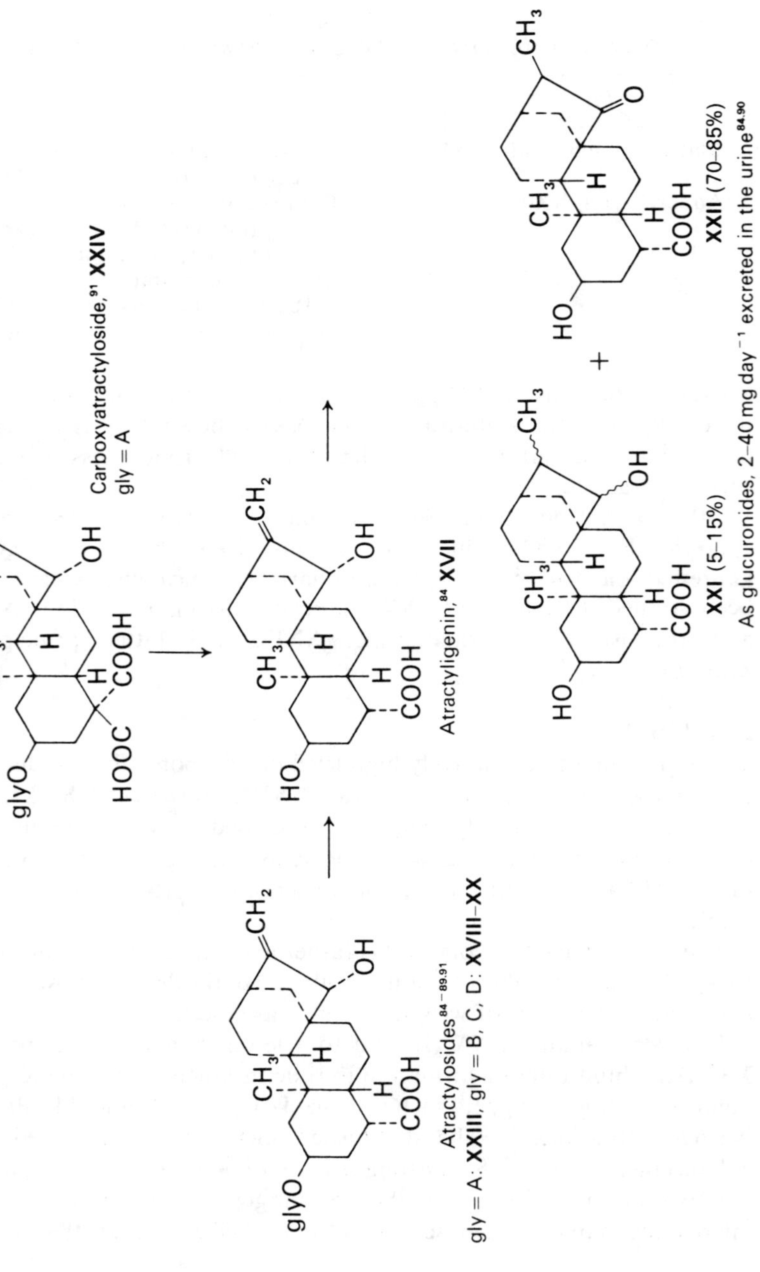

Atractylosides[84–89,91]
gly = A: **XXIII**, gly = B, C, D: **XVIII–XX**

Carboxyatractyloside,[91] **XXIV**
gly = A

Atractyligenin,[84] **XVII**

XXI (5–15%)

XXII (70–85%)

As glucuronides, 2–40 mg day^{-1} excreted in the urine[84,90]

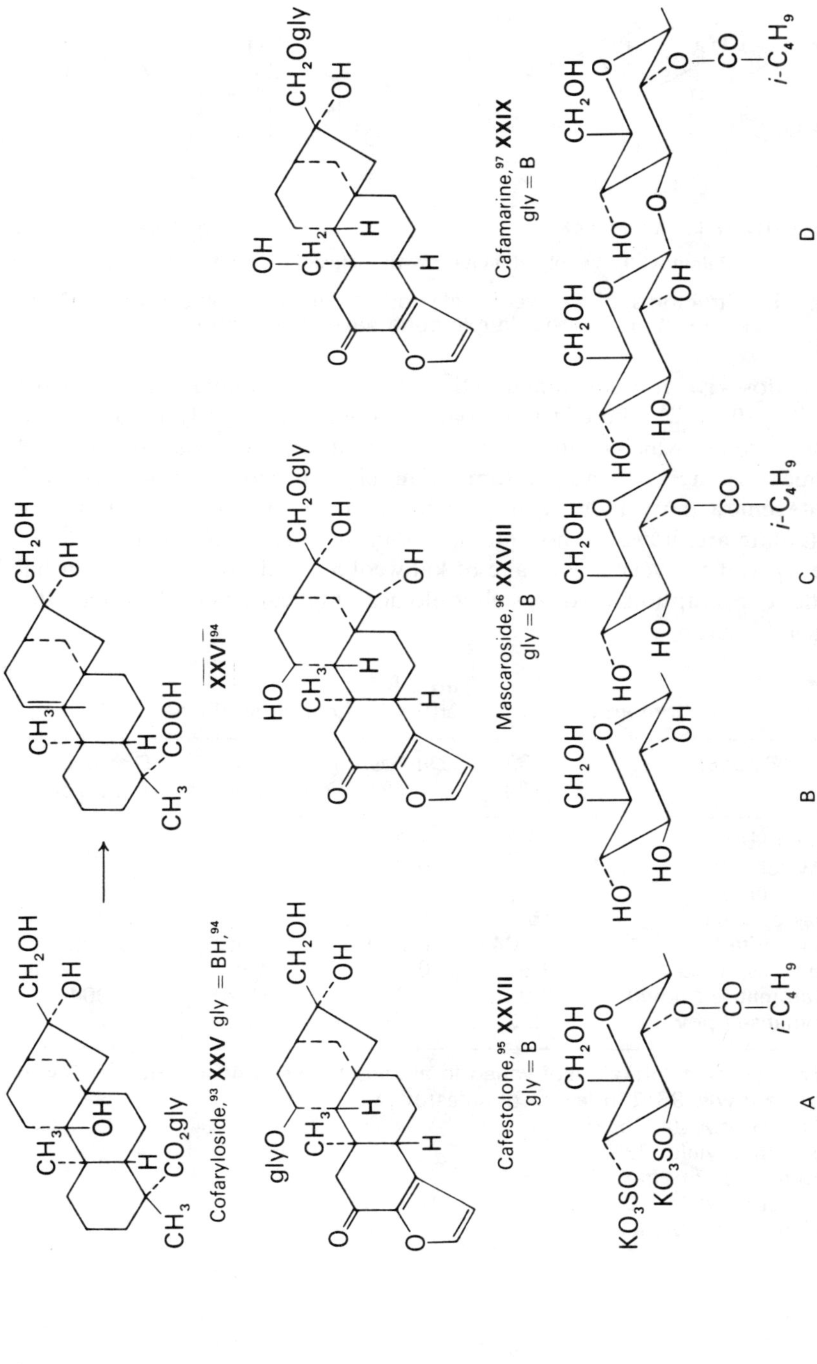

Fig. 4. Structures of glycosides of *Coffea* and of *Atractylis gummifera* L.

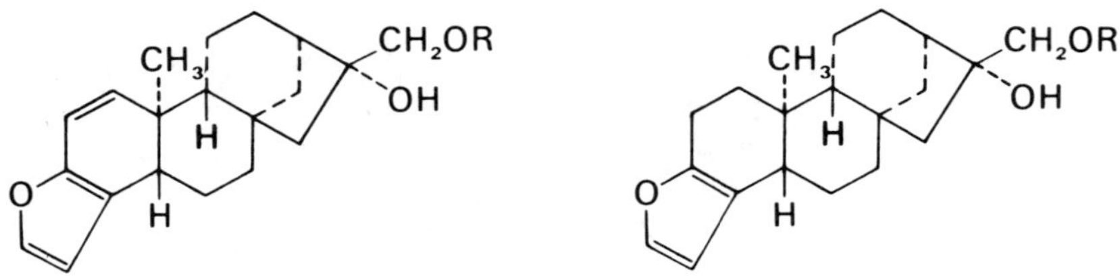

R = H: kahweol,[a] **XXX** Cafestol, **XXXI**

R = Palmitoyl: kahweol palmitate, **XXXII**; cafestol palmitate, **XXXIII**

Fig. 5. Structures of kahweol, cafestol and their palmitic esters. [a]Other positions for the double bond have been proposed.

i.e. at doses roughly equivalent to those of the pure palmitates tested, had a similar effect.[103] The effect of roasted coffee was slightly reduced, and instant coffee, which contains practically no diterpenes, was less active.[104] Some substances which act as tumour inhibitors in rodents may also exert a mutagenic activity under *in vitro* conditions: coffee oil, cafestol and cafestol palmitate are, however, non-mutagenic in the *Salmonella* bioassay.[105] In the light of the very low intake of kahweol and cafestol through normal coffee consumption, no direct physiological implications of their activities are apparent.

Table 5
Content of oil and diterpenes of coffee products

Product	*Oil* (%)	*Diterpenes*[a] (%)	*Oil* (*mg per cup*)	*Diterpenes*[a] (*mg per cup*)
Raw arabica	16	1·3		
Raw robusta	11	0·2		
Raw blend[b]	15	1·1		
Roasted blend[b]	15	1·1		
Brew (filter)[c]	0·04	0·0002	0·9[d]	0·005[d]
Brew (espresso)[e]	2·5	0·2	40[d]	3·2[d]
Unaromatised powder	0·02	0·0002	0·4[f]	0·004[f]
Aromatised powder	0·3	0·02	7[f]	0·4[f]

[a]The ratio cafestol:kahweol varied in arabica between 40:60 and 70:30; in robusta it was 98:2 in favour of cafestol.
[b]Arabica:robusta, 75:25.
[c]Extraction yield: 23%.
[d]From 10 g of coffee.
[e]Extraction yield: 16%.
[f]From 2 g of powder.

Table 6
Trigonelline, **XXXIV**, nicotinic acid, **XXXV**, and nicotinamide, **XXXVI**, content per cup (mg)

Product	*Number of samples*	***XXXIV***	***XXXV***	***XXXVI***	*Reference*
Brew[a]	6	40–55	0·03–0·05	0·005–0·008	109
Iced coffee[b]	1	20	3	0·2	109
Instant coffee[c]	13	20–35	NA	NA	108
Instant coffee[c]	22	5–15	0·2–1·2	0·07–0·2	109

NA = Not analysed.
[a] 10 g of coffee.
[b] Iced brew from 10 g of dark roasted coffee.
[c] 2 g of powder.

2.5. Trigonelline and Niacin

Trigonelline is transformed on roasting into several volatile products, mainly alkylpyridines,[106] and into niacin or vitamin PP.[107] The content per cup of residual trigonelline, determined by high performance liquid chromatography in instant coffees available in the United Kingdom,[108] and of residual trigonelline and of niacin, determined microbiologically as nicotinic acid and nicotinamide in roasted and instant coffees available in Japan,[109] is given in Table 6.

Trigonelline (Fig. 6) has a very low toxicity: doses of up to 1 g kg^{-1}, administered to rabbits and cats, failed to show any toxicity, and the LD_{50} (mouse) i.p. is higher than 3 g kg^{-1}. At doses of 40 mg kg^{-1}, well above the quantities ingested in normal coffee consumption, trigonelline produces a reduction of the excitability of the central nervous system, measured by the seizure threshold method, a reduction of bile secretion and an increase in intestinal motility.[100]

As was mentioned in Volume 1,[107] the niacin content of a cup, particularly from dark roasted coffee, may contribute up to 15% of the

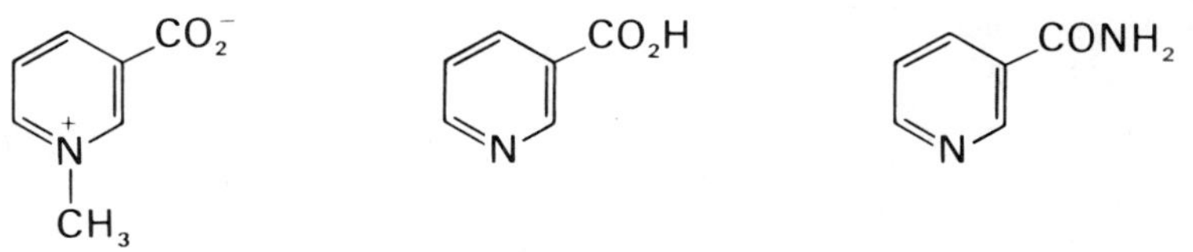

Trigonelline, **XXXIV** Nicotinic acid, **XXXV** Nicotinamide, **XXXVI**

Fig. 6. Structures of trigonelline and derivatives.

recommended daily intake of this vitamin, and coffee consumption may help in the prevention of pellagra, which still exists in maize-consuming cultures.[110,111] (See also Chapter 5 of this volume.)

Enzymes that demethylate trigonelline to nicotinic acid have recently been identified in various organisms, and in particular in hog and beef liver,[112] and trigonelline may be the mitotic cycle hormone that regulates cell proliferation.[113] Its reutilisation by the organism deserves further study.

2.6. Volatile Substances

Comprehensive lists of the volatile constituents of coffee (Fig. 7) record several hundreds of acyclic, isocyclic and heterocyclic substances[114–116] contributing to the overall flavour of the brew and whose pharmacological activity in the coffee context, particularly in behavioural motivation, is still unclear.[117]

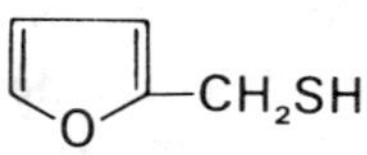

Furfurylmercaptan, **XXXVII**
0·01–0·05 ng g^{-1} in water: freshly roasted coffee note
1–10 ng g^{-1} in water: stale coffee note

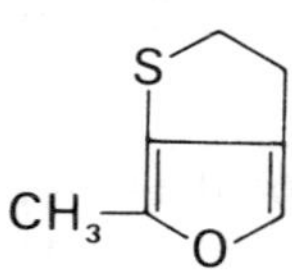

Kahweofuran, **XXXVIII**

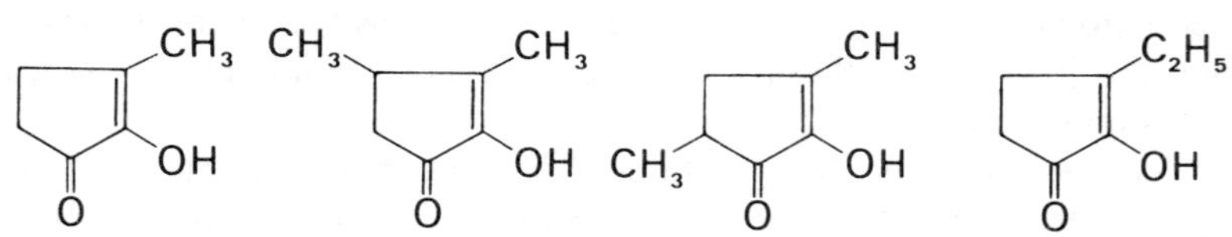

Cyclotenes, **XXXIX–XLII**

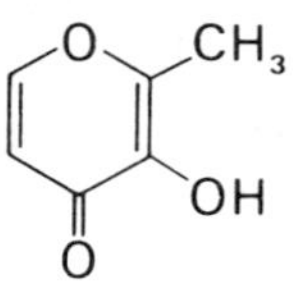

Maltol, **XLIII**

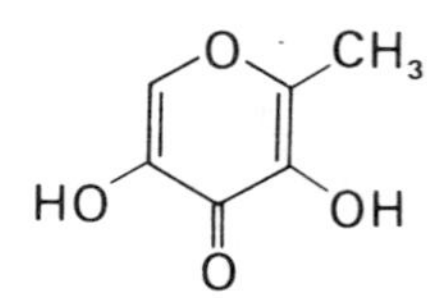

5-Hydroxymaltol, **XLIV**

Fig. 7. Structures of some volatile compounds found in roast coffee.

With the possible exception of furfurylmercaptan, **XXXVII**, patented in 1926 as the aroma of coffee,[118] and related sulphides and disulphides, such as kahweofuran (**XXXVIII**), that at certain dilutions have a coffee-like aroma,[119] no single component can account for the typical roasted coffee flavour. Many substances contribute specific notes to the aroma: cyclic compounds, such as the cyclotenes[120] (**XXXIX–XLII**), maltol (**XLIII**) and 5-hydroxymaltol (**XLIV**), have a sweet caramel flavour, and some aliphatic aldehydes show fresh woody notes,[121] and aliphatic diketones buttery notes.[122]

Application of multivariate statistics allows the correlation of information obtained by chemical analysis of the aroma with its sensory characteristics.[123–125] The branched C-4 and C-5 aldehydes, formed by thermal decomposition during roasting of valine, isoleucine and leucine, and volatile sulphur compounds are more important in the flavour of robusta than in that of arabica.[126]

Pyridine, N-substituted pyrroles, furans, etc., are associated with the organoleptic descriptors of roastiness such as bitterness, body, strength and intensity of flavour, which increase with roasting, and inversely associated with acidity and cereal-like flavour notes.[127] No specific volatile component or class has, however, yet been associated with the more vague descriptors of quality, such as fineness, freshness, richness, balance and purity of flavour.[127,128]

Volatile substances responsible for the putrid smell of defective raw beans ('stinkers' or 'quakers'), which give a tainted impure taste to the cup, have been identified as short-chain aliphatic esters and acids.[129]

Volatile components, usually present in coffee at levels much lower than those encountered in other common foods, have been linked with pharmacological actions as, for instance, salicylates, which may play a role in urticaria and asthma; however, a cup of instant coffee contains only negligible levels of 1–2 mg (in nine samples analysed)[130] of methyl salicylate, **XLV**.

Trace amounts of N-nitrosopyrrolidine, **XLVI**, in the order of 1–2 ng per cup were tentatively identified in five out of ten instant coffees analysed by gas chromatography and by high performance liquid chromatography,

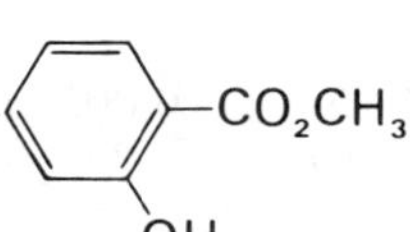

Methylsalicylate, **XLV**

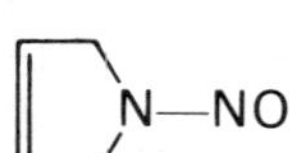

N-Nitrosopyrrolidine, **XLVI**

using an extremely sensitive indirect thermal energy analyser as detector. Here again, contrary to other foods, 'the levels ... were too low to be confirmed by GC–MS. The results ... should not be considered as unequivocal proof of the presence of N-nitrosoamines' in coffee.[131]

Coffee aroma is mutagenic in the *Salmonella* bioassay in the absence of S-9 mix.[132] A weak mutagenicity has been associated with formaldehyde,[133] **XLVII**, present in coffee at levels of 1 mg per cup of brew and of 1–2 mg per cup of instant coffee,[134] and with several dicarbonylic substances common in many food aromas; in particular, with glyoxal, **XLVIII** (found in instant coffee at levels of 200 μg per cup), methylglyoxal, **IL** (200–300 μg per cup) and diacetyl, **L** (160 μg per cup).[135] The cyclic compounds maltol,[135] **XLIII**, up to 400 μg per cup of brew,[136] and 5-hydroxymaltol,[135] **XLIV**, less than 15 μg per cup of brew,[136] have also been implicated. The mutagenicity of methylglyoxal, possibly the most active of these substances, is enhanced by hydrogen peroxide,[137] suppressed by cysteine[138] and sulphite,[139] and inhibited by the presence of S-9 mix.[140] The pharmacological implications of these findings will be discussed in detail later in this volume,[141] especially in the light of metabolic deactivation.[132,142]

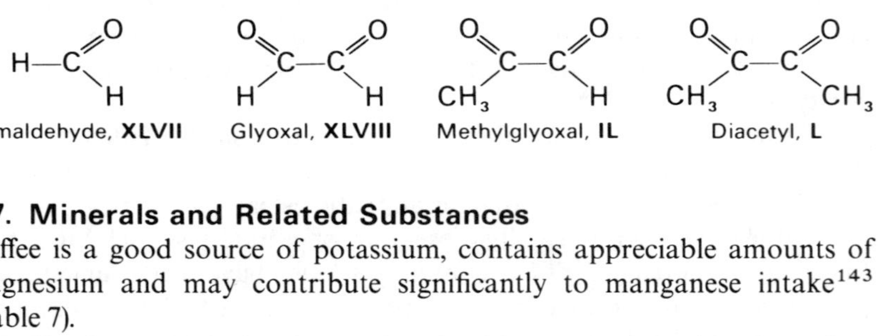

Formaldehyde, **XLVII** Glyoxal, **XLVIII** Methylglyoxal, **IL** Diacetyl, **L**

2.7. Minerals and Related Substances

Coffee is a good source of potassium, contains appreciable amounts of magnesium and may contribute significantly to manganese intake[143] (Table 7).

Coffee intake has been associated with a concentration-dependent reduction of non-haem iron absorption in human subjects if it is consumed

Table 7
Dietary minerals from coffee (mg per cup)

Product	*Potassium*	*Magnesium*	*Manganese*
Brew[a]	130–200	10–20	0·002–0·22
Instant infusion[b]	65–155	5–10	0·004–0·03

[a] From 10 g of roasted and ground coffee.
[b] 2 g of powder.

with a meal,[144] but the responsible mechanism and dietary implications have not yet been elucidated.

Phytate, the mono- to dodeca- anion of *myo*-inositol hexaphosphoric acid (1,2,3,4,5,6-cyclohexanehexolphosphoric acid), **LI**, which occurs naturally in coffee, can chelate minerals, and in particular zinc, with a reduction of their bioavailability. In particular, a phytate:zinc molar ratio of 6–10:1 has been shown to depress zinc bioavailability in the rat. Somewhat differing analytical results and conclusions were reached in two recent publications,[145,146] and the subject deserves further research.

2.8. Partially Identified Constituents

A psychoactive substance with receptor binding activity resembling that of opiate antagonists has been identified in *in vitro* experiments with rat brain membrane preparations in both regular and decaffeinated instant coffee.[147] It is heat stable, soluble in ether, resists enzyme digestion with papain and has a molecular weight of 1000–3500; its concentration in a cup of coffee is in the order of five times its half maximal effective dose.

Recent experiments with rats and mice[148] have shown that the efficiency of coffee in antagonising morphine *in vivo* is, however, too low to be of any practical importance.

3. EXOGENOUS COMPOUNDS

3.1. Polycyclic Aromatic Hydrocarbons

The presence of benzo[*a*]pyrene (B*a*P), **LII**, in coffee soot, first reported in 1938,[149] was confirmed in 1958[150,151] with the identification, beside quantities in the order of 3–4 ng g^{-1} of the skin carcinogen, of several other weakly carcinogenic polycyclic aromatic hydrocarbons in both dark roasted coffee and in the roasting fumes, the most important being phenanthrene, fluoranthene, pyrene, chrysene and anthracene.[152] Since

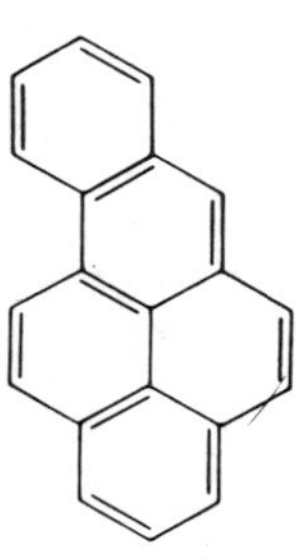

Benzo[*a*]pyrene (B*a*P), **LII**

then many authors have studied the occurrence of B*a*P, which is the most carcinogenic of the polycyclic aromatic hydrocarbons detected and is, therefore, used as an indicator of total contamination in coffee.

Quantities of B*a*P varying between less than 0·01 ng g^{-1} (Ref. 153) up to 2·8 ng g^{-1} (Ref. 154) have been detected in raw coffee beans; the figures are generally lower than those found for other common foodstuffs, and in particular vegetables, fruits and edible oils,[155] which are less protected than coffee beans from environmental contamination. The B*a*P content of roasted coffee appears to be independent of roasting time, temperature and degree, within normal roasting conditions;[153,156] its reduction[157] or even disappearance[154] during roasting may come from the mechanical removal of the silverskin as chaff. The amounts of B*a*P found in sound roasted coffee varies between less than 0·01[153] and 3·0[158] ng g^{-1}, with an average for commercial samples in the order of 0·3 ng g^{-1}.[153] Amounts of B*a*P as high as 23–24 ng g^{-1} have, however, been detected in burnt beans and in chaff,[157] which may explain the increase in B*a*P content with roasting observed by one author.[159] Accumulations up to 300 ng g^{-1} of B*a*P content in roaster soot, calamine and chaff have been described.[159]

Table 8
Coffee samples analysed for benzo[*a*]pyrene contamination (ng g^{-1})

Product	*No. of samples*	*BaP*	*Reference*
Raw beans	1	2·8	154
Raw beans	6	0·1–0·5	157
Roasted beans	1	0·1	154
Roasted beans	63	0·1–0·5	157
Roasted beans	13	0·1–3·0	158
Roasted beans	22	0·1–0·6	163
Roasted beans	55	<0·1–0·5	161
Roasted beans	59	0·1–0·7	164
Burnt chaff	1	22·7	157
Coffee concentrate	2	0·1	163
Brew	1	0·003	162
Brew	3	0·000 3–0·1	161
Brew	7	0·002–0·007	164
Instant coffee	6	0·02–0·06	162
Instant coffee	16	0·01–0·5	157
Instant coffee	20	0·1–0·7	158
Instant coffee	2	0·1	163
Instant coffee	4	<0·2	164

It had been stated that caffeine has a solubilising effect on polycyclic aromatic hydrocarbons during brewing and extracting.[160] However, it has been shown recently[161] that the solubilisation of B*a*P by caffeine is minimal in both arabica and robusta blends, and that, even in roasted coffee samples containing an elevated level of B*a*P of 2 ng g^{-1}, its extraction yield is no more than 1%.[161,164] Figures obtained by analytically sound methods[161–164] indicate amounts of B*a*P lower than 0·15 ng per cup, and intakes are negligible in relation to the quantities consumed from other sources;[155] it has been shown that coffee contributes less than 0·5% of the estimated average intake from foods and drinks.[157] Sample purification by liquid–liquid fractionation, followed by ion exchange,[156] has been improved by the use of high performance liquid chromatography[159] or by capillary gas chromatography,[163] and detection by fluorimetry or mass spectrometry has a sensitivity of 0·01 ng g^{-1}. Contamination of roasted coffee by B*a*P can be avoided by applying good manufacturing practices, viz. by careful roasting to avoid scorching of the beans, and regular removal of calamine from the exit port of the roaster and of sublimed matter deposited on the cooler. Reincorporation of chaff into ground roasted coffee is not recommended. The results of 282 published analyses are given in Table 8.

3.2. Pesticides

Coffee beans are protected within the cherry from direct application of pesticides, and coffee is consumed after further processing, particularly roasting, which eliminates almost quantitatively the contaminants,[165,166] and brewing or extracting, which further reduces the content in the cup to insignificant amounts (Table 9).

Some of the pesticides encountered as contaminants of coffee have, however, a zero tolerance level in the consuming countries, and the question is a regulatory one rather than a health issue.[166] It may then be concluded with the FDA in the USA that 'current levels of residues do not pose a hazard to the consumers'.[170]

3.3. Mycotoxins

When the moisture content of raw coffee exceeds 14%, a limit reached at 80% relative humidity after about two weeks, spoilage due to fungal growth starts.[171,172] This major defect can be easily detected organoleptically in the cup when the number of microorganisms present reaches 10^4 per gram.[173] Many moulds from the genera *Aspergillus*, *Penicillium*, *Alternaria* and *Cladosporium*, which can produce mycotoxins, have been detected on green coffee beans,[174,175] and ochratoxins A and B,

Table 9
Pesticides found in coffee products (ng g^{-1})

Product	Number of samples	α,β,δ-BHC	Lindane	DDT group
Raw beans	6	6 (1·5–335)	6 (0·5–10)	6 (1·5–21)
Raw beans	150	37 (tr–40)	15 (tr–290)	31 (tr–700)
Raw beans	19	8 (2·1–15)	15 (5–70)	14 (3·5–156 6)
Roasted beans	6	6 (tr–2·5)	6 (tr–0·5)	6 (1–5)
Roasted beans	19	13 (1·2–124)	1 (11)	3 (3–54)
Roasted beans	24	7 (1·2–2)	11 (1·5–3)	8 (3–5)
Instant coffee	6	1 (1·2)	1 (3)	0

Product	Aldrin, Dieldrin	Others	Reference
Raw beans	5 (0·5–1)	1 (6)	167
Raw beans	1 (tr)	9 (tr–160)	168
Raw beans	4 (4·3–5·0)	2 (1·5–6·5)	166
Roasted beans	5 (tr–1·5)	1 (2)	167
Roasted beans	0	0	166
Roasted beans	4 (12)	0	169
Instant coffee	0	0	169

tr = Trace.

xanthomegnin, viomellein and vioxanthin have been produced by isolates of *Aspergillus ochraceus*[176] and aflatoxin B_1 by isolates of *A. flavus*[177] from green coffee beans.

Caffeine may be a fungistatic agent with inhibition of growth and production of sterigmatocystin by *A. versicolor* at a concentration of 1·0 mg ml^{-1}; it also delays production of citrinin by *Penicillium citrinum*; production of patulin by *P. urticae* is already drastically reduced at a concentration of 0·5 mg ml^{-1}, but growth of the organism is only delayed.[178] Concentrations of caffeine of 1·0 mg ml^{-1} inhibit both growth and production of aflatoxin B_1 by *A. parasiticus*.[179] Caffeine at concentrations of 1·0–10·0 mg ml^{-1} has been said to produce both a small decrease in the growth of *A. ochraceus*, with up to 98% reduction of ochratoxin A production,[180] and a delay in the growth, with little effect on toxin production.[181] The growth of most strains of *A. flavus* isolated from peanuts, spices and rice is inhibited by caffeine at a concentration of 5·0 mg ml^{-1}; both aflatoxigenic and non-aflatoxigenic caffeine-resistant strains have, however, been isolated from coffee beans.[177] Inoculation with cultures of *A. parasiticus* in decaffeinated coffee beans with a caffeine concentration of 0·6–1·2 mg g^{-1} kept at a moisture level of around 40% resulted in the formation of both aflatoxins B_1 and G_1, whereas regular beans with a caffeine concentration of 14·7 mg g^{-1} kept at the same moisture level showed growth of mould without toxin formation; decaffeination in this experiment increased the aflatoxigenic potential of coffee.[182]

Aflatoxin B_1, **LIII**, ochratoxin A, **LIV**, and, rarely, sterigmatocystin, **LV**, have been identified in raw coffee beans from lots varying from visually free from mould to heavily mouldy (Fig. 8; Table 10).

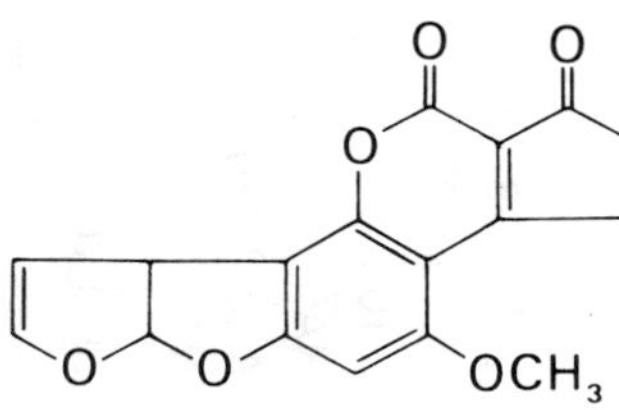

Aflatoxin B_1, **LIII**

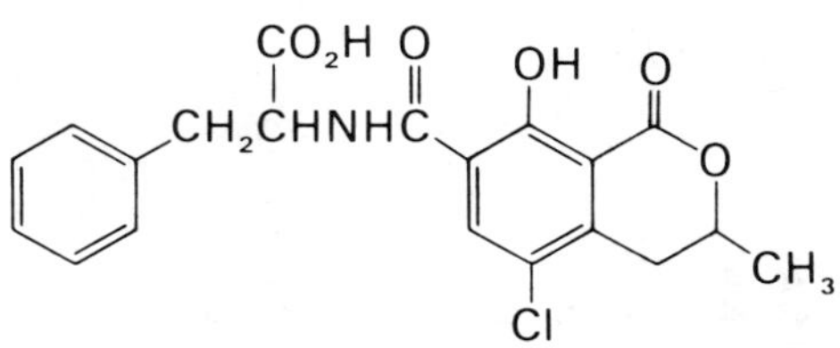

Ochratoxin A, **LIV**

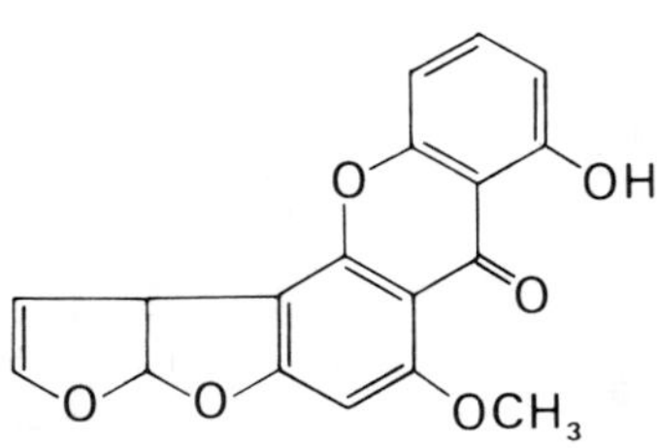

Sterigmatocystin, **LV**

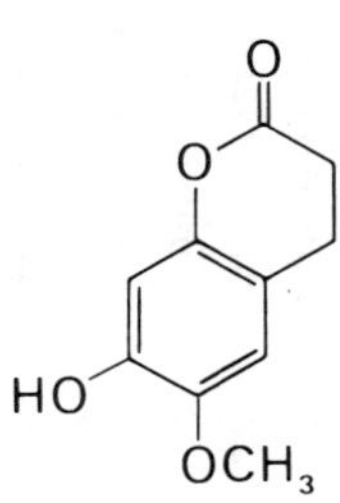

Scopoletin, **LVI**

Fig. 8. Structures of some mycotoxins in coffee, and of scopoletin.

Table 10
Coffee samples analysed for mycotoxin contamination

Number of samples	*Quality of beans*	*Number of samples (amount found in $ng\,g^{-1}$)*			*Reference*
		Aflatoxin B_1	*Ochratoxin A*	*Sterigmatocystin*	
2	Heavily mouldy	0	NA[a]	1 (1 140)	183
68	Commercial[b]	NA[a]	3 (20–80)	NA[a]	184
58	Heavily mouldy	0	NA[a]	NA[a]	185
267	Hand cleaned from mouldy lots	NA[a]	19 (20–360)	NA[a]	185
15	Mouldy	NA[a]	NA[a]	0	185
502	Commercial[b]	0	0	0	186
1	Heavily mouldy	NA[a]	NA[a]	1 (12 000)	187
640	Commercial[b]	0	NA[a]	NA[a]	187
22	Commercial[b]	0	NA[a]	NA[a]	187
201	Commercial[b]	2 (3, 12)	2 (24, 96)	0	187
2	Mouldy	0	2 (10, 90)	NA[a]	188
40	Commercial[b]	NA[a]	9 (0·5–23)	NA[a]	189
65	Commercial[b]	28 (2·6 ± 3·2)	NA[a]	NA[a]	190
22	Commercial[b]	0	4 (9·9–46)	NA[a]	191
8	Commercial[b]	0	NA[a]	NA[a]	192
1 913		30/1 522	39/1 102	2/721	

[a] NA = Not analysed.
[b] Commercial = visually free from mould.

Aflatoxin B_1, the most potent hepatocarcinogen known, has been identified in 2% of the samples analysed at levels varying between less than 1 and 13 ng g^{-1}. Recently, improved separation (high performance liquid chromatography) and identification (mass spectrometry, radioimmunoassay) techniques have shown that contamination with aflatoxin B_1 may be more widespread than was previously thought, although the levels encountered are generally quite low.[190]

Ochratoxin A, a non-carcinogenic nephrotoxic compound, has been identified in 3·5% of the samples analysed, at levels varying between 0·5 and 360 ng g^{-1}. Separation by high performance liquid chromatography coupled with a spectrofluorometric detector increases sensitivity 100-fold, lowering the detection limit down to 0·2 ng g^{-1}, with respect to the thin layer chromatographic methods used before:[189] 22·5% of the samples analysed showed some contamination; however, most of the contaminated samples contained 5 ng g^{-1} or less of toxin.

Sterigmatocystin, also a carcinogen, has only been described twice in coffee (in less than 0·3% of all the samples analysed), both times at high levels of 1100 and 12 000 ng g^{-1} in mouldy raw beans kept for many months in poor storage conditions that had become visually unfit for consumption; it does not seem to represent a major risk of contamination. No other mycotoxin has been identified in coffee samples.

The three toxins are partially degraded during roasting of coffee, with ochratoxin A the most heat labile and sterigmatocystin the most stable. Of the ochratoxin added to green beans, 80–90% was destroyed after 5 min at 200 ± 5°C during a heat treatment which simulated roasting.[184,191] No ochratoxin A was detectable after dark roasting of two green-bean samples which contained 3·8 and 23·0 ng g^{-1} of toxin, respectively, showing a destruction during roasting of at least 90%.[189]

The level of aflatoxin B_1 added to green beans decreased by 50% and 80% after 5 and 20 min, respectively, at 200 ± 5°C.[187] No aflatoxin B_1 could be found after roasting at 220°C green beans infected at a level of 10 ng g^{-1}.[193]

The level of sterigmatocystin added to green beans decreased by 70% after 20 min at 200 ± 5°C; however, little or no loss occurred before that.[185]

It can be concluded that the probability of finding any of these three mycotoxins in the coffee beverage is low, although there is one study reporting, without analytical details, the presence of 12 ng of aflatoxin in a cup of coffee prepared from contaminated beans.[190]

The presence of a natural fluorescent constituent of coffee, scopoletin, **LVI**, hinders a rapid screening of green coffee beans for the presence of aflatoxin B_1, and some of the early reports of toxin contamination may have been partly due to the presence of interferents.[194,195]

As has been described in a study of an instant coffee production line,[196] lactic acid bacteria were the predominant microflora with lactobacilli as the major group and *L. plantarum* the dominant species. 'Challenge' tests with pathogenic bacterial strains in instant coffee infusions demonstrated that these microorganisms could neither grow nor survive in the extract. Therefore it may be concluded that soluble coffee does not present a microbiological hazard to the consumer.

4. SUMMARY AND CONCLUSIONS

The complex chemical composition and pharmacological action of coffee have only partially been elucidated and 'we continue to enjoy poorly characterised infusions of roasted and ground coffee beans, and, indeed, still scrutinise the health implications of this practice'.[197]

Coffee contains, besides caffeine—by far its main pharmacologically active and best known component, trace quantities of related alkaloids.

The chlorogenic acids are, after caffeine, the most studied class of constituents present; their pharmacological properties are linked with their antioxidant character and depend strongly on the experimental conditions in which they are tested.

Glycosides of atractyligenin and related compounds, which inhibit oxidative phosphorylation by blocking the transport of ATP generated within the mitochondrial membrane in exchange for ADP outside, are pharmacologically important. The activity of the main glycoside identified and of atractyligenin is, however, 100 and 150 times, respectively, weaker than that of carboxyatractyloside, the active principle of *Atractylis gummifera* L., and no physiological effect has been shown in coffee.

The tumour-inhibiting activity of kahweol and cafestol and of their esters present in coffee oil is an interesting pharmacological action; its practical consequences in normal coffee consumption are, however, negligible, in view of their low concentration in the cup.

The pharmacology of trigonelline is now being rediscovered, and the vitamin PP activity formed during roasting might be of some importance in the now rare cases of vitamin malnutrition.

The aroma contains some of the most intense aroma principles identified in nature, such as furfurylmercaptan, but its organoleptic nature is still mysterious. In such a large number of constituents, a few, present also in many food aromas, show mutagenic activity in some *in vitro* tests: their role in explaining toxic effects attributed to coffee *in vivo* or in epidemiological studies appears doubtful, and certainly no reported epidemiological association with cancer can be related to the results in the *in vitro* mutagenicity tests.

The importance of the reduction of iron and zinc bioavailability is still unclear, and the activity of unidentified opiate antagonists in the brain seems now much weaker than was previously thought.

Contaminants and, in particular, B*a*P, pesticides and mycotoxins have sometimes been associated with coffee: here again their importance should not be dramatised as they do not present an undue toxicological hazard with the good manufacturing practices normally encountered in coffee production.

REFERENCES

1. Runge, F., *Neueste phytochemische Entdeckungen*, G. Reimers, Berlin, 1820, 1, 144.
2. Smith, A. W., in *Coffee, Vol. 1: Chemistry*, Eds R. J. Clarke and R. Macrae, Elsevier Applied Science Publishers, London, 1985, 34–36.
3. Macrae, R., in *Coffee, Vol. 1: Chemistry*, Eds R. J. Clarke and R. Macrae, Elsevier Applied Science Publishers, London, 1985, 123–124.
4. Lachance, M. P., *J. Food Safety*, 1982, **4**, 71–112.
5. Anon., *Food Technol.*, April 1983, 87–91.
6. Roberts, H. R. and Barone, J. J., *Food Technol.*, Sept. 1983, 32–39; Von Borstel, R. W., *Food Technol.*, Sept. 1983, 40–43, 47; Leviton, A., *Food Technol.*, Sept. 1983, 44–47; Miles, C. I., *Food Technol.*, Sept. 1983, 48–50.
7. Curatolo, P. W. and Robertson, D., *Ann. Int. Med.*, 1983, **98**, 641–653.
8. Dews, P. B. (Ed.), *Caffeine—Perspectives from Recent Research*, Springer, Berlin, 1984.
9. Bäettig, K., in *Coffee: Botany, Biochemistry and Production of Beans and Beverage*, Eds M. N. Clifford and K. C. Willson, Croom Helm, London, 1985, 394–439.
10. Elias, P. S., *Proc. 11th Coll. ASIC*, 1985, 93–112.
11. Pozniak, P. C., *J. Am. Diet. Assoc.*, 1985, **85**, 1127–1133.
12. Lecos, C., *FDA Consumer*, March 1984, 14–16.
13. Stahl, W. H. (Ed.), *Compilation of Odour and Taste Threshold Data*, DS48, Am. Soc. Testing and Materials, Philadelphia, PA, 1973, 77, 79, 80, 82.
14. Voilley, A., Sauvageot, F. and Durand, D., *Proc. 8th Coll. ASIC*, 1977, 251–259.
15. Wan Chung, C., Thesis, Techn. Univ. Munich, 1979, 28–29.
16. Lehmann, G. and Martinod, P., *Hoppe-Seyler's Z. Physiol. Chem.*, 1965, **341**, 155–156.
17. Franzke, C., Grunert, K. S., Hildebrandt, U. and Griehl, H., *Hoppe-Seyler's Z. Physiol. Chem.*, 1967, **348**, 1725–1726.

18. Franzke, C., Grunert, K. S., Hildebrandt, U. and Griehl, H., *Pharmazie*, 1968, **23**, 502–503.
19. Citroreksoko, P. S., Petermann, J., Wanner, H. and Baumann, T. W., *Proc. 8th Coll. ASIC*, 1977, 143–145.
20. Kappeler, A. W. and Baumann, T. W., *Proc. 11th Coll. ASIC*, 1985, 273–279.
21. Rall, T. W., in *The Pharmacological Basis of Therapeutics*, 6th edn, Eds A. Goodman and L. S. Gilman, Macmillan, New York, 1980, 592–607.
22. Kihlman, B. A., *Mutat. Res.*, 1977, **39**, 297–316.
23. Trugo, L. C. and Macrae, R., *Food Chem.*, 1984, **15**, 219–227.
24. Maier, H. G., *Café Cacao Thé*, 1987, **31**, 49–58.
25. Clifford, M. N., in *Coffee, Vol. 1: Chemistry*, Eds R. J. Clarke and R. Macrae, Elsevier Applied Science Publishers, London, 1985, 194.
26. Maier, H. G. and Grimsehl, A., *Kaffee u. Tee Markt*, 1982, **32**(23), 3–6.
27. Clifford, M. N., in *Coffee, Vol. 1: Chemistry*, Eds R. J. Clarke and R. Macrae, Elsevier Applied Science Publishers, London, 1985, 195–197.
28. Clifford, M. N., in *Coffee, Vol. 1: Chemistry*, Eds R. J. Clarke and R. Macrae, Elsevier Applied Science Publishers, London, 1985, 156–157.
29. Czok, G. and Lang, K., *Arzneimittel-Forsch.*, 1961, **11**, 335.
30. Cohen, S. and Booth, G. H., *New Engl. J. Med.*, 1975, **293**, 897–899.
31. Clifford, M. N., in *Coffee, Vol. 1: Chemistry*, Eds R. J. Clarke and R. Macrae, Elsevier Applied Science Publishers, London, 1985, 193.
32. Dart, S. K. and Nursten, H. E., in *Coffee, Vol. 1: Chemistry*, Eds R. J. Clarke and R. Macrae, Elsevier Applied Science Publishers, London, 1985, 242–243.
33. Malorny, G. and Walter, W., *Proc. 3rd Coll. ASIC*, 1968, 375–380.
34. Konig, W. and Walter, W., *Proc. 9th Coll. ASIC*, 1980, 145–149.
35. Lehmann, G., *Ernährung-Umsch.*, 1971, **18**, 43–47.
36. Anon., *Nutr. Rev.*, 1976, **34**, 167–169.
37. Michaud, J., Lesca, M. F. and Roudge, A. M., *Bull. Soc. Pharm. Bordeaux*, 1971, **110**, 66–72.
38. Czok, G. and Schulze, P. J., *Proc. 6th Coll. ASIC*, 1973, 343–351.
39. Westendorf, J. and Czok, G., *Z. Ernährungwiss.*, 1978, **17**, 26–36.
40. Westendorf, J. and Czok, G., *Z. Ernährungwiss.*, 1983, **22**, 255–270.
41. Czok, G., Walter, W., Knocke, K. and Degener, H., *Z. Ernährungwiss.*, 1974, **13**, 108–112.
42. Latza, R. and Ammon, R., *Ernährung-Umsch.*, 1977, **24**, 169–171.
43. Booth, A. N., Robbins, D. J., Masri, M. S. and Deeds, F., *Nature (London)*, 1960, **187**, 691.
44. Indahl, S. R. and Scheline, R. R., *Xenobiotica*, 1973, **3**, 549–556.
45. Peppercorn, M. A. and Goldman, P., *J. Bacteriol.*, 1971, **108**, 996–1000.
46. Booth, A. W., Emerson, D. H., Jones, F. T. and Deeds, F., *J. Biol. Chem.*, 1957, **229**, 51–59.
47. Jacobson, E. A., Newmark, H., Baptista, J. and Bruce, W. R., *Nutr. Rep. Int.*, 1983, **28**, 1409–1417.
48. Czok, G., *Wiss. Veröff. Deutsche Gesell. Ernährung*, 1970, **17**, 142–151.
49. Horman, I. and Viani, R., *Proc. 5th Coll. ASIC*, 1971, 102–111.
50. Czok, G. and Lang, K., *Arzneimittel-Forsch.*, 1961, **11**, 545–549.
51. Czok, G., Midani, W. and Finke, R. I., *Z. Ernährungwiss. Suppl.*, 1972, **14**, 68–77.

52. Somogyi, J. C. and Naegeli, U., *Int. J. Vit. Nutr. Res.*, 1976, **46**, 149–153.
53. Stalder, R., *Proc. 9th Coll. ASIC*, 1980, 427–432.
54. Horman, I. and Brambilla, E., *Int. J. Vit. Nutr. Res.*, 1981, **51**, 187.
55. Horman, I., Brambilla, E. and Stalder, R., *Int. J. Vit. Nutr. Res.*, 1981, **51**, 385–390.
56. Hayakawa, F., *Bull. Shiga Tanki-Daigaku*, 1983, **24**, 59–63.
57. Stalder, R. and Marchesini, M., *Int. J. Vit. Nutr. Res.*, 1981, **51**, 188.
58. Stich, H. F., Rosin, M. P., Wu, C. H. and Powrie, W. D., *Mutat. Res.*, 1981, **90**, 201–212.
59. Tyson, C. A. and Martell, A. E., *J. Am. Chem. Soc.*, 1972, **94**, 939–945.
60. Stich, H. F., Rosin, M. P., Wu, C. H. and Powrie, W. D., *Cancer Lett.*, 1981, **14**, 251–260.
61. Hanham, A. F., Dunn, B. D. and Stich, H. F., *Mutat. Res.*, 1983, **116**, 333–339.
62. Fujita, Y., Wakabayashi, K., Nagao, M. and Sugimura, T., *Mutat. Res.*, 1985, **144**, 227–230.
63. Challis, B. C. and Bartlett, C. D., *Nature (London)*, 1975, **254**, 532–533.
64. Lathia, D. and Frentzen, U., *Food Cosmet. Toxicol.*, 1980, **18**, 463–465.
65. Walker, E. A., Pignatelli, B. and Friesen, M., *J. Sci. Food Agric.*, 1982, **33**, 81–88.
66. Stich, H. F., Rosin, M. P. and Bryson, L., *Mutat. Res.*, 1982, **95**, 119–128.
67. Kimura, Y., Okuda, H., Okuda, T., Hatano, T., Agata, I. and Arichi, S., *Planta Med.*, 1984, 50, 473–477.
68. Fukuhara, Y., Yoshida, D. and Goto, F., *Agric. Biol. Chem.*, 1981, **45**, 1061–1066.
69. Wood, A. W., Huang, M. T., Chang, R. L., Newmark, H. L., Lehr, R. E., Yagi, H., Sayer, J. M., Jerina, D. M. and Conney, A. H., *Proc. Natl Acad. Sci. USA*, 1982, **79**, 5513–5517.
70. Huang, M. T., Chang, R. L., Wood, A. W., Newmark, H. L., Sayer, J. M., Yagi, H., Jerina, D. M. and Conney, A. H., *Carcinogenesis*, 1985, **6**, 237–242.
71. Hirono, I., Hosaka, S., Haga, M., Sakata, M., Mori, H., Tanaka, T. and Hikino, H., *J. Food Safety*, 1980, **4**, 205–211.
72. Bories, G. and Durand, G., *Proc. 5th Coll. ASIC*, 1971, 403–405.
73. Wattenberg, L. H., *Cancer Res. (Suppl.)*, 1983, **43**, 2448s–2453s.
74. Auf'Mkolk, M., Ingbar, J. C., Amir, S. M., Winterhoff, H., Sourgens, H., Hesch, R. D. and Ingbar, S. H., *Endocrinol.*, 1984, **115**, 527–533.
75. Auf'Mkolk, M., Amir, S. M., Kubota, K. U. and Ingbar, S. H., *Endocrinol.*, 1985, **116**, 1677–1685.
76. Czok, G. and Lang, K., *Arzneimittel-Forsch.*, 1961, **11**, 448–450.
77. Valette, G. and Morin, H., *Proc. 4th Coll. ASIC*, 1969, 248–253.
78. Kronig, D. and Kunkel, H., *Nachrichtentechnisch. Gesell. VDI*, May 1973, 185–193.
79. Bariana, D. S., Krupey, J., Scarpati, L. M., Freedman, S. O. and Sehon, A. H., *Nature (London)*, 1965, **207**, 1155.
80. Lehrer, S. B., Karr, R. M. and Salavaggio, J. E., *Clin. Allergy*, 1981, **11**, 357–366.
81. Zuskin, E., Valic, F. and Kanceliak, B., *Thorax*, 1981, **36**, 9–13.
82. Layton, L. L., Panzani, R., Greene, F. C. and Corse, J. W., *Int. Arch. Allergy Appl. Immunol.*, 1965, **28**, 116–127.
83. Layton, L. L., Panzani, R. and Corse, J. W., *J. Allergy Clin. Immunol.*, 1966, **38**, 268–275.

84. Ludwig, H., Obermann, H. and Spiteller, G., *Chem. Ber.*, 1974, **107**, 2409–2411.
85. Obermann, H. and Spiteller, G., *Chem. Ber.*, 1976, **109**, 3450–3461.
86. Richter, H. and Spiteller, G., *Chem. Ber.*, 1978, **111**, 3506–3509.
87. Maier, H. G. and Wewetzer, H., *Z. Lebensm. Unters. Forsch.*, 1978, **16**, 105–107.
88. Aeschbach, R., Kusy, A. and Meier, H. G., *Z. Lebensm. Unters. Forsch.*, 1982, **175**, 337–341.
89. Maetzel, U. and Meier, H. G., *Z. Lebensm. Unters. Forsch.*, 1983, **176**, 281–284.
90. Obermann, H., Spiteller, G. and Hoyer, G. A., *Chem. Ber.*, 1973, **106**, 3506–3518.
91. Santi, R. and Luciani, S. (Eds), *Atractyloside—Chemistry, Biochemistry and Toxicology*, Piccin, Padova, 1978.
92. Vignais, P. M., Vignais, P. V. and Defaye, G., *Atractyloside—Chemistry, Biochemistry and Toxicology*, Piccin, Padova, 1978, 51.
93. Richter, H., Obermann, H. and Spiteller, G., *Chem. Ber.*, 1977, **110**, 1963–1970.
94. Obermann, H. and Spiteller, G., *Chem. Ber.*, 1975, **108**, 1093–1100.
95. Richter, H. and Spiteller, G., *Chem. Ber.*, 1979, **112**, 1088–1092.
96. Ducruix, A., Hamonniere, M., Pascard, E. and Poisson, J., *Café Cacao Thé*, 1975, **19**, 57–58.
97. Poisson, J., *Proc. 8th Coll. ASIC*, 1977, 33–57.
98. Folstar, P., in *Coffee, Vol. 1: Chemistry*, Eds R. J. Clarke and R. Macrae, Elsevier Applied Science Publishers, London, 1985, 210–213.
99. Davis, G. E., unpublished results.
100. Czok, G., *Proc. 2nd Coll. ASIC*, 1966, 239–244.
101. Lam, L. K. T., Sparnins, V. L. and Wattenberg, L. W., *Cancer Res.*, 1982, **42**, 1193–1198.
102. Wattenberg, L. W., *Cancer Res. (Suppl.)*, 1983, **43**, 2448s–2453s.
103. Wattenberg, L. W., *Banbury Rep.*, 1984, **17** (Coffee Health), 137–145.
104. Sparnins, V. L., Lam, L. K. T. and Wattenberg, L. W., *Proc. Am. Assoc. Cancer Res.*, 1981, **22**, 114.
105. Aeschbacher, H. U. and Wolleb, U., personal communication.
106. Viani, R. and Horman, I., *J. Food Sci.*, 1974, **39**, 1216–1217.
107. Macrae, R., in *Coffee, Vol. 1: Chemistry*, Eds R. J. Clarke and R. Macrae, Elsevier Applied Science Publishers, London, 1985, 127–136.
108. Trugo, L. C., Macrae, R. and Dick, J., *J. Sci. Food Agric.*, 1983, **84**, 300–306.
109. Taguchi, H., Sakaguchi, M. and Shimabayashi, Y., *Agric. Biol. Chem.*, 1985, **49**, 3467–3471.
110. Adrian, J., Xabregas, J., Pena, J., Morais de Carvalho, J. and Gomes, N., *Proc. 5th Coll. ASIC*, 1971, 371–374.
111. Davis, R. G., *Chem. Senses Flavor*, 1978, **3**, 423–429.
112. Taguchi, H. and Shimabayashi, Y., *Biochem. Biophys. Res. Cancer*, 1983, **113**, 569–574.
113. Lynn, D. G., Nakanishi, K., Patt, S. L., Occolowitz, J. L., Almeida, S. and Evans, L. S., *J. Am. Chem. Soc.*, 1978, **100**, 7759–7760.
114. Maier, H. G., *Kaffee*, Paul Parey, Berlin, 1981, 51–59.
115. Van Straten, S. and Maarse, H. (Eds), *Volatile Compounds in Food—Qualitative Data*, 5th edn, Inst. CIVO Anal. TNO, Zeist (NL), 1983, 72.1–10.
116. Maarse, H. (Ed.), *Volatile Compounds in Food—Quantitative Data, Vol. 2*, Inst. CIVO Anal. TNO, Zeist (NL), 1984, 72.1–11.

117. Baettig, K., in *Coffee: Botany, Biochemistry and Production of Beans and Beverage*, Eds M. N. Clifford and K. C. Willson, Croom Helm, London, 1985, 398–399.
118. Reichstein, T. and Staudinger, H., *Brit. Pat.*, 1926, 26060.
119. Tressl, R. and Silwar, R., *J. Agric. Food Chem.*, 1981, **29**, 1078–1082.
120. Gianturco, M. A., Giammarino, A. S. and Pitcher, R. G., *Tetrahedron*, 1963, **19**, 2051–2059.
121. Yamanishi, T., in *Flavour Research—Recent Advances*, Eds R. Teranishi, R. A. Flath and H. Sugisawa, Marcel Dekker, New York, 1981, 231–304.
122. Kung, J. F. T., *J. Agric. Food Chem.*, 1974, **22**, 494–496.
123. Pictet, G. and Vuataz, L., *Proc. 7th Coll. ASIC*, 1975, 189–200.
124. Moll, H. R. and Pictet, G., *Proc. 9th Coll. ASIC*, 1980, 87–98.
125. Shibamoto, T., in *The Quality of Foods and Beverages, Vol. 2*, Eds G. Charalambous and G. Inglett, Academic Press, New York, 1981, 301–334.
126. Liardon, R. and Ott, U., *Lebensm. Wiss. Technol.*, 1984, **17**, 32–38.
127. Liardon, R., Ott, U. and Daget, N., in *Analysis of Volatiles*, Ed. P. Schreier, Walter de Gruyter, Berlin, 1984, 447–459.
128. Liardon, R. and Spadone, J. C., *Proc. 11th Coll. ASIC*, 1985, 181–196.
129. Guyot, B., Cros, E. and Vincent, J. C., *Proc. 10th Coll. ASIC*, 1982, 253–269.
130. Swain, A. R., Dutton, S. P. and Truswell, A. S., *J. Am. Diet. Assoc.*, 1985, **85**, 950–960.
131. Sen, N. P. and Seaman, S., *J. Am. Offic. Assoc. Chem.*, 1981, **64**, 1238–1242.
132. Aeschbacher, H. U., *TIPS*, 1984, **5**, 512–513.
133. Levin, D. E., Hollstein, M., Christman, M. F., Schwiers, E. A. and Ames, B. N., *Proc. Natl Acad. Sci. USA*, 1982, **79**, 7445–7449.
134. Hayashi, T., Reece, C. A. and Shibamoto, T., *J. Assoc. Offic. Analyt. Chem.*, 1986, **69**, 101–105.
135. Bjeldanes, L. F. and Chew, H., *Mutat. Res.*, 1979, **67**, 367–371.
136. Maier, H. G., *Kaffee*, Paul Parey, Berlin, 1981, 57.
137. Fujita, Y., Wakabayashi, K., Nagao, M. and Sugimura, T., *Mutat. Res.*, 1985, **144**, 227–230.
138. Fujita, Y., Wakabayashi, K., Nagao, M. and Sugimura, T., *Mutat. Res.*, 1985, **142**, 145–148.
139. Suwa, Y., Nagao, M., Kosugi, A. and Sugimura, T., *Mutat. Res.*, 1982, **102**, 383–391.
140. Kasai, H., Kumeno, K., Yamazumi, Z., Nishimura, S., Nagao, M., Fujita, Y., Sugimura, T., Nukaya, H. and Kosuge, T., *Gann*, 1982, **73**, 681–683.
141. Aeschbacher, H. U., Chapter 8, this volume.
142. De Flora, S., Camoirano, A., Zanacchi, P. and Bennicelli, C., *Mutat. Res.*, 1984, **134**, 159–165.
143. Gillies, M. E. and Birkbeck, J. A., *Am. J. Clin. Nutr.*, 1983, **38**, 936–942.
144. Morck, T. A., Lynch, S. R. and Cook, J. D., *Am. J. Clin. Nutr.*, 1983, **37**, 416–420.
145. McKenzie, J. M., *Nutr. Rep. Int.*, 1984, **29**, 387–395.
146. Harland, B. F. and Oberleas, D., *J. Food Sci.*, 1985, **50**, 832–833, 842.
147. Boublik, J. H., Quinn, M. J., Clements, J. A., Herington, A. C., Wynne, K. N. and Funder, J. W., *Nature (London)*, 1983, **301**, 246–248.
148. Strubelt, O., Kashube, M. and Zetler, G., *Experientia*, 1986, **42**, 35–37.

149. Roffo, A. H., *Bol. Inst. Med. Exper. Estud. Trat. Cancer*, 1938, **15**, 741–762.
150. Kuratsune, M. and Hueper, W. C., *J. Natl Cancer Inst.*, 1958, **20**, 37–51.
151. Kuratsune, M. and Hueper, W. C., *J. Natl Cancer Inst.*, 1960, **24**, 463–469.
152. Maier, H. G., *Kaffee*, Paul Parey, Berlin, 1981, 58–59.
153. Chassevent, F. and Heros, M., *Café Cacao Thé*, 1963, **7**, 349–358.
154. Morgante, A., *Riv. Merceol.*, 1980, **19**, 215–222.
155. Soos, K., *Arch. Toxicol. Suppl.*, 1980, **4**, 446–448.
156. Grimmer, G. and Hildebrandt, A., *Deutsche Lebensm. Rdsch.*, 1966, **62**, 19–21.
157. Van der Stegen, G. H. D. and Overbruggen, G. J. J., *Proc. 10th Coll. ASIC*, 1982, 347–354.
158. Levi, C. P., *Proc. 9th Coll. ASIC*, 1980, 125–133.
159. Guyot, B., Lahmi, S. and Vincent, J., *Café Cacao Thé*, 1982, **26**, 199–206.
160. Fritz, W., *Deutsche Lebensm. Rdsch.*, 1969, **65**, 83–85.
161. De Kruijf, W., Schouten, T. and Van der Stegen, G. H. D., *J. Agric. Food Chem.*, 1987, **35**, 545–549; *Café Cacao Thé*, 1987, **31**, 151–154.
162. Strobel, R. G. K., *Proc. 6th Coll. ASIC*, 1973, 128–134.
163. Ruschenburg, U. and Jahr, D., *Café Cacao Thé*, 1986, **30**, 3–10.
164. Hischenhuber, C. and Stijve, T., *Deutsche Lebensm. Rdsch.*, 1987, **83**, 1–4.
165. Nagayama, T., Kan, K., Tamura, Y., Maki, T. and Nishima, T., *Kenkyu Nenpo-Tokyo-Toritsu Eisei Kenkyusho*, 1984, **35**, 210–218.
166. Cetinkaya, M., Von Duszeln, J., Thiemann, W. and Silwar, R., *Z. Lebensm. Unters. Forsch.*, 1984, **179**, 5–8.
167. Blumenthal, A. and Cerny, M., *Mitt. Lebensm. Hyg.*, 1976, **67**, 515–520.
168. Hearne, S. A., *Harvest of Unknowns*, NRCD, New York, 1984, 17 (FDA survey 1978–83).
169. Anon., *Natur*, 1983, **3**, 85–87.
170. Anon., *FDA Compliance Program Evaluation FY 77/78, Pesticides in Imported Coffee Beans*, Washington, 25 January 1978.
171. Betancourt, L. E. and Frank, H. K., *Deutsche Lebensm. Rdsch.*, 1983, **79**, 366–369.
172. Corte Dos Santos, A., Hahn, D., Cahagnier, B., Drapron, R., Guilbot, A., Lefebvre, J., Multon, J. L., Poisson, J. and Trentesaux, E., *Proc. 5th Coll. ASIC*, 1971, 304–315.
173. Poisson, J., Cahagnier, B., Multon, J. L., Hahn, D. and Corte Dos Santos, A., *Proc. 7th Coll. ASIC*, 1975, 311–321.
174. Betancourt, L. E. and Frank, H. K., *Deutsche Lebensm. Rdsch.*, 1983, **79**, 404–407.
175. Mislivec, P. B., *J. Food Prot.*, 1981, **44**, 723–726.
176. Stack, M. E., Mislivec, P. B., Denizel, T., Gibson, R. and Pohland, A. E., *J. Food Prot.*, 1983, **46**, 965–968.
177. Tsubouchi, H., Yamamoto, K., Terada, H., Hisada, K. and Sakabe, Y., *Proc. Jpn Assoc. Mycotoxicol.*, 1984, **20**, 10–16.
178. Buchanan, R. L., Harry, M. A. and Gealt, M. A., *J. Food Sci.*, 1983, **48**, 1226–1228.
179. Buchanan, R. L. and Fletcher, A. M., *J. Food Sci.*, 1978, **43**, 654–655.
180. Buchanan, R. L., Tice, G. and Marino, D., *J. Food Sci.*, 1981, **47**, 319–321.
181. Tsubouchi, H., Yamamoto, K., Hisada, K. and Sakabe, Y., *Proc. Jpn Assoc. Mycotoxicol.*, 1984, **19**, 16–21.

182. Nartowicz, V. B., Buchanan, R. L. and Segall, S., *J. Food Sci.*, 1979, **44**, 446–448.
183. Purchase, I. F. H. and Pretorius, M. E., *J. Assoc. Offic. Analyt. Chem.*, 1973, **56**, 225–226.
184. Levi, C. P., Trenk, H. L. and Mohr, H. K., *J. Assoc. Offic. Analyt. Chem.*, 1974, **57**, 866–870.
185. Levi, C. P., Trenk, H. L. and Yeransian, J. A., *Proc. 7th Coll. ASIC*, 1975, 287–301.
186. De Palo, D., Gabucci, G. and Valussi, S., *Proc. 8th Coll. ASIC*, 1977, 539–543.
187. Levi, C., *J. Assoc. Offic. Analyt. Chem.*, 1980, **63**, 1282–1285.
188. Fritz, W., Buhtig, C., Dowath, R. and Engst, R., *Z. Gesamte Hyg. Ihre Grenzgeb.*, 1979, **25**, 929–932.
189. Cantafora, A., Grossi, M., Miraglia, M. and Benelli, L., *Riv. Soc. Ital. Sci. Alim.*, 1983, **12**, 103–108.
190. Tsuboi, S., Kawamura, K. and Iwamura, N., *JCMR Ann.*, 1983, **3**, 109–115.
191. Gallaz, L. and Stalder, R., *Chem. Mikrobiol. Technol. Lebensm.*, 1976, **4**, 147–149.
192. Goto, T., Manabe, M. and Matsuura, S., *Rep. Natl Food Res. Inst.*, 1983, **42**, 83–86.
193. Haberle, V., Balenovic, J. and Briski, B., *Hrana i Ishrana*, 1978, **19**, 451–460.
194. Thier, H. P., Bricout, J., Viani, R., Reymond, D. and Egli, R. H., *Z. Lebensm. Unters. Forsch.*, 1968, **137**, 1–4.
195. Takahashi, H., Ueno, I., Haraikawa, K. and Ueno, Y., *Chiba-ken Eisei Kenkyusho Kenkyu Hokoku*, 1983, **7**, 16–20.
196. Vanos, V. and Bindschedler, O., *Food Microbiol.*, 1985, **2**, 187–197.
197. Goldman, P., *New Engl. J. Med.*, 1984, **310**, 783–785.

Chapter 2

The Metabolism of Coffee Constituents

M. J. ARNAUD
Nestlé Research Centre, Nestec Ltd, Lausanne, Switzerland

1. INTRODUCTION

Like seeds of other plants, the coffee bean contains all of the plant cell materials constituting its cell walls: cellulose, hemicelluloses, pectins, lignin, gums, proteins, minerals, pigments, fats, waxes and oils. Within the cells, the organelles, including nucleus and mitochondria, and the cytoplasm contain proteins, lipids, nucleic acids, carbohydrates, chlorophyll, carotenoids and a large variety of enzymes and plant hormones.

All of these constituents are transformed during the roasting of the coffee bean, resulting in a tremendous variety of identified and unknown compounds which can be extracted and found in the coffee brews. Some constituents of coffee beans can be destroyed during roasting, giving either new compounds present in the brew or volatile substances lost in waste gases. See Volume 1 for detailed information.

The composition of the beverage depends on the roasting process, the different coffee species used and whether instant or brewed coffees are considered. These differences affect only the quantitative aspect of the products ingested. This chapter, devoted to the metabolism in humans of products found in coffee, will be too restricted if animal studies are not taken into account (see Chapter 7) and too long if the minor constituents of coffee such as well known nutrients (amino acids, sucrose, glucose and fructose) are also described. This review will therefore describe in detail substances that characterise coffee and for which information is available; well known nutrients will only be mentioned with appropriate references.

Among other nutrients found in the coffee drink, minerals constitute an important part of the extract, and several reports have evaluated positive or negative effects of the coffee beverage on mineral balance (see also this volume, Chapter 5).

2. MINERALS

Minerals constitute *c*. 14% of the water-soluble components of roasted arabica coffee.[1] The main components are potassium, magnesium, phosphorus, sulphur, chlorine, silicon, calcium, sodium, iron and trace elements such as copper, bromine, fluorine, manganese, molybdenum, rubidium and zinc. Coffee intake has been considered as a source of minerals in the New Zealand diet.[2] The estimation of the daily mineral intake showed that coffee is a good source of potassium, magnesium and manganese. The other elements are in too low a concentration to have any nutritional significance, and in fact their contribution is influenced more by the tap water used in the preparation of the drink. In contrast with this study, coffee was shown to reduce the absorption of non-haemic iron in humans.[3–5] This effect, observed in healthy subjects, corresponds to a very small quantitative reduction in iron absorption, the nutritional significance and biochemical mechanisms of which are unclear.[6,7]

3. CARBOHYDRATES

3.1. Polysaccharides

These are the major constituents of roasted coffee, and they have been reported as complex and poorly characterised carbohydrates.[8] The water-soluble carbohydrates found in the beverage are derived from arabino-galactan, galactomannan and pectic substances found in the green bean. Although galactomannans are used commercially as food stabilisers, the intestinal absorption of the modified polysaccharides has not been demonstrated.

3.2. Mono- and Disaccharides

Monosaccharides and disaccharides are usually destroyed during roasting although arabinose, glucose, mannose and galactose with traces of fructose, rhamnose and sucrose are observed.[9,10] When ingested, these mono- and disaccharides are absorbed from the intestine at a faster rate for galactose and glucose, as compared with fructose, mannose and arabinose.

The metabolism of galactose,[11] fructose, mannose, rhamnose and arabinose has already been reported.[12]

3.3. Phytate

The content of the phytate anion, the hexaphosphoric ester of *myo*-inositol in different coffees, has been analysed, and a lower concentration has been found in instant coffee (1·1 mg per cup) than in percolated coffee (from 4 to 18 mg per cup).[13] When ingested, phytate was shown to be hydrolysed in conventional rats fed a high- (22% hydrolysis) or a low- (56% hydrolysis) calcium diet, whereas negligible phytate hydrolysis occurred in germ-free rats. Bacteria were responsible for the hydrolysis of phytate, and this confirms the observation of the capacity of human faeces to hydrolyse phytate.[14] [U-^{14}C]Phytate metabolism was studied after oral administration to male rats adapted either to a low- or a high-calcium diet. The high-calcium diet increased the recovery of radioactivity in faeces (54% of the dose), with 25% of the dose collected as $^{14}CO_2$. With the low-calcium diet, radioactive phytate was quantitatively absorbed and extensively oxidised as CO_2 (60% of the dose). Only 2–4% of the dose was recovered in urine within 48 h. It was concluded that high calcium intake inhibits the utilisation of phytate, which releases *myo*-inositol on hydrolysis to be used as a gluconeogenic compound.[15]

4. PROTEINS AND AMINO ACIDS

'Protein' content in roasted coffee amounts to about 10% when estimated by acid hydrolysis. During roasting, some protein nitrogen is lost as a result of browning reactions between proteins and carbohydrates producing heterocyclic aroma volatiles. These 'proteins' have not been fully characterised, but the amino acids present in roasted coffee have been identified. The most important of the latter are glutamic acid, leucine, valine, aspartic acid, proline, glycine, phenylalanine, alanine, isoleucine and tyrosine[1] (see also Volume 1, Chapter 4). We can expect the 'proteins' to be extracted with water, and when ingested with coffee hydrolysed in the gastrointestinal tract, giving free amino acids which are absorbed.

5. LIPIDS

5.1. Fatty Acids

Lipids are minor constituents of the coffee beverage, the majority being left with the grounds, though some of the lipids present yield a range of volatiles on roasting. Some of these aroma volatiles produced during roasting are trapped by the oil expelled at the surface of the bean. Palmitic acid and linoleic acid are the most important fatty acids of the crude lipids, which amount to less than 1% of the soluble powder of instant coffee.[8] In roasted coffee, lipids are present as triglycerides along with unsaponifiables such as terpenes, sterols and waxes.

5.2. Waxes

While waxes are known to be poorly absorbed, or not at all, the components of coffee wax, hydroxytryptamine derivatives with arachic, behenic, lignoceric and stearic acids, have been shown to decompose partly during roasting, the residue remaining in the spent grounds. These products do not seem to be present in the beverage, but see Volume 1, Chapter 6.

5.3. Sterols

The most important sterols reported in roasted coffee beans are sitosterol, stigmasterol, campesterol and cycloartenol, which will be present in trace amounts in the beverage (Fig. 1; see also Volume 1, Chapter 6). The absorption and metabolism of dietary plant sterols have been reported in animals and humans. In rabbits fed a diet containing 2% plant sterols, campesterol and β-sitosterol were found in the blood and tissues whereas stigmasterol could not be detected in any tissue studied.[16] In the rat,

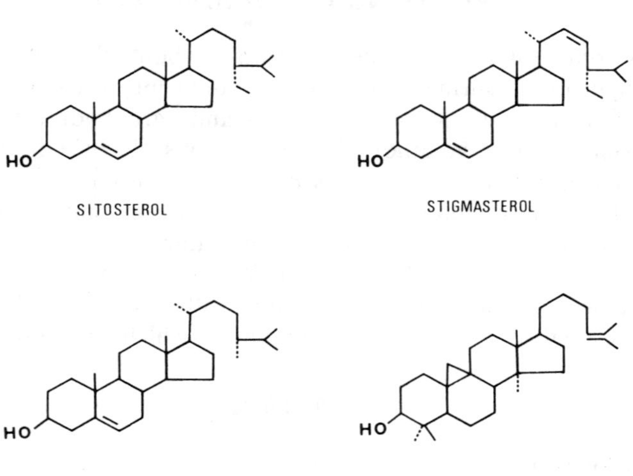

Fig. 1. Structures of the most important sterols present in coffee.

quantitative data showed that the extent of the absorption amounted to only 3–8% of a dose of 50 mg.[17] These results were extrapolated to humans, in whom labelled β-sitosterol was used as a non-absorbable marker of dietary lipid absorption. It was postulated that the poor faecal recovery observed in man was due to the formation of volatile products in the colon.[18] The small fraction of absorbed plant sterols is catabolised to a variety of products such as bile acids. The intake of sterols from coffee can be neglected compared with the 250 mg of plant sterols which might be present daily in the typical American diet.[19] These plant sterols contribute to approximately 20% of the total sterols in the diet of an adult human being.[20]

5.4. Terpenes

Diterpenes (such as kahweol, cafestol and their esters; see Volume 1, Chapter 6) are extensively destroyed during roasting and may be the origin of volatile (mono) terpenes, naphthalenes and quinolines.

The metabolism of ^{14}C-*d*-limonene, a monoterpene, has been investigated in the rat. After oral administration by stomach tube, 60% of the radioactivity was recovered in urine, 2% from expired CO_2 and 5% from faeces. It was shown that 25% of the dose was excreted in bile within 24 h.[21] The level of radioactivity in blood and tissue was negligible 48 h after the administration, and it was suggested that loss of ^{14}C occurred through volatile compounds other than CO_2. More than 10 metabolites were observed: four of them were identified and also found in rabbit urine. The rabbit was characterised by having two metabolites not found in the rat. Species differences in the metabolism of limonene were reported for the rat, hamster, guinea pig, rabbit, dog and man.[22] In 1981, limonene epoxides were shown to be the intermediates in the production of glycols with rat liver microsomes.[23] In the rabbit, the biotransformations of other terpenoids such as α- and β-pinene, pinane, 3-carene, carane, myrcene and *p*-cymene were reported, and included allylic oxidation, epoxidation, stereoselective gem-dimethylhydroxylation and these metabolic transformations lead to the identification of several new compounds.[24] Data from acute poisoning with pine oil containing α- and β-pinene, carene and limonene suggested that monoterpenes are poorly reabsorbed in the gastrointestinal tract. Part of the reabsorbed monoterpenes was exhaled via the lungs, possibly excreted in the bile and slowly metabolised and then excreted via the kidneys. As shown in Fig. 2, the metabolism of monoterpenes in man occurs via reduction, hydration, hydroxylation and acetylation.

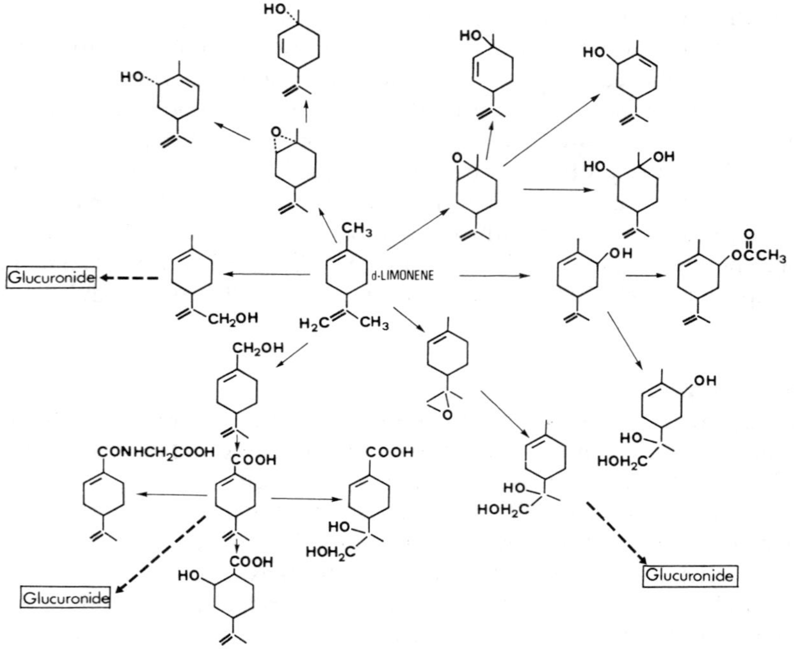

Fig. 2. Proposed metabolic pathways of *d*-limonene in human and several animal species *in vivo*, and also in microsomal preparations. From references 21, 22 and 23.

5.4.1. Naphthalene

Considerable attention has been focused on the metabolism of aromatic hydrocarbons and particularly naphthalene. Large doses of naphthalene incorporated into the diet have been administered to rats, and the isolation of urinary metabolites led to the identification of eight methylthio derivatives,[25] later shown to be formed during enterohepatic circulation of premercapturic acid pathway metabolites of naphthalene, in a process mediated by the intestinal microflora.[26] The main metabolites were naphthalene 1,2-dihydrodiol and 1-naphthol and their glucuronide conjugates. Several reports indicate that epoxides are formed during the metabolism of naphthalene and that a covalently bound product(s) was obtained with either microsomal or hepatocyte preparations. Differences in toxicity between mouse and rat were observed and explained by a more severe depletion of the non-protein sulphydryl level in the case of the mouse, as a result of a 10-fold higher covalent binding and metabolism.[27]

5.4.2. Quinoline

Quinoline and its derivatives are widely found in chemicals and drugs, and are also formed in the body from tryptophan metabolism. *In vitro* tests showed that quinoline binds to nucleic acids, and it was suggested that the reactive intermediate was an epoxy derivative.[28] The metabolites produced *in vitro* from microsomes have been identified as quinoline-1-oxide, 5,6-dihydroxy-5,6-dihydroquinoline and 3-hydroxyquinoline.[29] However, unidentified water-soluble metabolites have been detected which were responsible for the mutagenic activity.[30] Among these polar products, glutathione, glucuronic and sulphate conjugates of different quinoline derivatives have been suggested.

Compared with these model compounds such as naphthalene and quinoline, the derivatives detected in coffee are present only in trace amounts and may have a quite different intestinal absorption and metabolic disposition.

5.5. Kahweol, Cafestol and Diterpene Glycosides

In spite of some decomposition during roasting, the presence of kahweol is observed and characterises arabica coffees where the concentration is five times higher than in robusta coffees. In contrast, cafestol is less affected by roasting. Water-soluble diterpene glycosides have also been identified. Although it was shown that total atractyligenin decreased by about 35% on roasting[1] with a concomitant 5- to 10-fold rise of free atractyligenin, the glycoside conjugates still present were water-soluble and found in the coffee beverage. Atractyligenin (Fig. 3) was found in the urine of coffee drinkers,

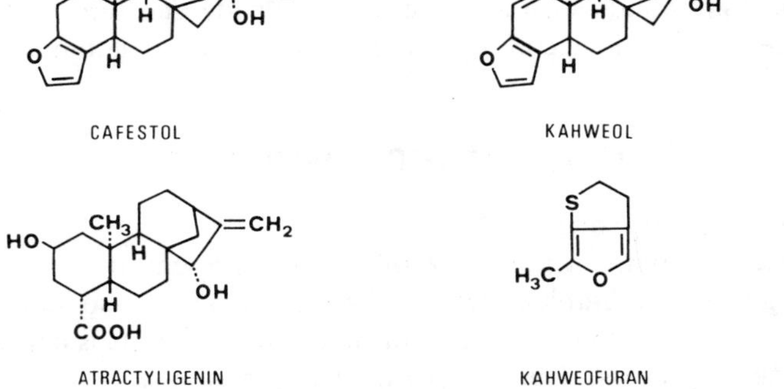

Fig. 3. Products that characterise coffee and whose metabolic fate has not been studied.

with two derivatives identified as the glucuronide of dihydroatractyligenin and 2β-hydroxy-15-oxoatractylan-4α-carboxylic acid;[31,32] the metabolic fate of cafestol and kahweol have not been reported (Fig. 3).

6. ALIPHATIC COMPOUNDS

Free aliphatic acids are generated during roasting from complex carbohydrates (see also Volume 1, Chapter 8). These main acids are identified as lactic, pyruvic, oxalic, glycolic, tartaric and citric acids, in amounts up to about 1·5% in roasted coffee. They are extracted completely and are present in the beverage. These products, found endogenously in the human body, when ingested are absorbed and then subjected to oxidative metabolism or excreted in the urine. During roasting, a number of other carboxylic acids are produced along with carbonyl compounds in very small amounts. Glyoxal, methylglyoxal and diacetyl have been identified in roasted coffee. Recently, methylglyoxal and hydrogen peroxide were shown to be responsible for most of the mutagenicity of instant coffee.[33] *In vivo*, glyoxal, methylglyoxal and diacetyl must be absorbed from the gastrointestinal tract and then oxidised. Their detailed metabolism and tissue distribution after oral ingestion in animals including man have not been reported. Acetaldehyde, formaldehyde, propanol, acetone and methylethylketone with a multitude of volatile aliphatic compounds are in part responsible for the aroma, and are present in trace amounts in the beverage. With labelled formaldehyde, extensive oxidation was demonstrated with more than 80% of the dose recovered as $^{14}CO_2$ in the expired air. The rest of the radioactivity was recovered in urine as methionine, serine and an adduct formed from cysteine and formaldehyde.[34] Sulphur aliphatic compounds such as dimethylsulphoxide and dimethylsulphone have been detected in coffee, and their metabolism in humans is well known.[35]

7. AROMATIC COMPOUNDS

7.1. Benzo(*a*)pyrene

The polycyclic hydrocarbon benzo(*a*)pyrene is present at a concentration below 1 $\mu g\ kg^{-1}$ in soluble coffee. 2,3-Benzofluorene has also been reported, but the presence of these products in significant amounts is dependent on the roasting process. The comparative absorption and metabolism of naphthalene and benzo(*a*)pyrene in the rat jejunum *in situ* were studied and showed a slow absorption of the latter in contrast with naphthalene.

Whereas naphthalene appeared unchanged in the portal blood, benzo(*a*)-pyrene was extensivley metabolised by mucosal cells. Labile metabolites were observed but not identified, and the water-soluble metabolites were glucuronides as shown by β-glucuronidase treatment.[36] More recently, chromatographic fractionation of urinary and faecal metabolites of ^{14}C-labelled benzo(*a*)pyrene excreted from germ-free and conventional rats was reported.[102] The higher excretion of acidic conjugates in faeces of germ-free rats showed the influence of intestinal flora. The results also demonstrated the quantitative importance of mercapturic acid pathways. However, these analyses, showing a complex pattern of benzo(*a*)pyrene metabolism, failed to identify new metabolites.

7.2. Chlorogenic Acids

An important series of phenolic compounds other than pyrocatechol and pyrogallol are the phenols derived from caffeic and ferulic acids, present as depside esters of quinic acid. The most important, the esters of caffeic and quinic acids, are the chlorogenic acids. Chlorogenic acid (5-caffeoylquinic acid, according to IUPAC nomenclature) and neochlorogenic acid (3-caffeoylquinic acid) are the most important in quantity. The total content of chlorogenic acid isomers in instant coffee, expressed as a percentage of dry matter, ranges from 4 to 10%[37] (see Volume 1, Chapter 5).

Quinic acid is released during roasting, and chlorogenic acids have been shown to be as much as 90% destroyed in some instances. In commercial coffee extracts, concentrations of 20 to 50 g kg^{-1} quinic acid were determined.

7.2.1. Caffeic Acid

In rats and also in humans, chlorogenic acid was shown to be hydrolysed into caffeic and quinic acids.[38] Booth *et al.*[39] demonstrated in 1957 that caffeic acid was the precursor of *m*-hydroxyhippuric acid in human urine. More than ten additional metabolites were identified in human and rat urine, and Booth *et al.* proposed the metabolic pathways of caffeic acid shown in Fig. 4. Caffeic, dihydrocaffeic, ferulic, dihydroferulic, vanillic and *m*-coumaric acids, and also glucuroconjugates of *m*-coumaric acid, feruloylglycine, vanilloylglycine and *m*-hydroxyhippuric acid, were identified in urine. Glucuroconjugates of *m*-coumaric acid and *m*-hydroxyhippuric acid seemed to be the main metabolites in humans.

When administered i.p. to rats, labelled cinnamic acid [3-^{14}C] and ferulic acid [2-^{14}C] were not incorporated into tissues, but most of the radioactivity was excreted in urine (48 and 68%, respectively) and faeces (25

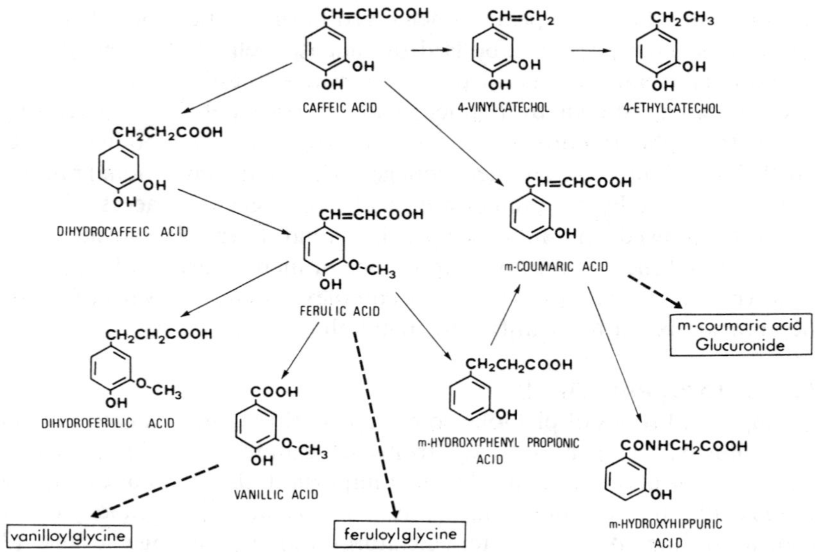

Fig. 4. Proposed metabolic pathways of caffeic acid. From references 39–45.

and 17%). In addition to unchanged cinnamic acid, hippuric acid was identified as the main metabolite along with benzoic acid. Unchanged ferulic acid was estimated as 25% of the administered dose, and *m*-hydroxyphenylpropionic acid was also identified.[40] However, in contrast with Booth *et al.*,[39] Teuchy and Van Sumere were unable to show, with labelled ferulic acid, the presence of dihydroferulic acid, feruloylglycine and vanilloylglycine.[41] Radiolabelled cinnamic acids were used to study their choleretic effects and administered intraduodenally to rats. It was shown that 20% of the dose of ferulic acid, isoferulic acid, 3,4-dimethoxycinnamic acid, 3,4,5-trimethoxycinnamic acid, 8% of *p*-methoxycinnamic acid and less than 5% of caffeic acid, *m*-coumaric acid and *p*-coumaric acid were recovered in biliary secretion.[42]

In humans, after the oral administration of caffeic acid, the O-methylated phenolic acids (ferulic, vanillic, etc.) appeared rapidly in the urine, whereas *m*-hydroxyphenyl derivatives appeared later. By suppressing the gut flora with neomycin, it was shown that the dehydroxylation reaction was due solely to the action of intestinal bacteria.[43]

These results were confirmed by comparing the metabolism of caffeic acid in germ-free[44] and various gnotobiotic[45] rats. However, discrepancies were observed in the ability of bacteria cultures to transform caffeic acid into ferulic acid, dihydrocaffeic acid, dihydroferulic acid, *m*-hydroxypropionic acid, vinyl catechol and ethyl catechol and the metabolites produced by these bacteria in the host.

Recently, the oral administration of 1 g of caffeic acid or ferulic acid was performed in adult volunteers, to quantitate and identify urinary metabolites in order to estimate the dietary intake of plant phenolics in the human population. An accurate determination of this intake is of importance in the correlation between phenolic constituents of the diet and the lower risk of gastrointestinal cancer. This hypothesis was sustained by animal experiments showing reduced incidence of neoplasia experimentally induced by benzo(*a*)pyrene when caffeic and ferulic acids were added to the diet.[46]

The highest level of phenolic metabolites from dietary caffeic acid appeared in the urine samples collected in the first 4 h and then decreased rapidly. In addition to caffeic acid, vanillic, ferulic and isoferulic acids were identified and quantified. No isoferulic acid and dihydroferulic acid could be detected as metabolites of ferulic acid. As no phenolic metabolite can be detected in faecal samples, the total metabolite recovery of only 10% of the 1 g of caffeic acid dose can be questioned.[47]

7.2.2. Quinic Acid

The metabolic transformation of quinic acid to give the aromatic ring structure was first reported as early as 1863 by Lautemann.[48] A century later, Cotran *et al.*[49] found that the metabolic reaction leading to the excretion of hippuric acid in urine occurred in man and guinea pig after oral administration but not after parenteral injection in the guinea pig and after neomycin treatment in man.

The role of gut bacteria in the 'aromatisation' of quinic acid was confirmed by Adamson *et al.*[50] The fate of quinic acid was investigated in 22 species of animals including man. In man, oral quinic acid was 'aromatised' to an extent of 60% of the dose and excreted in the urine as hippuric acid. The suppression of the 'aromatisation' of oral quinic acid in the rhesus monkey pretreated with neomycin supports the view that quinic acid was 'aromatised' by the gut flora in this species as in man.

In addition to quinic acid, the 'aromatisation' of shikimic acid,[51] another product found in coffee,[52] has been reported.

7.3. Eugenol

Eugenol is another aromatic compound found in coffee (Fig. 5), with a human exposure from the total diet estimated at 0·6 mg per capita per day.

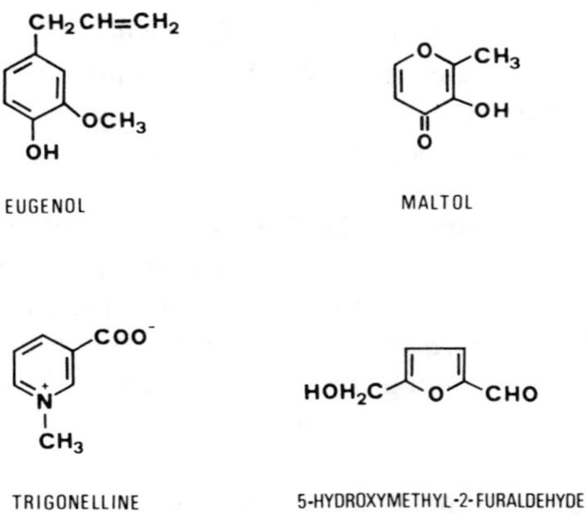

Fig. 5. Structures of typical aromatic compounds found in coffee (maltol, 5-hydroxymethyl-2-furaldehyde, trigonelline and eugenol).

Oral administration to rats of a high dose of eugenol increased the specific activity of liver microsomal uridine diphospho-D-glucuronyltransferase. This observation indicates a rapid formation and excretion of a glucuronide of eugenol, shown to be one of the most effective glucosidurogenic foreign compounds.[53] The metabolism of ^{14}C-eugenol has been reported in rats at a range of doses close to human dietary exposure and also at doses similar to those used in toxicology. Independent of the dose size, faecal excretion amounted to only 10% of the dose, whereas 75–80% was recovered in urine. Half of the urinary metabolites were conjugated eugenols. Eugenol sulphate predominates at low dose with the glucuronide as the major metabolite at higher doses. Sulphate and glucuronide conjugates of 4-hydroxy-3-methoxypropylbenzene and 3,4-dihydroxypropylbenzene were also found whereas 10% of the dose corresponded to unidentified metabolites.[54] A detailed paper dealing with the metabolism of eugenol methyl ether and isoeugenol methyl ether has been published by Solheim and Scheline[55] in which more than 20 metabolites were identified.

7.4. Guaiacol

Guaiacol has been used to esterify drugs such as ibuprofen, improving its gastrointestinal tolerance. The time course of plasma concentrations of guaiacol in man showed a peak 2 h after oral administration, followed by a plateau explained by delayed absorption.[56]

Guaiacol glyceryl ether metabolism has been studied in man because this product has been used as an expectorant, and β-(2-methoxyphenoxy)lactic acid has been identified as a major metabolite.[57]

8. HETEROCYCLIC COMPOUNDS

Heterocyclic compounds are present in trace amounts as volatile constituents of the aroma (see Volume 1, Chapter 7). Some of the most important are furfuryl alcohol, furan carboxylic acid, maltol, 5-hydroxymethyl-2-furaldehyde and furaneol.

8.1. Maltol

The metabolism of maltol (Fig. 5), which has been marketed as a food flavour enhancing agent, has been studied in dogs after intravenous administration of a single dose of $10\,mg\,kg^{-1}$. These experiments showed that maltol is rapidly and extensively metabolised and excreted as sulphate and glucuronide conjugates. The imprecision of the methods used was suggested as explaining why only 58% of the dose was recovered in urine over the 3 days following the administration, and no metabolite was found in the faeces.[58]

8.2. 2-Methylfuran

The metabolic activation of 2-methylfuran by the hepatic and pulmonary microsomal system to 4-oxo-2-pentenal which binds covalently to proteins has been studied *in vitro*. *In vivo*, i.p. injection of doses from 50 to $200\,mg\,kg^{-1}$ into rats resulted in a higher covalent binding to both protein and DNA in the liver and the kidney, the maximal binding being observed 4 h after the administration. Conjugation reactions led to depressed levels of liver glutathione. The mediation of reactive metabolites of 2-methylfuran in the covalent binding reactions was assessed by an increase of the binding on phenobarbital pretreatment of rats.[59]

8.3. 5-Hydroxymethyl-2-furaldehyde

Hydroxymethylfuraldehydes are produced during advanced Maillard reactions through the dehydration of 3-deoxyhexosones. 5-Hydroxymethyl-2-furaldehyde (Fig. 5) has been described as a reactive substance, and this view is supported by metabolic experiments in humans, where only half of the dose administered parenterally was recovered as 5-hydroxymethyl-2-furoic acid and furan-2,5-dicarboxylic acid.[60] The synthesis of uniformly ^{14}C-labelled hydroxymethylfuraldehyde[61] and its subsequent use in metabolic experiments in the rat showed a rapid elimination of radioactivity in the urine with a recovery of 85% after 8 h and 95–100% 24 h after the oral administration. Hydroxymethylfuraldehyde was completely converted into 5-hydroxymethyl-2-furoic acid and N-(5-hydroxymethyl-2-fureoyl)glycine. After 24 h, no accumulation of radioactivity could be seen in the body of the animal.[62]

8.4. Maillard Reaction Products

Melanoidins, the products of the Maillard reaction giving the characteristic colour of the coffee extract, correspond to many identified and unknown compounds with complex chemical structures. Only the metabolism of model compounds has been investigated in the rat.

Tritium-labelled 'early' and 'advanced' Maillard reaction products of casein and glucose were given in a single meal to rats. Faecal excretion of the radioactivity of untreated casein compared with 'early' and 'advanced' Maillard casein amounted to 4%, 14% and 40–50% of the ingested dose, respectively, whereas cumulated urinary excretion corresponded to 10%, 15% and 22% on the third day after the ingestion. Thus, total excretion was the most important for 'advanced' Maillard casein (60% of the dose), lower for 'early' Maillard casein (35%) and less than 15% for untreated casein. Urine analysis showed the presence of ε-deoxyfructosyllysine, but the other metabolites were not studied and identified. Premelanoidins prepared from protein–glucose interaction were mainly excreted in faeces (60–70%), and 23–30% of the dose was found in urine. Melanoidins were prepared from glycine and labelled glucose and isolated by chromatography. After oral administration, 90–95% of this melanoidin fraction was excreted in faeces, and only 1% was recovered in expired CO_2, in urine and also in the carcase of the animal, 24 h after the ingestion.[63] The melanoidins present in coffee must have a similar metabolic fate. All these experiments demonstrated that these Maillard reaction products were partially absorbed by the intestines depending on the treatment, but they did not yield information on the metabolic reactions involved in their elimination.

8.5. Pyridines

Nitrogen and sulphur heterocyclics formed during roasting are pyrroles, pyrazines, pyridines, quinoxalines, oxazoles, thiazoles, thiophenes and dithiolanes.[1] N-Methylation of pyridine has been shown by several investigators, along with ring hydroxylation followed by glucuronic acid conjugation. The metabolic N-oxidation of pyridine and 3-methylpyridine has been studied after intraperitoneal administration in various animal species.[64] Pyridine-N-oxide was a quantitatively important metabolite and ranged from 10% in rats to about 40% in mice and guinea pigs. In the case of 3-methylpyridine, 3-methylpyridine-N-oxide accounted for less than 7% of the dose.

8.6. Other Heterocyclic Compounds

In instant coffee, products such as furfurylmercaptan, furfurylmethylsulphide and furfurylmethyldisulphide are present at the ppm level. Kahweofuran (Fig. 3) is a heterocyclic compound present at the same level and is characteristic of coffee. The list of heterocyclic compounds is increasing continually as these compounds, which are responsible for flavour and stale odour, are being studied intensively.

9. ALKALOIDS

Caffeine, the main physiologically active component of coffee, is quantitatively the most important alkaloid. The caffeine content of roasted beans amounts to *c.* 1% in arabica coffee and *c.* 2% in robusta coffee. In instant coffee, it varies from 2 to 5% according to the blend and yield, giving an absolute content of 40–180 mg per cup of coffee. The other methylxanthines, theobromine and theophylline, are present in low concentration in roasted coffee.

9.1. Trigonelline

Trigonelline (Fig. 5) is present in green coffee at a concentration of *c.* 1%. Although it is rapidly decomposed on roasting, a concentration of 0·1% is still present in a high-roast coffee, along with its breakdown products, nicotinic acid, methyl nicotinate, pyridine and 3-methylpyridine.[1] The metabolism of pyridine and 3-methylpyridine has been discussed previously. Since the beginning of this century, the metabolism of nicotinic acid in mammals has been investigated by many research groups. These studies have led to the discovery of the following metabolites: nicotinuric acid, nicotinamide, trigonelline, N-methylnicotinamide, N-methyl-2-pyridone-5-carboxamide, N-methyl-4-pyridone-3-carboxamide, nicotinamide-N-oxide, 6-hydroxynicotinic acid and 6-hydroxynicotinamide, in addition to the unchanged product.[65–71] Trigonelline is thus a metabolite of nicotinic acid, and with the labelled product it has been shown that approximately 3–5% of nicotinic acid is converted in the rat into trigonelline.[72]

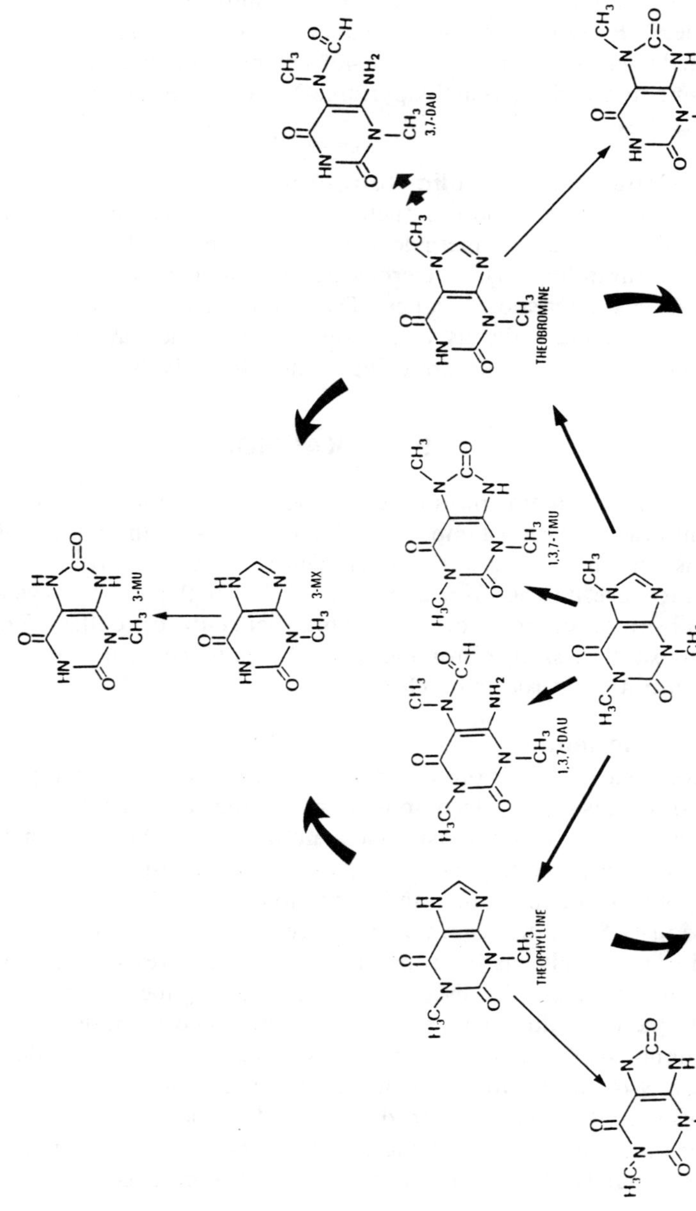
3,7-DAU
3,7-DMU
THEOBROMINE
1,3,7-TMU
3-MU
3-MX
CAFFEINE
1,3,7-DAU
THEOPHYLLINE
1,3-DMU

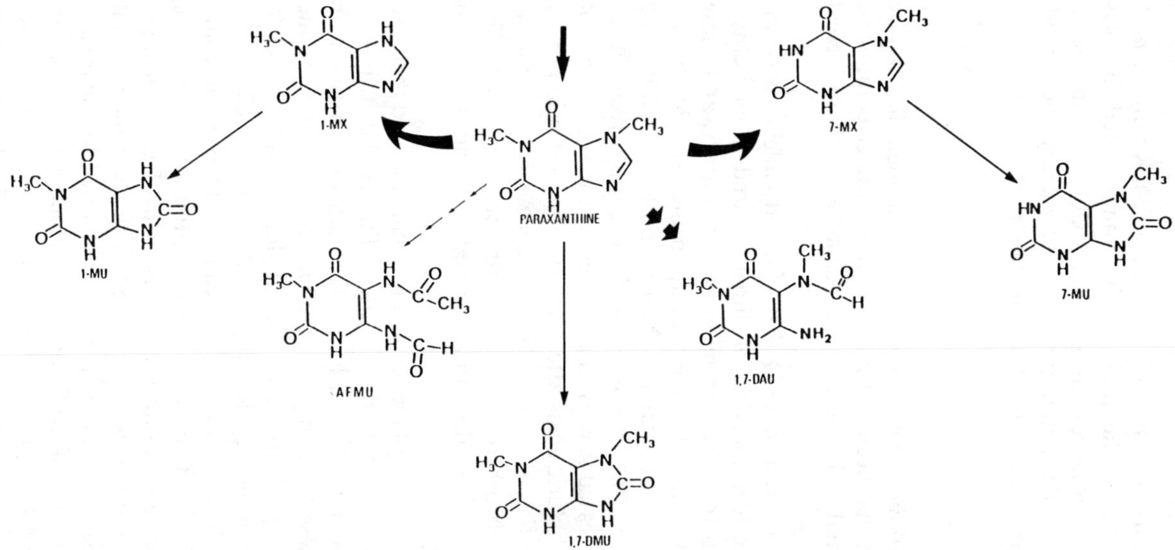

Fig. 6. Metabolic pathways of caffeine in man (AFMU: 5-acetylamino-6-formylamino-3-methyluracil; DAU: uracil derivatives; DMU: dimethyluric acid; MU: monomethyluric acid; MX: methylxanthine; TMU: trimethyluric acid).

In 1942, Handler and Dann[73] administered 80 mg trigonelline per day in the diet of rats, obtaining 56–86% recoveries of the unchanged product in the urine. More recently, using [Me-^{14}C]trigonelline, McKennis *et al.*[74] demonstrated that 92–97% of the radioactivity administered i.p. was recovered in the urine, 1% in faeces and 0·1% in $^{14}CO_2$. In conclusion, trigonelline seems to be essentially eliminated in urine, mainly unchanged, but its possible transformation into other metabolites has not been investigated in detail.

9.2. Caffeine

Since its discovery in 1820 by Runge[75] and its chemical identification by Medicus in 1875,[76] caffeine has been and still is the most studied compound present in coffee, presumably because of its remarkable physiological properties.

It had already been observed in 1850 that ingested caffeine was not recovered in urine.[77] The identification of dimethyl- and monomethylxanthines was reported before the beginning of this century, and these studies have recently been reviewed.[78] The formation and excretion of uric acid derivatives was demonstrated later by Myers and Hanzal, following extensive studies between 1929[79] and 1946.[80] The improvement of chromatographic techniques led to recoveries in human urine of 65–75% of methylxanthines and methyluric acids after the oral administration of 1 g of caffeine.[81] These studies suggested that caffeine on ingestion is completely and rapidly absorbed from the gastrointestinal tract. The complete bioavailability of orally administered caffeine has been demonstrated recently.[82]

The synthesis of methyl- or ring-labelled forms of caffeine, and their use in metabolic experiments in animals, demonstrated the rapid distribution of caffeine throughout the body[83] and led to the discovery of two new unknown metabolites. These two compounds did not have an intact purine ring and were subsequently identified as trimethylallantoin[84,85] and 6-amino-5[N-methylformylamino]1,3-dimethyluracil.[86] Five uracil derivatives have now been identified, produced from caffeine, theobromine and paraxanthine[87–89] (see Fig. 6). Except for the formation of 1-methyluric acid from 1-methylxanthine where xanthine oxidase is involved,[90] all of the other metabolic transformations were observed with liver microsomal enzymes. Although almost all of the metabolites recovered in urine have been identified, the mechanisms of the formation of uracil derivatives,[91] in particular 5-acetylamino-6-formylamino-3-methyluracil,[88] are not yet understood. Another interesting pathway was reported in the premature infant, where theophylline, a metabolite of caffeine, was shown to be methylated back to caffeine.[92] This methylation process has also been suggested from pharmacokinetics data to occur in adult subjects,[93] and was recently confirmed using labelled theophylline.[94] In the animal model, sulphur-containing metabolites of caffeine[95] could be metabolites initially present in bile, subjected to enterohepatic recycling and transformed by the intestinal microflora. Until recently, it was believed that caffeine metabolites in human and animal species were similar, and that differences were only quantitative. However, with the discovery of 5-acetylamino-6-formylamino-3-methyluracil, observed until now only in human subjects,[78] and more recently of paraxanthine glucuronide,[96] shown to be the most important urinary metabolite of caffeine in the NMRI mouse,[97] species-specific metabolic pathways were demonstrated.

In clinical research, a cup of coffee or a similar dose of caffeine is now used more and more as a safe and non-invasive test to detect liver disease[98] and to characterise acetylator phenotype.[99] More sophisticated breath tests using either stable or radioactive caffeine have been applied to human subjects showing, in addition to liver disease,[100] metabolic induction due to smoking.[101]

10. CONCLUSIONS

New techniques involving synthetic compounds labelled with radioactive or stable isotopes have made possible the study of the distribution (by whole animal body autoradiography), the identification (by radiochromatography and mass spectrometry) of the metabolites and accurate metabolic balance of the isotope. We may therefore in the future see a further increase in our knowledge of the metabolism of coffee constituents, both in animals and in humans.

For caffeine, already a well-worked field, progress may be limited, but marked advances should be seen in the metabolism of the myriad of other compounds and compound classes, not only present in coffee but also often found in other plants and foods.

ACKNOWLEDGEMENTS

The author thanks Jacqueline Jaquier for the review of the literature, Dr Ian Horman for correcting the manuscript, and Catherine Isom for manuscript preparation.

REFERENCES

1. Spiller, M. A., in *The Methylxanthine Beverages and Foods: Chemistry, Consumption and Health Effects, Prog. Clin. Biol. Res.*, 1984, **158**, 91–147.
2. Gillies, M. E. and Birkbeck, J. A., *Am. J. Clin. Nutr.*, 1983, **38**, 936–942.
3. Derman, D., Sayers, M., Lynch, S. R., Charlton, R. W., Bothwell, T. H. and Mayet, F., *Br. J. Nutr.*, 1977, **38**, 261–269.
4. Layrisse, M., Martinez-Torres, C., Renzi, M., Velez, F. and Gonzalez, M., *Am. J. Clin. Nutr.*, 1976, **29**, 8–18.
5. Morck, T. A., Lynch, S. R. and Cook, J. D., *Am. J. Clin. Nutr.*, 1983, **37**, 416–420.
6. Kojima, N., Wallace, D. and Bates, G. W., *Am. J. Clin. Nutr.*, 1981, **34**, 1392–1401.
7. Layrisse, M., Martinez-Torres, C. and Renzi, M., *Am. J. Clin. Nutr.*, 1976, **29**, 274–279.
8. Clifford, M. N., in *Coffee: Botany, Biochemistry and Production of Beans and Beverage*, Eds M. N. Clifford and K. C. Willson, Croom Helm, London, 1985, 305–374.
9. Trugo, L. C. and Macrae, R., in *10th Int. Coll. ASIC*, 1982, 187–192.
10. Pictet, G. A., in *7th Int. Coll. ASIC*, 1975, 189–200.
11. Williams, C. A. and Macdonald, I., *Wld Rev. Nutr. Diet*, 1982, **39**, 23–52.
12. Jeanes, A. and Hodge, J. (Eds), *Physiological Effects of Food Carbohydrates*, ACS Symposium Series 15, Washington, DC, 1975.
13. Harland, B. F. and Oberleas, D., *J. Food Sci.*, 1985, **50**, 832–833, 842.
14. Wise, A. and Gilburt, D. J., *Appl. Environ. Microbiol.*, 1982, **43**, 753–756.
15. Nahapetian, A. and Young, V., *J. Nutr.*, 1980, **110**, 1458–1472.
16. Bhattacharyya, A. K. and Lopez, L. A., *Biochim. Biophys. Acta*, 1979, **574**, 146–153.
17. Vahouny, G. V., Connor, W. E., Subramaniam, S., Lin, D. S. and Gallo, L. L., *Am. J. Clin. Nutr.*, 1983, **37**, 805–809.
18. Newton, D. F. and Mansbach, C. M., *Clin. Chim. Acta*, 1978, **89**, 331–339.
19. Connor, W. E., *J. Am. Diet. Assoc.*, 1968, **52**, 202–208.
20. Subbiah, M. T. R., *Am. J. Clin. Nutr.*, 1973, **26**, 219–225.
21. Igimi, H., Nishimura, M., Kodama, R. and Ide, H., *Xenobiotica*, 1974, **4**, 77–84.
22. Kodama, R., Yano, T., Furukawa, K., Noda, K. and Ide, H., *Xenobiotica*, 1976, **6**, 377–389.
23. Watabe, T., Hiratsuka, A., Ozawa, N. and Isobe, M., *Xenobiotica*, 1981, **11**, 333–344.
24. Ishida, T., Asakawa, Y., Takemoto, T. and Aratani, T., *J. Pharm. Sci.*, 1981, **70**, 406–415.
25. Stillwell, W. G., Bouwsma, O. J., Thenot, J.-P., Horning, M. G., Griffin, G. W., Ishikawa, K. and Takaku, M., *Res. Commun. Chem. Pathol. Pharmacol.*, 1978, **20**, 509–530.
26. Bakke, J., Struble, C., Gustafsson, J.-A. and Gustafsson, B., *Proc. Natl Acad. Sci. USA*, 1985, **82**, 668–671.
27. O'Brien, K. A. F., Smith, L. L. and Cohen, G. M., *Chem.–Biol. Interactions*, 1985, **55**, 109–122.

28. Tada, M., Takahashi, K., Kawazoe, Y. and Ito, N., *Chem.–Biol. Interactions*, 1980, **29**, 257–266.
29. Tada, M., Takahashi, K. and Kawazoe, Y., *Chem. Pharm. Bull.*, 1982, **30**, 3834–3837.
30. Hollstein, M., Talcott, R. and Wei, E., *J. Natl Cancer Inst.*, 1978, **60**, 405–410.
31. Obermann, H., Spiteller, G. and Hoyer, G. A., *Chem. Ber.*, 1973, **106**, 3506–3518.
32. Ludwig, H., Obermann, H. and Spiteller, G., *Chem. Ber.*, 1974, **107**, 2409–2411.
33. Fujita, Y., Wakabayashi, K., Nagao, M. and Sugimura, T., *Mutat. Res.*, 1985, **144**, 227–230.
34. Neely, W. B., *Biochem. Pharmacol.*, 1964, **13**, 1137–1142.
35. Hucker, H. B., Miller, J. K., Hochberg, A., Brobyn, R. D., Riordan, F. H. and Calesnick, B., *J. Pharm. Exp. Ther.*, 1967, **155**, 309–317.
36. Bock, K. W., Clausbruch, U. G. von and Winne, D., *Med. Biol.*, 1979, **57**, 262–264.
37. Trugo, L. C. and Macrae, R., *Analyst*, 1984, **109**, 263–266.
38. Czok, G., Walter, W., Knoche, K. and Degener, H., *Z. Ernährungswiss.*, 1974, **13**(3), 108–112.
39. Booth, A. N., Emerson, O. H., Jones, F. T. and Deeds, F., *J. Biol. Chem.*, 1957, **229**, 51–59.
40. Teuchy, H. and Van Sumere, C. F., *Soc. Belge Biochimie*, 1968, 953–954.
41. Teuchy, H. and Van Sumere, C. F., *Arch. Int. Physiol. Biochim.*, 1971, **79**, 589–618.
42. Westendorf, J. and Czok, G., *Z. Ernährungswiss.*, 1983, **22**, 255–270.
43. Shaw, K. N. F., Gutenstein, M. and Jepson, J. B., in *Proc. 7th Int. Cong. Biochem.*, Moscow, 1961, ed. N. M. Sissakian, Pergamon Press, Oxford, 1963, **9**, 427.
44. Scheline, R. R. and Midtvedt, T., *Experientia*, 1970, **26**(10), 1068–1069.
45. Peppercorn, M. A. and Goldman, P., *Proc. Natl Acad. Sci. USA*, 1972, **69**(6), 1413–1415.
46. Wattenberg, L. W., Coccia, J. B. and Lam, L. K., *Cancer Res.*, 1980, **40**, 2820–2823.
47. Jacobson, E. A., Newmark, H., Baptista, J. and Bruce, W. R., *Nutr. Rep. Int.*, 1983, **28**(6), 1409–1417.
48. Lautemann, E., *Justus Liebigs Ann. Chem.*, 1863, **125**, 9–13.
49. Cotran, R., Kendrick, M. I. and Kass, E. H., *Proc. Soc. Exp. Biol. Med.*, 1960, **104**, 424–426.
50. Adamson, R. H., Bridges, J. W., Evans, M. E. and Williams, R. T., *Biochem. J.*, 1970, **116**, 437–443.
51. Asatoor, A. M., *Biochem. Biophys. Acta*, 1965, **100**, 290–292.
52. Bohm, B. A., *Chem. Rev.*, 1965, **65**, 435–466.
53. Yuasa, A., *Jap. J. Vet. Sci.*, 1974, **36**, 427–432.
54. Sutton, J. D., Sangster, S. A. and Caldwell, J., *Biochem. Pharmacol.*, 1985, **34**, 465–466.
55. Solheim, E. and Scheline, R. R., *Xenobiotica*, 1976, **6**, 137–150.
56. Catanese, B., Barillari, G., Iorio, E. and Pini, L., *Boll. Chim. Farm.*, 1982, **121**, 567–572.

57. Vandenheuvel, W. J. A., Smith, J. L. and Silber, R. H., *J. Pharm. Sci.*, 1972, **61**, 1997–1998.
58. Rennhard, H. H., *J. Agric. Food Chem.*, 1971, **19**, 152–154.
59. Ravindranath, V., McMenamin, M. G., Dees, J. H. and Boyd, M. R., *Toxicol. Appl. Pharmacol.*, 1986, **85**, 78–91.
60. Jellum, E., Borresen, H. C. and Eldjarn, L., *Clin. Chim. Acta*, 1973, **47**, 191–201.
61. Germond, J. E. and Arnaud, M. J., *J. Lab. Comp. Radiopharm.*, 1987, **24**, 343–348.
62. Germond, J. E., Philippossian, G., Richli, U., Bracco, I. and Arnaud, M. J., *J. Toxicol. Environ. Health*, 1987, **22**, 79–89.
63. Finot, P. A. and Magnenat, E., *Prog. Food Nutr. Sci.*, 1981, **5**, 193–207.
64. Gorrod, J. W. and Damani, L. A., *Eur. J. Drug Met. Pharmacokinetics*, 1980, **5**, 53–57.
65. Komori, Y. and Sendju, Y., *J. Biochem., Tokyo*, 1926, **6**, 163–169.
66. Ackermann, D., *Z. Biol.*, 1913, **59**, 17–22.
67. Huff, J. W. and Perlzweig, W. A., *J. Biol. Chem.*, 1943, **150**, 395–400.
68. Knox, W. E. and Grossmann, W. I., *J. Biol. Chem.*, 1946, **166**, 391–392.
69. Wuchang, M. L. and Johnson, B. C., *J. Biol. Chem.*, 1959, **234**, 1817–1821.
70. Bonavita, V., Narrd, S. A. and Kaplan, N. O., *J. Biol. Chem.*, 1961, **236**, 936–939.
71. Lee, Y. U., Gholson, R. K. and Raica, N., *J. Biol. Chem.*, 1956, **244**, 3277–3282.
72. Mason, J. B. and Kodicek, E., *Biochem. J.*, 1970, **120**, 515–521.
73. Handler, P. and Dann, W. J., *J. Biol. Chem.*, 1942, **146**, 357–368.
74. McKennis, H., Bowman, E. R., Horvath, A. and Bederka, J. P., *Nature (London)*, 1964, **202**, 699–700.
75. Runge, F., in *Neueste Phytochemische Entdeckungen*, G. Reiners, Berlin, 1820, 204.
76. Medicus, L., *Liebigs. Ann. Chem.*, 1875, **175**, 230–251.
77. Lehmann, C. G., *Lehrbuch der physiologischen Chemie*, 2nd edn, Leipzig, 1850, 367.
78. Arnaud, M. J., in *Caffeine*, Ed. P. B. Dews, Springer-Verlag, Berlin, 1984, 3–38.
79. Myers, V. C. and Hanzal, R. F., *Am. J. Physiol.*, 1929, **90**, 458–459.
80. Myers, V. C. and Hanzal, R. F., *J. Biol. Chem.*, 1946, **162**, 309–323.
81. Cornish, H. H. and Christman, A. A., *J. Biol. Chem.*, 1957, **228**, 315–323.
82. Blanchard, J. and Sawers, S. J. A., *Eur. J. Clin. Pharmacol.*, 1983, **24**, 93–98.
83. Burg, A. W. and Werner, E., *Biochem. Pharmacol.*, 1972, **21**, 923–936.
84. Rao, G. S., Khanna, K. L. and Cornish, H. H., *Experientia*, 1973, **19**, 953–955.
85. Arnaud, M. J., Ben Zvi, Z., Yaari, A. and Gorodischer, R., *Res. Commun. Chem. Pathol. Pharmacol.*, 1986, **52**, 407–410.
86. Arnaud, M. J. and Welsch, C., *J. Agric. Food Chem.*, 1979, **27**, 524–527.
87. Callahan, M. M., Robertson, R. S., Arnaud, M. J., Branfman, A. R., McComish, M. F. and Yesair, D. W., *Drug Metab. Dispos.*, 1982, **10**, 417–423.
88. Tang, B. K., Grant, D. M. and Kalow, W., *Drug. Metab. Dispos.*, 1983, **11**, 218–220.
89. Arnaud, M. J. and Welsch, C., *Experientia*, 1979, **35**, 34.
90. Birkett, D. J. and Grygiel, J. J., in *Drug Metabolism in the Immature Human*, Eds L. F. Soyka and G. P. Redmond, Raven Press, New York, 1981, 229–239.
91. Ferrero, J. L. and Neims, A. H., *Life Sci.*, 1983, **33**, 1173–1178.

92. Bory, C., Baltassat, P., Porthault, M., Bethenod, M., Frederich, A. and Aranda, J. V., *Lancet*, 1978, 1204–1205.
93. Tang-Liu, D. D.-S. and Riegelman, S., *Res. Commun. Chem. Pathol. Pharmacol.*, 1981, **34**, 371–380.
94. Caldwell, J. and Arnaud, M. J., unpublished observations.
95. Kamei, K., Matsuda, M. and Momose, A., *Chem. Pharm. Bull. Tokyo*, 1975, **23**, 683–685.
96. Arnaud, M. J., Richli, U. and Philippossian, G., *Experientia*, 1986, **42**, 696.
97. Arnaud, M. J., *Drug Metab. Dispos.*, 1985, **13**, 471–478.
98. Jost, G., Wahlländer, A., Mandach, U. v. and Preisig, R., *Hepathology*, 1987, **7**, 338–344.
99. Grant, D. M., Tang, B. K. and Kalow, W., *Clin. Pharmacol. Ther.*, 1983, **33**, 591–602.
100. Renner, E., Wietholtz, H., Huguenin, P., Arnaud, M. J. and Preisig, R., *Hepathology*, 1984, **4**, 38–46.
101. Wietholtz, H., Voegelin, M., Arnaud, M. J., Bircher, J. and Preisig, R., *Eur. J. Clin. Pharmacol.*, 1981, **21**, 53–59.
102. Egestad, B., Pettersson, P., Sjövall, J., Rafter, J., Hyvönen, K. and Gustafsson, J. A., *Biomed. Chromatogr.*, 1987, **2**, 120–134.

Chapter 3

Epidemiological Studies on the Effects of Coffee Drinking

S. HEYDEN
Duke University Medical Center, Durham, North Carolina, USA

1. GENERAL INTRODUCTION

Coffee and caffeine have always maintained a position high on the list for suspicion of causing ill health, ranging from myocardial infarction, arrhythmias, hypertension, gout and anxiety to fibrocystic breast disease, various cancers and birth defects, osteoporosis and hyperlipidaemia. There is no other known agent in man's environment which has been as frequently associated with such a variety of chronic degenerative diseases as coffee.

Acute administration of caffeine in volunteers has demonstrated temporary increases in blood pressure, serum catecholamine levels, plasma renin activity, serum free fatty acid levels, urine production and gastric acid secretion.[1] Chronic caffeine consumption lacks these physiological responses. Older caffeine users showed no increase in blood pressure or heart rate,[2] and 'continuous heavy coffee ingestion does not involve a risk of the development of hypertension',[3] Chronic caffeine consumption has no effect on plasma catecholamine levels,[3] plasma renin activity, blood cholesterol or glucose levels, or urine production.[1]

The more recent literature appears to negate most of the assertions previously alluded to, notably in regard to myocardial infarction,[1] kidney and bladder cancers,[1] pancreatic cancer,[4] anxiety,[5] fibrocystic breast disease,[1,6] maternal coffee consumption and pregnancy outcome,[1,7,8] and hyperlipidaemia.[9] These assertions are illustrated in Fig. 1. In fact, a growing number of positive effects of coffee are being reported. Among others, coffee is shown to be an effective bronchodilator in young patients with asthma,[10] a good source of potassium[11] and a booster of pain-free walking time for patients with chronic stable angina.[12,13] Introduction of either one or two cups of coffee increased the exercise duration until onset of angina by 8% and 12%, respectively, whereas decaffeinated coffee had no effect. Caffeine 'may be effective as a topical treatment of atopic dermatitis and as systemic therapy for neonatal apnea',[1] and was recently

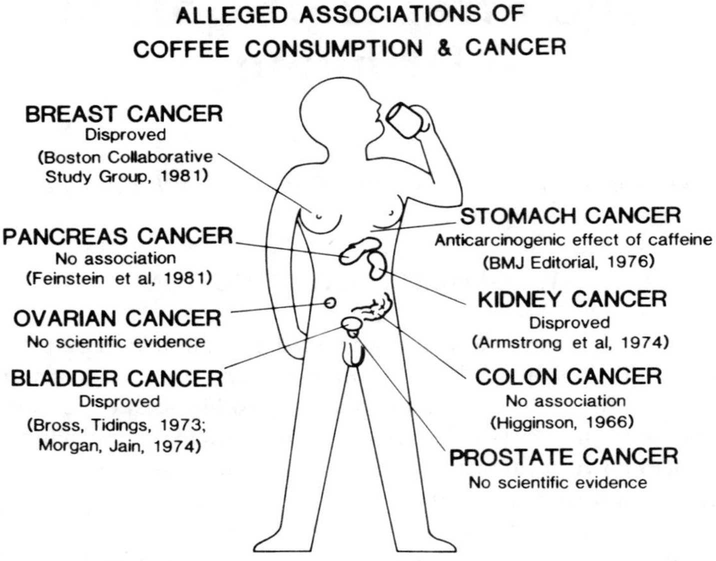

Fig. 1. Alleged associations of coffee consumption and cancer.

praised as 'an analgesic adjuvant'.[14] The positive effect of coffee in the prevention of postprandial hypotension, particularly after breakfast and mainly in the elderly, has been shown experimentally.[15] Patients with 'autonomic failure are advised to drink two cups of coffee (approximately 200–250 mg of caffeine)'. The only remaining uncertainty is the possibility of inducement of arrhythmias in susceptible patients with either two cups of coffee or their equivalent as caffeine.[16] Even this finding has been challenged: 'What is not yet appreciated is that ventricular premature beats are innocuous in the overwhelming majority of persons. They no more augur against sudden death than a sneeze portends pneumonia.'[17]

2. STUDY OF COFFEE CONSUMPTION AND MORTALITY RATES AMONG 10 000 HYPERTENSIVES*

2.1. Introduction

Because of the vast medical literature which produced a negative image for the effect of coffee on the health of regular coffee consumers, and the obvious contradictions created by the more recent publications, a large prospective study is needed to investigate the effect of coffee and caffeine on the hardest end-point of any study, mortality from all causes. Several epidemiological studies have already shown an absence of any effect of coffee consumption on mortality in total communities,[18–20] but a representative prospective study of an important segment of the general population—hypertensive men and women—has never been undertaken. As many as 60 million Americans may have 'mild' to moderate or severe degrees of hypertension. Caffeine is a widely used stimulant, which assumes a major position in research aimed at elucidating the effects it may have on the health and disease of hypertensives.

This section will focus on the mortality data from the largest American hypertension intervention trial, the Hypertension Detection and Follow-up Program,[21] in which one of us (SH) participated as Co-Principal Investigator from 1973 to 1979. The purpose of this report is to examine mortality rates over 4 years from all causes as well as specific causes among more than 10 000 hypertensive patients, and to relate these rates to coffee consumption during the lifetime. If this study in hypertensive patients were to reveal a higher all-cause as well as cardiovascular and cerebrovascular death rate among heavy or moderate coffee drinkers compared with non-users of caffeine, one would have an important argument for limiting coffee consumption among hypertensive patients.

2.2. Methodology

The Hypertension Detection and Follow-up Program (HDFP) was a community-based, collaborative trial of antihypertensive treatment over a 5-year period (1973–79).[21] Participants between the ages of 30 and 69 years were recruited from 14 population groups across the United States. A total of 10 064 individuals with diastolic blood pressure of 90 mm Hg or above were enrolled, after appropriate screening at home and clinics. The type of treatment for hypertension is irrelevant as far as this report is concerned.

At the first home visit, at the end of year 1, the HDFP participants were

* Prepared in conjunction with Dr J. B. Martin, University of Texas.

classified on the basis of their caffeine intake from both beverages and medication according to answers given to standardised questions. Caffeine consumption data were collected at the year 2 and year 5 follow-up visits using a slightly modified questionnaire. Modifications included a distinction between regular coffee and decaffeinated coffee, the addition of a cola consumption question, and a deletion of the question on medication. These data were used to evaluate the validity of the baseline caffeine consumption data.

The total daily milligrams of caffeine consumed for all selected sources were converted (by JBM) to the equivalent of cups of coffee consumed per day. The conversion averaged the daily caffeine consumption at 317 mg (275 mg from coffee, 26 mg from tea and 16 mg from medication). The 10 064 men and women were stratified into four caffeine consumption groups (by JBM; see Table 1).

Among the potentially confounding factors, age, race, sex and smoking habits were found to be associated with both caffeine consumption and the outcome variable death. The analyses of the data were stratified (by JBM) according to age, sex and race categories to control for confounding. Smoking status classified the cohort into two groups: current smokers (37%) and non-smokers. Ex-smokers (2%) were combined with persons reporting never having smoked (61%). For the 592 cases of death which occurred between year 1 and year 5, death certificates were coded by a nosologist's utilisation of the Eighth Revision of the International Classification of Diseases Adapted (ICDA) Code.

Table 1
Equivalency of caffeine consumption per day and distribution of study cohort by level of caffeine consumption[a]

Equivalency	*Level of caffeine consumption*			
	None	*Low*	*Medium*	*High*
Total caffeine consumed per day[b] (mg)	0	0·1–214	214–428	428
Number of cups of coffee consumed per day equivalent	0	0·1–2	2·1–4	>4
Distribution				
Participants (*n*)	1 425	4 164	2 650	1 825
Distribution (%)	14	41	26	18

[a] Level of caffeine consumption was fairly stable over the 4-year study period.
[b] Beverages and medication caffeine sources.

Rate ratios were calculated (by JBM) in order to compare mortality rates within the study population by level of caffeine intake. Rate ratios were summarised over sex, race and 10-year age bands. Standardised mortality ratios (SMR) were calculated to compare mortality rates of hypertensives by level of caffeine consumption with the mortality rates of the general population. Death rates for the United States in 1975 were used as the standard population rates.

2.3. Results

Table 2 demonstrates no association between caffeine consumption and all-cause mortality. Using the non-caffeine consumption group as the referent group, the rate ratio for all-cause mortality was 0·82 for low-, 0·82 for medium- and 0·90 for high-caffeine consumption group.

There was no association for the three cause-specific mortality groupings and caffeine consumption. It should be noted that, in the category of cerebrovascular deaths, neither the apparently low rate of 0·61 for medium- nor the slightly increased rate of 1·30 for high-caffeine consumer groups are statistically significantly different from the no-caffeine consumer group.

In an attempt to investigate cancer mortality in more detail, the 102 site-specific cancer deaths which had occurred over the four-year observation period were divided in the four caffeine consumption groups. Analyses by 12 different cancer sites showed no accumulation of cases at any one site, and cancer mortality was distributed uniformly in all four groups with an incidence varying between 0·8% in medium-caffeine consumers, 1·1% in the low-caffeine consumption group and 1% in the abstainers.

The two habits, cigarette smoking and coffee drinking, were strongly associated in this study population. The proportion of current smokers increased with the four levels of caffeine consumption.

Table 2

Mortality rates over 4-year follow-up by level of caffeine intake (rate ratios[a] totalled across subgroups; 95% confidence interval)

Mortality (n = *592*)	*Caffeine consumption group*			
	None	*Low*	*Medium*	*High*
All-cause	1·00	0·82	0·82	0·90
Cerebrovascular	1·00	0·73	0·61	1·30
All other cardiovascular	1·00	0·93	0·81	0·80
Cancer	1·00	0·98	0·95	1·05

[a] Standardised by race and sex, and across five 10-year age groups.

As expected, smokers had a significantly higher mortality than non-smokers. However, the mortality from cerebrovascular diseases, other cardiovascular diseases (mainly coronary heart disease) and cancer was not increased among smokers with medium or heavy intake of caffeine as compared with smokers with no-caffeine or low-caffeine consumption.

2.4. Discussion

The most important result of this study was the lack of association between caffeine consumption and mortality among hypertensive patients. The major strength of this study rests with the uniquely comprehensive follow-up of deaths (99·9%) that HDFP was able to achieve. The consistency of the findings in all-cause and cause-specific mortality including cancer corroborates evidence provided by prospective epidemiological studies with the same results, as far as all-cause mortality is concerned, in Framingham,[20] in Evans County[19] and in Gothenburg, Sweden.[22,23] However, in studying cancer mortality and habitual coffee intake, the present report represents probably the first prospective study following such a large number of people between 30 and 69 years of age for four years. In the literature, one frequently finds reports linking coffee consumption with a variety of cancer sites, beginning with stomach, colon and prostatic cancer, followed by kidney and bladder cancer, up to the more recent past alleging an association with pancreatic, ovarian and breast cancer (Fig. 1). In view of the present report, it is reassuring that several cancer sites have already been eliminated in later epidemiological studies as well as in several experimental observations in animal models.

There was never any real concern among scientists about coffee consumption in normotensive persons, but this report for the first time addressed the age-old notion shared by some health officials and paramedical personnel that coffee may increase blood pressure and that, consequently, hypertensives should drink less coffee or even abstain from it. 'It now seems quite certain that although the acute ingestion of caffeine alters many haemodynamic variables, the chronic ingestion of caffeine has little or no effect on blood pressure, heart rate, plasma catecholamine levels and plasma renin activity in normal subjects.'[1] Obviously, in the HDFP population we are dealing with habitual caffeine consumers, and 'it seems unlikely that the habitual consumption of caffeine would significantly alter cardiac output or stroke volume'.[1] The tolerance to caffeine's effect on haemodynamic variables has frequently been overlooked in previous short-term experiments with caffeine-naïve volunteers.

The level of caffeine consumption by the study cohort was fairly stable throughout the four years. The average caffeine consumption from cola was 19 mg per day and remained unchanged from the 2- to 5-year follow-up home visits. If an adjustment was made to the year 1 data for a possible inclusion of decaffeinated coffee, the average daily caffeine consumption from coffee decreased slightly from 228 mg in year 1 to 201 mg in year 5 and from tea decreased from 26 mg in year 1 to 24 mg in year 5.

In conclusion, there is no evidence from this large study to support the hypothesis that the level of caffeine consumption is related to death rates from all causes, strokes, heart disease or cancer in hypertensive patients.*

3. COFFEE CONSUMPTION AND FIBROCYSTIC BREAST DISEASE†

3.1. Introduction

In the past 20 years no health warning has had a more measurable impact on the coffee-consuming public than the anecdotal report[24] that linked methylxanthine intake with a condition commonly called 'fibrocystic breast disease'. Over the last 2 years clinicians have exhibited a growing appreciation of a widespread and essentially harmless condition[25,26] prevailing in a clinically recognisable form in 50% of American women and in a histologically detectable form in 90%.[25]

Previous reports linking coffee consumption with myocardial infarction,[27] bladder cancer[28] and pancreatic cancer[29] found instant outlet in the lay press, but they were not followed by a decrease in coffee consumption. Nevertheless, many physicians have been surprised at the readiness with which women have accepted advice, mainly from women's journals, to reduce caffeine intake from various sources: soft drinks, chocolate and cookies, certain drugs, tea and coffee. The results of voluntary abstention from these sources of methylxanthine have been equivocal: some women have temporarily been free of breast discomfort or experienced spontaneous disappearance of nodules, but others had no changes. The purpose of this report is to analyse the natural history of fibrocystic disease in women whose intake of methylxanthine is constant. The first part reviews

*A more detailed analysis of the data presented has been submitted for publication under the title 'Mortality Patterns among Hypertensives by Reported Level of Caffeine Consumption', by Jeanne B. Martin, J. Fred Annegers, J. David Curb, Siegfried Heyden, Christopher Howson, Eun Sul Lee and Marshall Lee, on behalf of the Hypertension Detection and Follow-up Cooperative Group.

†Prepared in conjunction with Dr J. G. Fodor, St John's Memorial University of Newfoundland, Canada.

the evolution of changes among 358 women who were considered to have fibrocystic breasts and were followed up every 6 months for 5 years. This study was completed in 1979, the year Minton's report[24] on caffeine and benign breast disease was published. No previous article had ever suggested an association between coffee and diseases of the breast. The second part is a prospective study carried out in 1982 of 72 women with palpable breast cysts. Their caffeine intake was recorded on a monthly basis and purposely kept constant.

3.2. Patients and Methods

3.2.1 Study No. 1

This was a retrospective chart analysis of 358 patients, examined at a time when there was no concern about a potential association between coffee consumption and breast disease.

From 1974 to 1979 women between 35 and 74 years of age were examined for benign and malignant breast disease. In all, 10 000 women, whose physicians had attested to absence of palpable lesions, were recruited into the Breast Cancer Demonstration Project at Duke Medical Center. The chosen patients were re-examined every 6 months until surgery was performed when the diagnosis of fibrocystic breast disease was confirmed histologically.

The clinical findings and menstrual status were abstracted from the charts at each recorded visit; in some women as few as three, in others as many as ten, follow-up visits were made over the 5-year period. The records of women with menstrual activity were separated from those of menopausal women. The average age at the end of the study of the first group was 49 years and of the second was 59 years. Findings were classified as follows:

- Always positive: clinical examination revealed the same or a similar description of findings in the same quadrant of one or both breasts throughout the observation period.
- Unidirectional change: patients were described as having undergone a single or unidirectional change, that is from normal to abnormal in the same breast or vice versa, or from a single lesion to multiple lesions, or from a single node to diffuse nodularity or vice versa. The essential criterion was that, once these changes had occurred, they remained unchanged until surgery.
- Fluctuations: findings changed back and forth from one examination to the next, for example from negative first to positive and then back

to negative, or from a single mass to multiple masses and back to a single one, or from diffuse nodularity to a single nodule and then again back to diffuse nodularity.

The clinicians had an opportunity to read the reports of the previous examinations, confirming or updating earlier results. Study forms required uniform reporting on location, size, mobility and texture of lesions. The women were instructed not to come for their clinic visits if they were premenstrual, so it was unnecessary to relate clinical findings to the individual's menstrual cycle.

3.2.2. Study No. 2

The purpose of the second study was to evaluate the clinical course of fibrocystic breast disease over a 6-month period, recording the women's usual methylxanthine intake in an employed population. In 72 women fibrocystic breasts were diagnosed by clinical examination during a cancer screening programme at Cannon Mills in Kannapolis, NC; the women agreed to return for monthly follow-up examination. Mean age was 48 years (range from 24 to 65 years). The examiner who performed the breast examination was unaware of previous findings. For each nodule the position, size, texture and mobility were recorded. A questionnaire on methylxanthine consumption from beverages (15 questions), foods (nine questions) and caffeine-containing drugs (six questions) was completed at each visit. Data collected included method of preparation, serving size and frequency of consumption. Methylxanthine content was obtained from several sources.[30,31]

The number of nodules that disappeared was considered separately from those that disappeared but recurred over the 6 months. Cysts were counted and the numerical change during the observation periods was recorded. With the clock-face image as a guide and division of the breast into 12 sections (hours), 'no change' in the position of a given nodule was assumed when its location was recorded either in the same position or ± 1 hour from the location previously noted.

3.3. Results

3.3.1. Study No. 1

The most important observation was that 41% of the premenopausal women and 51% of the menopausal women showed fluctuations (Table 3). When unidirectional changes are combined with these fluctuations, a clear picture of the 'waxing and waning' conditions emerges. Only 29% of

Table 3
Study 1: Clinical findings over 5 years in 153 premenopausal and 205 menopausal women with fibrocystic breasts

Findings	*Premenopausal, No. (%)*	*Menopausal, No. (%)*
Always positive	44 (29)	36 (18)
Unidirectional change	46 (30)	64 (31)
Fluctuations	63 (41)	105 (51)

women in the premenopausal and 18% in the menopausal group had findings similar from one examination to the next throughout the study period.

Overall, 31% had changes recorded as 'unidirectional', either improving or worsening over the 5-year period, and 47% had fluctuating changes. These findings suggest that more than 70% of women with histologically proven fibrocystic breast disease show change with breast examinations 6 to 12 months apart. Stable clinical findings in fibrocystic breasts over a long period appear to be the exception rather than the rule, at least in this retrospective study. It is highly unlikely that these women made any major change in methylxanthine intake during the study period.

3.3.2. Study No. 2

This study attempted to achieve more precise measurements of change than those recorded retrospectively, by using the clock-face diagram to locate breast nodules.

Table 4
Study 2: Clinical findings over 6 months in 72 women with fibrocystic breasts[a]

Findings	*Left breast, No. (%)*	*Right breast, No. (%)*	*Both breasts, No. (%)*
Cysts present at baseline	50 (69)	58 (81)	108 (75)
Cysts disappeared	9 (13)	12 (17)	21 (15)
Cysts disappeared then reappeared	20 (28)	28 (39)	48 (33)
Cysts changed in number only	32 (44)	19 (26)	51 (35)
Cysts changed in position and number	65 (90)	60 (83)	125 (87)

[a] Reproduced in part with permission from S. Heyden and L. H. Muhlbaier, *Surgery*, 1984, **96**, 479–483.

During the study period, 90% of left breasts and 83% of right breasts showed changes in position or number of nodules (Table 4). All nodules disappeared from 13% of the right breasts and from 17% of the left. In 28% of left breasts and 39% of right breasts, the nodules disappeared only to reappear during the 6 months. The total methylxanthine intake did not change significantly, although caffeine consumption from coffee increased by 24 mg per day between the first and last examination (Table 5).

3.4. Discussion

The results of these two studies offer no support for the claim that caffeine is associated with benign breast disease.

Clinically, the important point in the prospective study was the spontaneous disappearance of breast nodules without any relation to the methylxanthine consumption. Equally relevant was the observation that only 13% of breasts examined did not undergo nodular changes over the 6 months. Another indicator is the number of cysts counted from one examination to the next and changes in their position within the breasts.

All these measurements demonstrate a high degree of unreliability of clinical findings over such a brief observation period.

For the purpose of the second study, it was important that the

Table 5

Study 2: Daily methylxanthine consumption (mg) in 72 women with fibrocystic breasts[a,b]

Source	*First month mean ± SE*	*Last month mean ± SE*	*Maximum*
Coffee	121·35 ± 18·77	145·17 ± 19·73	915·0
Tea	64·48 ± 8·56	44·96 ± 6·88	36·0
Soft drinks	29·02 ± 4·31	28·48 ± 4·73	157·7
Drugs	3·84 ± 1·68	2·91 ± 1·34	64·0
Candies	0·23 ± 0·07	0·20 ± 0·05	3·3
Food	0·08 ± 0·13	0·18 ± 0·05	7·3
Totals[c]	230·06 ± 19·51	225·80 ± 22·25	960·0

[a] The result of self-reported intake of methylxanthines.

[b] Reproduced in part with permission from S. Heyden and L. H. Muhlbaier, *Surgery*, 1984, **96**, 479–483.

[c] Average individual methylxanthine consumption (total from all sources). Overall methylxanthine consumption decreased by only 4·3 mg per day between the first and the last examination (NS).

methylxanthine intake of the study population did not change. The change in total consumption between the first and last examination was a decrease of only 4·3 mg per day.

Thus, for a condition in which stability of clinical findings is the exception, it would be difficult to associate certain habits (i.e. methylxanthine consumption or abstention from it) with the presence or absence of benign breast disease. A number of women might, by chance, have either breast lesions while consuming methylxanthines, or no lesions while abstaining from them.

3.5. Review of Literature

Patients with fibrocystic breast disease are advised to abstain from coffee, cola drinks, tea and chocolate.[24] This recommendation is based on a preliminary study of 47 women with fibrocystic breasts: 13 of 20 who stopped consuming beverages containing methylxanthines reportedly experienced complete disappearance of all palpable breast nodules within 1 to 6 months. No long-term follow-up of this group has been reported.

Later the same group found[32] that in 37 women with fibrocystic breast disease the condition resolved when they stopped using methylxanthine-containing products.

From our findings in the retrospective study, it could be expected that 70% of premenopausal or even 80% of menopausal women would show 'waxing and waning' of their condition and, of these, a considerable number by chance might have negative clinical findings after initial positive ones. The nature of 'waxing and waning', which is the predominant feature of this chronic condition, precludes any conclusion that there is a causal association between giving up the habit and becoming asymptomatic.

Following these publications,[24,32] five studies[33–37] were reported, by other investigators, on the same subject.

One[33] reported on 72 women with fibrocystic breast disease on a methylxanthine-free diet and 68 patients who did not change their dietary habits. Among a subsample from whom breast fluid specimens were obtained, no relation was found between changes in breast-fluid caffeine levels and improvement of fibrocystic breast disease. 'Absolute changes in clinically palpable breast findings in the abstaining group were minor.' In addition, 'comparison of before–after mammograms offered little support for the methylxanthine hypothesis'.

The second study[34] was of 323 women with palpable benign lesions of the breast and 1458 women free of palpable nodules. No difference in coffee and tea consumption between the two groups was noted. 'These findings do not support the recent suggestion of Minton *et al.* that methylxanthine ingestion is causally related to benign breast disease or breast cancer.'

Methylxanthine consumption among 210 women with fibrocystic breast lesions and 241 breast cancer patients was the object of a third investigation.[35] In comparing the patients' daily intake with that of controls, the authors found no significant differences in the consumption of coffee or tea. For both benign and malignant breast disease, the risk was slightly higher for those who drank one to three cups per day than it was in those who drank seven or more.

In a case-control study[36] of 854 women in whom benign breast disease was diagnosed histologically, 755 matched surgical controls and 723 matched neighbourhood controls, no association was noted between coffee consumption and benign breast disease. The estimated mean intake of methylxanthines was similar: 302, 312 and 313 mg for cases, surgical controls and neighbourhood controls, respectively. No evidence of a dose response was noted.

In contrast, only one hospital-based case-control study[37] showed a positive association with coffee consumption. In 634 women with fibrocystic disease and 1066 control women, the authors reported as unclear why an association between symptoms and caffeine consumption was found only among the controls: 'we of course do not know which controls may have undiagnosed fibrocystic disease'.

Because of the uncertainty expressed by the authors concerning the status of the control women, their results are of doubtful value, especially in view of the widespread prevalence of fibrocystic breast changes which, according to autopsy reports, involve 90% of women in our society. On the basis of the reports mentioned, the Council on Scientific Affairs of the American Medical Association concluded: 'There is currently no scientific basis for associating methylxanthine consumption with fibrocystic disease of the breast. Indeed, it has been suggested that lumpy, fibrous breast tissue in women is normal and represents a response to physiological hormonal variation.'[38]

Minton has taken his hypothesis one step further on the assumption 'that breast cancer occurs in patients with fibrocystic breast disease four to six times more often than in women with normal breasts'. In an animal experiment, he and his colleagues[39] showed that caffeine and a fat diet enhanced the development of mammary tumours in DMBA-treated rats. Indirect support for this new hypothesis is claimed from epidemiological studies supposed to have shown independently a correlation with breast cancer in countries where there is documented high caffeine and high fat consumption. 'Our analyses of their independent reports show that the countries reporting the highest incidence of breast and other cancers are countries where both high fat diets and high caffeine consumption have occurred simultaneously.'[39]

One must separate the issues addressed in this statement on coffee consumption. The basic premise that the risk of breast cancer is increased four to six times in women with fibrocystic breasts is no longer tenable. A recent editorial[36] found it high time to empty the 'fibrocystic disease' waste basket and thereby mitigate the anxiety of a large number of women who no longer need to be concerned about the risk of breast cancer. After surgical excision of non-cancerous breast tissue, approximately 70% of women with fibrocystic breasts could not be considered at high risk. But, as Hutter[26] reminds us, histopathological findings do not tell the whole story. A more meaningful picture is revealed by combining histological findings with age and family history of breast cancer. Women over 55 years with non-proliferative lesions have a reduced risk (0·3 times), whereas those with proliferative lesions have a higher risk (2·2 times). Women with cysts are not at increased risk unless they have a positive family history of breast cancer, in which case their risk rises threefold.

Before one can associate the development of breast cancer with 'high coffee and high fat' consumption, one has to account for the missing link, the non-existing causal association between coffee intake and benign breast conditions. Is it reasonable to associate a very common habit (coffee drinking) with such a widespread disorder when there is no biologically plausible connection between the two and when experimental evidence is lacking?

A dietary case-control study[40] of 818 breast-cancer patients compared methylxanthine intake from all sources with matched surgical and neighbourhood controls. In addition, analysis was done in relation to fat consumption. Neither methylxanthine consumption nor the amount or quality of fat showed any association with breast cancer.

A study[41] of 2651 women with newly diagnosed breast cancer failed to disclose any association between breast cancer and coffee consumption when data were compared with those on non-cancerous control women.

As others have previously disproved in pancreatic cancer,[42] bladder cancer[43] and renal cancer,[44] we conclude that in breast cancer coffee is not a carcinogen.

3.6. Summary and Acknowledgement

Caffeine and other methylxanthines have recently come under scrutiny as substances that may cause breast tissue to undergo fibrocystic changes.

This chapter describes a retrospective 5-year study of 358 women with fibrocystic breast disease and a 6-month prospective study of 72 women with this condition. Although methylxanthine consumption was constant, the clinical findings in semiannual or monthly examinations in both studies were inconsistent. On the basis of these results and a review of available studies, the authors believe there is no scientific basis for an association between the consumption of methylxanthines and the development of fibrocystic breast disease.

We wish to acknowledge the kind permission of the Canadian Medical Association to reproduce this section, which has appeared in the *Canadian Journal of Surgery*, Vol. 29, No. 3.

4. HABITUAL COFFEE CONSUMPTION AND CHOLESTEROL LEVELS*

4.1. Introduction

The concern among hypercholesterolaemic patients about coffee as a potential factor increasing cholesterol levels is shared by many physicians who have followed the recent medical literature on this subject. This review will attempt to arrive at a conclusion, answering two questions: (1) Is the relationship between coffee and cholesterol a consistent finding in each sex in all studies? (2) Is the apparent positive association between coffee consumption and serum cholesterol found in some studies biologically significant enough to warrant a health warning?

4.2. Worldwide Studies

Table 6 summarises the 21 presently available international studies on the association between coffee consumption and cholesterol levels. The studies can be classified into five groups with similar outcomes, hereafter designated A to E and discussed in the sections that follow.

4.2.1. Group A Studies

Five reports, from Tromso,[45] the United States[46–48] and Jerusalem,[49] show a positive association between cholesterol concentrations and coffee consumption in both sexes, and the results of a German study[50] appear to be age-dependent.

Over 72 000 men and women took part in these six studies, which reported unequivocally significant associations between coffee and total

* Prepared in conjunction with Dr K. A. Schneider and Dr J. Ruh, Duke University, and Dr J. G. Fodor, University of Newfoundland.

Table 6
Disparity of results in 21 studies of the association of cholesterol levels and coffee consumption among men and women

Study group	*Name or location of study*	*Number of participants*	*Correlation of increasing coffee consumption with serum cholesterol (P-values)*	
			Men	*Women*
A	Tromso	13 799	0·001	0·001
	Kaiser–Permanente	47 611	0·001	0·001
	US vegetarians	116	0·05	0·05
	HDFP	9 038	0·01	0·01
	Jerusalem	1 596	0·001	0·01
	Germany			
	Age 18–24	387	0·05	0·05
	Age 65–74	383	NS	NS
B	HANES, USA	5 098	NS	NS
	Minnesota LRC	632	0·11	NS
	Evans County	361		
	Smokers		0·01	0·01
	Non-smokers		NS	NS
	Southern Pacific	2 455	NS	NS
C	Tecumseh	3 394	NS	0·05
	Australia	4 757	NS	0·01
	California LRC	701	NS	0·01
	San Antonio	2 151	NS	0·01
D	Holland[a]	4 342	0·08	0·29
E	Norway (Bjelke)	444	'Significant'	No exam
	Honolulu	1 859	'Significant'	No exam
	Stanford, CA	77	0·05	No exam
	Framingham	1 992	NS	No exam
	Chicago	1 900	NS	No exam
	Philadelphia	34	NS	No exam

[a]All significant correlation coefficients are positive except for Dutch females.

cholesterol and a dose–response relationship. However, it is also relevant to note that HDL cholesterol levels were unrelated to coffee intake in most studies in which they had been measured.

Exceptions to this rule were the Tromso study[45] and the Jerusalem study.[49] In the former an increase in HDL concentrations was noted at low coffee intake in women, with a subsequent decrease at higher level of coffee intake. There was also an increase in HDL among men at low levels of coffee intake, but no decrease at high coffee consumption was seen.[45] In the Jerusalem study,[49] HDL levels were increased in women who drank coffee ($P<0.01$) but not in men.

The Kaiser–Permanente study[47] found a positive association between increased coffee intake and higher cholesterol levels, but HDL levels were unrelated to coffee intake in both sexes.

The authors of the Tromso study speculated on the difference in coffee brewing methods between various study locations as the reason for different outcomes as far as cholesterol levels were concerned. However, the investigators of the Kaiser–Permanente study rejected this explanation for the relationship between coffee and total serum cholesterol. In the latter study, no association was seen between tea-drinking and cholesterol. The disparity in findings between coffee and tea suggested to them that 'caffeine is not the substance involved in an apparent effect on cholesterol ... closer examination leads us to consider a noncausal interpretation [between coffee and cholesterol]'.[47]

In a study of 116 vegetarians, total and LDL cholesterol were associated with coffee intake.[48] HDL levels were unrelated to the level of coffee intake.

The Hypertension Detection and Follow-up Program (HDFP)[46] analysed cholesterol levels in four coffee consumption groups: less than one cup per day, between one and four cups per day, between five and eight cups per day, and equal to or more than nine cups per day. Age-adjusted mean serum cholesterol levels increased, respectively, from 228 to 230 to 233 to 235 mg dl^{-1} in the four subsets. These differences were statistically significant as determined by analysis of covariance ($P<0.006$).

The German study[50] represented randomly selected residents from three cities. A statistically significant positive correlation between the number of cups of coffee consumed per day and cholesterol levels was seen only in men and women between 18 and 24 years, but not in the age group 65–74 years.[50]

4.2.2. Group B Studies

Four studies have been analysed which reveal no association between coffee and cholesterol in both sexes.

The US National Health and Nutrition Survey (HANES)[51] examined four groups with levels of coffee consumption from low to high. Average serum cholesterol concentrations ranged in men from 202 to 200 mg dl^{-1}, and in women from 195 to 202 mg dl^{-1}. The results among the coffee subgroups were not significantly different.

The Minnesota LRC study[52] suggested 'that differences in dietary fat intake might confound the coffee lipid relationship reported by others'. After adjusting for body mass index, exercise, smoking, alcohol, and hormone use (in women), the cholesterol coffee association was borderline ($P = 0{\cdot}11$) for men and non-significant for women. HDL was unrelated to coffee intake in both sexes.

In the Evans County study,[53] no relationship was observed between coffee consumption and LDL, HDL or total cholesterol among non-smokers. However, in cigarette smokers, there was a significant interaction effect between coffee drinking and total and LDL cholesterol. The coffee–smoking–cholesterol interaction effect was confirmed in the Australian study[54] but not in the San Antonio Heart study.[55] In the Evans County study, HDL cholesterol was unrelated to coffee drinking among both non-smokers and smokers.

The Southern Pacific Railroad study[56] analysed correlations between lipid and lipoprotein levels and consumption of caffeine-containing beverages among 2175 men and 280 women. The authors found no significant variation in HDL cholesterol level associated with coffee intake.

4.2.3. Group C Studies

The results of the following four studies exemplify an apparent lack of association of the two variables in men but a positive one in women.

The investigators of the Tecumseh study[57] emphasised the correlation coefficients between frequency of coffee consumption and serum cholesterol levels for women as 'positive at a low level'.

The Australian authors[54] found a 'highly significant interaction between caffeine and smoking habits in their association with serum cholesterol ... confined to females'. There was no interaction between coffee, smoking and cholesterol among males and no increase of cholesterol levels with higher coffee intake.

The Southern California LRC study[58] reported a difference of $20\,\mathrm{mg\,dl^{-1}}$ in cholesterol levels between low-coffee- and high-coffee-consuming women. The study was adjusted for age, obesity, cigarettes per day, alcohol use per day, gonadal hormone use, regular exercise, daily intake of saturated, monounsaturated and polyunsaturated fat, polyunsaturated/saturated fats (P/S) ratio and use of cream and sugar in coffee. HDL levels were elevated slightly in women who drank two cups or more of coffee per day. The absence of an association between coffee and any plasma lipid or lipoprotein in men remained unexplained.

The San Antonio heart study,[55] with its combined Mexican, American and Anglo populations, confirmed a positive relationship between coffee intake and both total and LDL cholesterol only for women, after adjustment for age, dietary intake, obesity, ethnicity, alcohol consumption, cigarette smoking and use of hormones. HDL levels increased slightly with higher coffee intake in women. Men with heavy coffee drinking habits were found to consume significantly more total fat and saturated fats and cholesterol than light coffee drinkers. The more atherogenic diet in men explained the higher cholesterol levels in heavy coffee drinkers. HDL levels were unrelated to coffee intake in men.

4.2.4. Group D Studies

A study from Holland demonstrated a different outcome in men and women. This Dutch study[59] is also one of the few comparing cholesterol levels (in $mg\,dl^{-1}$) in various coffee-drinking groups with those in non-drinkers. For men, age-adjusted cholesterol levels rose from 218 in non-drinkers to 221 in light coffee drinkers to 225 in moderate and 229 in heavy coffee drinkers. The *P*-value for differences in means was 0·08. For women, differences in cholesterol levels ranged from 220 to 215 $mg\,dl^{-1}$, i.e. there was a downward trend with higher coffee intake although it was not statistically significant.

4.2.5. Group E Studies

In the remaining six investigations, cholesterol levels were not examined in women.

Another Norwegian study[60] showed a significant increase in cholesterol levels with higher coffee consumption in men.

The Honolulu Heart study[61] also compared coffee drinkers and abstainers; LDL levels among coffee-consuming Japanese men living in Hawaii were higher than in those who did not drink coffee. The difference was slight but statistically significant. HDL levels were unrelated to coffee consumption.

The Stanford study[62] found total and LDL cholesterol levels to be weakly (albeit statistically significantly) correlated with coffee intake after adjusting for age, cigarette use, adiposity, aerobic capacity, nutrient intake and stress. The number of cups consumed per day correlated positively with levels of apolipoprotein B. The HDL levels were unrelated to coffee intake.

Men in Framingham[63] displayed cholesterol levels of 233 $mg\,dl^{-1}$ in non-drinkers of coffee and of 235 $mg\,dl^{-1}$ in moderate and heavy coffee drinkers and again of 235 $mg\,dl^{-1}$ in heavy coffee drinkers. These differences were not statistically significant.

In the Chicago study of employed men,[64] stratified into four coffee-consuming groups, the differences were also not statistically significant; 245 mg dl^{-1} in the lightest coffee users and 253 mg dl^{-1} in the heaviest coffee drinkers.

In the earliest study of this subject, 34 healthy men in Philadelphia did not show any significant acute effect of caffeine on serum cholesterol.[65]

4.3. Discussion

The discrepancies in the results of these studies between the sexes are of some concern. In 21 studies of men, 11 failed to show a significant association between coffee intake and cholesterol. In 15 studies of women, 10 did show a positive, i.e. statistically significant, association. Half of the presently available data thus show no increase in cholesterol concentrations among men with higher consumption. These same studies, however, suggest an increase in cholesterol concentration in women with high coffee consumption. What is clearly needed in such a situation is a clinical experiment with randomised groups of men and women exposed to various levels of coffee consumption. Unfortunately, this would be confounded in the coffee–cholesterol issue by the great number of other variables which one must control because they could potentially affect serum cholesterol concentrations.

In answer to the second question, it is our opinion that no health warning is warranted. This judgement is prompted by the present review, and also appears justified in view of the fact that HDL cholesterol was not reduced among heavy coffee consumers in ten studies in which HDL levels were examined, and, in fact, increased in women in the Tromso, the Jerusalem, the California LRC and the San Antonio studies.

It is the Tromso study[45] where the greatest increment in cholesterol levels occurred for women. From the lowest to the highest coffee consumption group, an increase of 26 mg dl^{-1} was found. For men, an increase of 31 mg dl^{-1} was noted. Another Norwegian study[60] reported a 14 mg dl^{-1} difference between the lowest and highest coffee-consuming males. Cholesterol levels in these men were extremely high, ranging from 276 to 290 mg dl^{-1}. In men with such high average levels, a difference of 14 mg dl^{-1} may be of some importance within an overall dietary-drug treatment programme.

In the much larger samples of the Kaiser–Permanente study, cholesterol differences between women not drinking any coffee and women who drank more than six cups per day were 5·8 mg dl^{-1}, and for the men in these two categories 9·6 mg dl^{-1}. We agree with the conclusion from the Kaiser–

Permanente study that it remains premature to prohibit coffee as a means of reducing risk of coronary heart disease because 'the preponderance of current evidence fails to support an independent coffee–CHD relation'.[47]

It is difficult to reconcile the discrepancies in the results among the various studies from six industrialised countries. One important caveat is relevant in this review: none of the quoted studies was planned with an *a priori* hypothesis, in a prospective randomised fashion to test the hypothesis of a relationship between coffee drinking and cholesterol. The coffee–cholesterol association is a by-product of observational studies which were neither prospective nor randomised. It is inherent in the nature of cross-sectional studies that they can demonstrate an association but cannot prove causality between coffee consumption and cholesterol levels.

4.4. Summary of Findings

Twenty-one observational studies involving over 103 000 persons from six different countries have reported their findings on the association between coffee consumption and cholesterol levels. Results of these reports display a variety of trends in the association between coffee intake and serum cholesterol concentrations: there were six studies showing a significant association and four studies showing no association in both sexes. Another five reports found a significant association for only one sex. The remaining six investigations examined only men and were equally divided in reporting both significant and insignificant correlations between coffee and cholesterol. This unexplained incongruity of cross-sectional data makes a causal relationship between coffee and cholesterol rather unlikely. In addition, HDL cholesterol levels appeared unrelated to coffee intake in the ten studies in which it was measured. A critical assessment of the published data would lead to the conclusion that consumption of coffee is of no clinical importance in cholesterol metabolism.

REFERENCES

1. Curatolo, P. W. and Robertson, D., *Ann. Int. Med.*, 1983, **98**, 641–653.
2. Izzo, J. L., Ghosal, A., Kwong, T., Freeman, R. B. and Jaenike, J. R., *Am. J. Cardiol.*, 1983, **52**, 769–773.
3. Ammon, H. P. T., Bieck, P. R., Mandalaz, D. and Verspohl, E. J., *Br. J. Clin. Pharmacol.*, 1983, **15**, 701–706.
4. Wynder, E. L., Hall, N. E. L. and Polansky, M., *Cancer Res.*, 1983, **43**, 3900–3906.
5. Eaton, W. W. and McLeod, J., *Am. J. Publ. Health*, 1984, **74**, 66–68.

6. Heyden, S. and Muhlbaier, L. H., *Surgery*, 1984, **96**, 479–484.
7. Kurppa, K., Holmberg, P. C., Kuosma, E. and Saxen, L., *Am. J. Publ. Health*, 1983, **73**, 1397–1399.
8. Martin, J. C., *Neurobehav. Toxicol. Teratol.*, 1982, **4**, 421–427.
9. Heyden, S., Heiss, G., Manegold, C., Tyroler, H. A., Hames, C. G., Bartel, A. G. and Cooper, G., *Circulation*, 1979, **60**, 22–25.
10. Becker, A. B., Simons, K. J., Gillespie, C. A. and Simons, F. E. R., *N. Engl. J. Med.*, 1984, **310**, 743–746.
11. Gillies, M. E. and Birkbeck, J. A., *Am. J. Clin. Nutr.*, 1983, **38**, 936–942.
12. Piters, K. M., Coffee boosts pain-free walking time for patients with chronic stable angina. Presented to the Western Section, American Federation for Clinical Research, Carmel, CA, February 1984. *Med. World News*, 12 March 1984, 137.
13. Piters, K. M., Colombo, A. and Olson, H. G., *Am. J. Cardiol.*, 1985, **55**, 277–280.
14. Laska, E. M., Sunshine, A., Mueller, F., Elvers, W. B., Siegel, C. and Rubin, A., *J. Am. Med. Assoc.*, 1984, **251**, 1711–1733.
15. Onrot, J., Goldberg, M. R., Biaggioni, I., Hollister, A. S., Kincaid, D. and Robertson, D., *N. Engl. J. Med.*, 1985, **313**, 549–554.
16. Dobmeyer, D. I., Stine, R. A., Leier, C. V., Greenberg, R. and Schaal, S. F., *N. Engl. J. Med.*, 1983, **308**, 814–816.
17. Graboys, T. B. and Lown, B., *N. Engl. J. Med.*, 1983, **308**, 835–836.
18. Yano, K., Rhoads, G. C. and Kagan, A., *N. Engl. J. Med.*, 1977, **297**, 405–409.
19. Heyden, S., Tyroler, H. A., Heiss, G., Hames, C. G. and Bartel, A., *Arch. Int. Med.*, 1978, **138**, 1472–1475.
20. Dawber, T. R., Kannel, W. B. and Gordon, T., *N. Engl. J. Med.*, 1974, **291**, 871–874.
21. Hypertension Detection and Follow-up Program Cooperative Group, *J. Am. Med. Assoc.*, 1979, **242**, 2562–2571.
22. Tibblin, G., Wilhelmsen, L. and Werko, L., *Am. J. Cardiol.*, 1975, **35**, 514–522.
23. Wilhelmsen, L., Tibblin, G., Elmfeldt, D., Wedel, H. and Werko, L., *Acta Med. Scand.*, 1977, **201**, 547–552.
24. Minton, J. P., Foecking, M. K., Webster, D. J. T. *et al.*, *Surgery*, 1979, **86**, 105–109.
25. Love, S. M., Gelman, R. S. and Silen, W., *N. Engl. J. Med.*, 1982, **307**, 1010–1014.
26. Hutter, R. V. P., *N. Engl. J. Med.*, 1985, **312**, 179–181.
27. Boston Collaborative Drug Surveillance Program, 1972, **2**, 1278–1281.
28. Cole, P., *Lancet*, 1971, **1**, 1335–1337.
29. MacMahon, B., Yen, S., Trichopoulous, D. *et al.*, *N. Engl. J. Med.*, 1981, **304**, 630–633.
30. Burg, A. W., *Tea Coffee Trade J.*, 1975, **88**, 40–43.
31. Morgan, K. J., Stults, V. J. and Zabik, M. E., *Regul. Toxicol. Pharmacol.*, 1982, **2**, 296–307.
32. Minton, J. P., Abou-Issa, H., Reiches, N. *et al.*, *Surgery*, 1981, **90**, 299–304.
33. Ernster, V. L., Mason, L., Goodson, W. H. III *et al.*, *Surgery*, 1982, **91**, 263–267.
34. Marshall, J., Graham, S. and Swanson, M., *Am. J. Publ. Health*, 1982, **72**, 610–612.
35. Lawson, D. H., Jick, H. and Rothman, K. J., *Surgery*, 1981, **90**, 801–803.

36. Lubin, F., Ron, E., Wax, Y. *et al.*, *J. Am. Med. Assoc.*, 1985, **253**, 2388–2392.
37. Boyle, C. A., Berkowitz, G. S., LiVoisi, V. A. *et al.*, *JNCI*, 1984, **72**, 1015–1019.
38. Council on Scientific Affairs, American Medical Association, *J. Am. Med. Assoc.*, 1984, **252**, 803–806.
39. Minton, J. P., Abou-Issa, H., Foecking, M. K. *et al.*, *Cancer*, 1983, **51**, 1249–1253.
40. Lubin, F., Ron, E., Wax, Y. *et al.*, *JNCI*, 1985, **74**, 569–573.
41. Rosenberg, L., Miller, D. R., Helmrich, S. P. *et al.*, *Am. J. Epidemiol.*, 1985, **122**, 391–399.
42. Wynder, E. L., Hall, N. E. L. and Polansky, M., *Cancer Res.*, 1983, **43**, 3900–3906.
43. Curatolo, P. W. and Robertson, D., *Ann. Int. Med.*, 1983, **98**, 641–653.
44. Armstrong, B., Garrod, A. and Doll, R., *Br. J. Cancer*, 1976, **33**, 127–136.
45. Thelle, D. S., Arensen, E. and Forde, O. H. (the Tromso heart study), *N. Engl. J. Med.*, 1983, **308**, 1454–1457.
46. Davis, B. R., Curb, J. D., Borhani, N. O., Prineas, R. J. and Molteni, A., *Cardiovasc. Dis. Epidemiol. Newsletter AHA*, 1985, **37**, 48.
47. Klatsky, A. L., Petitti, D. B., Armstrong, M. A. and Friedman, G. D. (Kaiser–Permanente study), *Am. J. Cardiol.*, 1985, **55**, 577–578.
48. Sacks, F. M., Castelli, W. P., Donner, A. and Dass, E. H., *N. Engl. J. Med.*, 1975, **292**, 1148–1151.
49. Kark, J. D., Friedlander, Y., Kaufmann, N. A. and Stein, Y. (Jerusalem LRC study), *Br. Med. J.*, 1985, **291**, 699–704.
50. Arab, L., Kohlmeier, M., Schlierf, G. and Schettler, G. (West Germany), *N. Engl. J. Med.*, 1983, **309**, 1250.
51. Kovar, M. D., Fulwood, R. and Feinleib, M. (US National Health and Nutrition Survey), *N. Engl. J. Med.*, 1983, **309**, 1249.
52. Folsom, A. R., Jacobs, D. R., Luepker, R. V., Hannan, P., Taylor, H. L. and Blackburn, H. (Minnesota LRC study), *Cardiovasc. Dis. Epidemiol. Newsletter AHA*, 1984, **35**, 53.
53. Heyden, S., Heiss, G., Manegold, C., Tyroler, H. A., Hames, C. G., Bartel, A. G. and Cooper, G. (Evans County study), *Circulation*, 1979, **60**, 22–25.
54. Shirlow, M. J. and Mathers, C. D. (University of Sydney, Australia), *Int. J. Epidemiol.*, 1984, **13**, 422–430.
55. Haffner, S., Knapp, A., Stern, M., Hazuda, H. P., Rosenthal, M. and Franco, L. J. (San Antonio heart study), *Am. J. Epidemiol.*, 1985, **122**, 1–12.
56. Phillips, N. R., Havel, R. J. and Kane, J. P. (Southern Pacific Railroad study), *Arteriosclerosis*, 1981, **1**, 13–24.
57. Nichols, A. B., Ravenscroft, C., Lamphierar, D. E. and Ostrander, L. D. (Tecumseh study), *J. Am. Med. Assoc.*, 1976, **236**, 1948–1953.
58. Mathias, S., Garland, C., Barrett-Connor, E. and Wingard, D. L. (Southern California LRC study), *Am. J. Epidemiol.*, 1985, **115**, 896–905.
59. Hofman, A., Van Laar, A., Klein, F. and Valkenburg, H. A. (Zoetermeer Districts, Holland), *N. Engl. J. Med.*, 1983, **309**, 1249.
60. Bjelke, E. (Norway), *Lancet*, 1974, **1**, 1116–1117.
61. Rhoads, G. G., Kagan, A. and Yano, K. (Honolulu heart study), *Circulation*, 1976, **54** (Suppl. II), 53.
62. Williams, P. T., Wood, P. D., Vraniza, K. M., Albers, J. J., Garay, S. C. and Taylor, C. B. (Stanford University), *J. Am. Med. Assoc.*, 1985, **253**, 1407–1411.

63. Dawber, T. R., Kannel, W. B. and Gordon, R., *N. Engl. J. Med.*, 1974, **291**, 871–874.
64. Shekelle, R. G., Gale, M., Paul, O. and Stamler, J. (Chicago, Illinois), *N. Engl. J. Med.*, 1983, **309**, 1249–1250.
65. Bellet, S., Kershbaum, A. and Aspe, J. (Philadelphia), *Arch. Int. Med.*, 1965, **116**, 750–752.

Chapter 4

Physiological Effects of Coffee and its Components

H. MILON, R. GUIDOUX and J. A. ANTONIOLI

Nestlé Research Centre, Nestec Ltd, Lausanne, Switzerland

1. INTRODUCTION

Coffee brew contains hundreds of substances, many of which are common to many other natural products, and some of which are potentially active agents. Their physiological importance has only been poorly investigated, and very few effects, if any, can be attributed to them. Two exceptions, however, are worthy of mention: the role of volatile components on the olfactory sensory organs, and the digestive action of the still unidentified constituents of decaffeinated coffee.

Most of the well known physiological effects of coffee can be related to caffeine, which shares the properties of other methylxanthines.[1] These compounds stimulate the central nervous system, act on the kidney to produce diuresis, stimulate cardiac muscle and relax smooth muscle, notably bronchial muscle. Caffeine itself increases the capacity for muscular work and produces a slight stimulation of basal metabolic rate.

The practical implications of these properties in relation to coffee consumption are not always obvious. Nevertheless, effects are mostly observed on three physiological systems, namely the central nervous system, the circulatory system and the digestive system.

2. CENTRAL NERVOUS SYSTEM

2.1. Introduction

As mentioned in the general introduction to this chapter, many constituents of coffee are potentially biologically active. However, the activity of any substance depends on the amount ingested, and in this respect the number of substances present in coffee that are capable of producing effects on the central nervous system (CNS) is very small. This is why caffeine, the main compound of interest from the neurobiological point of view, will be the focus of this review.

The second important point to be mentioned at the beginning of this section is the notion of dose, especially relevant for caffeine where the CNS is concerned. Caffeine, like the other methylxanthine theophylline, is either consumed in food and beverages or given as a drug (it is often administered with other drugs). Although our main interest will be caffeine ingested in coffee, i.e. relatively small amounts, pharmacological studies have often used much higher doses and great care must be taken when extrapolating these results to real-life effects on man.

Coffee drinking (or intake as a flavouring agent) is not essential for man's nurture. It is drunk mainly for pleasure, for its stimulating effects and in social circumstances. Therefore the first physiological effects on the CNS concern two senses, namely olfaction and taste. These will be discussed first. Then behavioural effects in their broad sense will be reviewed, followed by biochemical effects (including those on the central respiratory system). Finally, the presently proposed mechanisms of action of caffeine on the CNS will be described.

Several excellent reviews have been published recently on the central effects of caffeine.[2–5] The other components of coffee are seldom treated, even in books or reviews dedicated to the beverage coffee, because they are much less important.

2.2. Behavioural Effects

2.2.1. Senses

The main reasons for consuming coffee are its popularly admitted stimulatory effects and the pleasure it provides.

The first contact our mind has with coffee is through the senses of olfaction and taste. These two senses are very important in the case of this beverage because it is strongly attractive to both of them. They provide quite different sensations and imply different substances.[6] It is worth mentioning that although many children enjoy the smell of coffee, most of them do not like drinking it before their teens. Even then many adults add milk and/or sugar, not only to soothe some believed gastric effects but mostly to tone down its taste. (See also chapter 6 in this volume.)

Many factors, chosen to produce a given smell and taste, influence the composition and hence the potential physiological effects of coffee: blend, year, date of production, roasting, and national and individual brewing habits. These variations affect the caffeine content of the coffee itself or of the cup. For example, a pure robusta blend may contain twice as much caffeine as a pure arabica blend, and the amount of ingested caffeine from the same blend depends strongly on the way the beverage is prepared.[7]

It is probably in this respect that substances other than caffeine affect mostly behaviour through coffee drinking. Whereas none of them, to date, has been shown to be able to affect the CNS, at least in the amounts present in coffee, those participating in its flavour and aroma certainly influence coffee consumption (see the introduction to this volume) and, therefore, the effects of caffeine. For example, although caffeine *per se* is quite bitter (detection threshold level of 0·65 mg per 100 ml), it only accounts for 10% of coffee bitterness and cannot contribute very much to the appealing taste of coffee. Moreover, when coffee is used in behaviour experiments, its good or bad taste, as it is perceived by the subjects, may well interact through cognitive factors in learning or conditioning mechanisms.

Finally, when enquiring why people drink coffee, the answers are very interesting from two points of view. First, there appear to be ethnic differences. As observed by Rozin and Cines[8,9] in several reports, whereas Jews enjoy particularly the taste and flavour of coffee other groups prefer it as a hot beverage for its stimulatory effects. In Italy it is closely related to family and social life, and in this population its consumption starts at a particularly early age. By contrast, Blacks have the lowest intake. Secondly, when questioned on the reasons for drinking coffee, it seems that whereas three-quarters of the subjects 'like the taste' heavy drinkers clearly look for its stimulant effects.[10]

2.2.2. Performance

2.2.2.1. Animal studies. Most of the published studies were performed on rats and mice. In some of them, golden hamsters,[11] rhesus monkeys[12] and baboons[13] were also used.

The effects of caffeine have principally been tested on learning capabilities in an attempt to discriminate short-term from long-term memory. Caffeine has been shown to improve performance in shock-motivated avoidance,[14] discriminative lever pressing[15] or maze learning.[16] Improved performances were usually observed at doses ranging from 1 to 10 mg kg^{-1}, somewhat lower than those stimulating locomotor activity.

2.2.2.2. Human studies. Several factors make the study of the effects of caffeine on human performance very difficult.

(a) At the doses used (1–4 mg kg^{-1}, i.e. 70–300 mg caffeine), the observed effects are necessarily extremely weak, hardly detectable and difficult to specify.

(b) Caffeine has often been added in protocols designed for studying other compounds (because it is a very easy-to-use molecule) or for testing its synergistic effects with other drugs. In this kind of study, it is hard to extract specific information.

(c) More important, the very wide individual variations of response (which will be discussed later) make it even more difficult to attribute specific positive or negative effects of caffeine on human performance.

Weiss and Laties[17] have reviewed the results obtained in human studies. Most of these are early investigations. Considering different types of performance, specific effects of caffeine have been evaluated, including those that are adverse.

(a) *Physical endurance and capacity.* Although the number of subjects involved in these studies was rather small, it seems that caffeine could increase the capability for hard physical work.[18]

(b) *Motor co-ordination and control.* In this domain, caffeine appears to have little and inconsistent positive effects.[19] On the other hand, it has a negative effect on hand steadiness.[20]

(c) *Monitoring.* This term includes the ability to deal with data produced by devices (mechanical, electronic). Contrary to what was observed for motor co-ordination, caffeine repeatedly palliated performance decrement, possibly by delaying the effects of boredom of the task.[21,22]

(d) *Learning.* Learning *per se* is not improved, but after caffeine administration the response tended to be faster and stronger.[20,23,24]

(e) *Intellectual performance.* This was tested by asking subjects to perform verbal and arithmetic tasks. Caffeine was shown to improve performance, probably by decreasing the effects of fatigue because the results were clearer during long sessions.[25] However, it does not permit the statement that intellectual performance was enhanced.

2.2.3. Activity

2.2.3.1. Animal studies. Since caffeine has often been identified as a stimulant of the CNS, many studies have used spontaneous locomotor activity measurements as a parameter. The interesting aspects of these studies are the dose–response results[26] and the amplitude of the effects. Several observations are common to the different species studied. First, the minimum dose necessary to observe an effect lies between 1·5 and 10 mg kg^{-1}, whereas the peak response appears at doses between 10 and 20 mg kg^{-1}. However, the dose above which no further increase of activity occurs varies from 20 to 100 mg kg^{-1} (Table 1). In fact, if the dose is increased still further, a decrease of locomotor activity is observed whereas brain caffeine levels remain dose-dependent. Finally, when some time has elapsed even after administration of a large dose, activity increases can be observed but only when brain caffeine levels have decreased sufficiently as a result of

Table 1

Effects of caffeine on spontaneous locomotor activity (from Hirsh[4])

Animal	*Route of administration*	*Threshold stimulant dose ($mg\,kg^{-1}$)*	*Peak stimulant dose ($mg\,kg^{-1}$)*	*Lowest supra-maximal dose ($mg\,kg^{-1}$)*	*Reference*
Mouse	p.o.	—	—	60	27
Rat	s.c.	5	20	40	28
				100	29
Mouse	p.o.	8	16	32	30
Mouse	s.c.	5	10	20	31
Rat	i.p.	—	—	50	32
Mouse	s.c.	2	16	32	33
Rat	i.p.	10	20	30	34
Mouse	p.o.	2·5	10	20	26
Mouse	i.p.	6	20	—	35
Mouse	p.o.	—	10	25	36

p.o. = per os.
s.c. = subcutaneous.
i.p. = intra-peritoneal.

elimination and metabolism.[34] These observations have been confirmed by other studies in mice.[33] Although the mechanisms of action of caffeine will be discussed in detail later, a speculative explanation could be the following. At low doses caffeine acts mainly as an adenosine antagonist whereas at high doses brain caffeine concentrations become large enough to inhibit phosphodiesterase activity. This biphasic effect would be due to a competition between the two mechanisms, provoking opposite effects on locomotor activity, each mechanism becoming predominant at different concentration ranges. This is further demonstrated by the results of Hirsh,[4] as illustrated in Fig. 1.

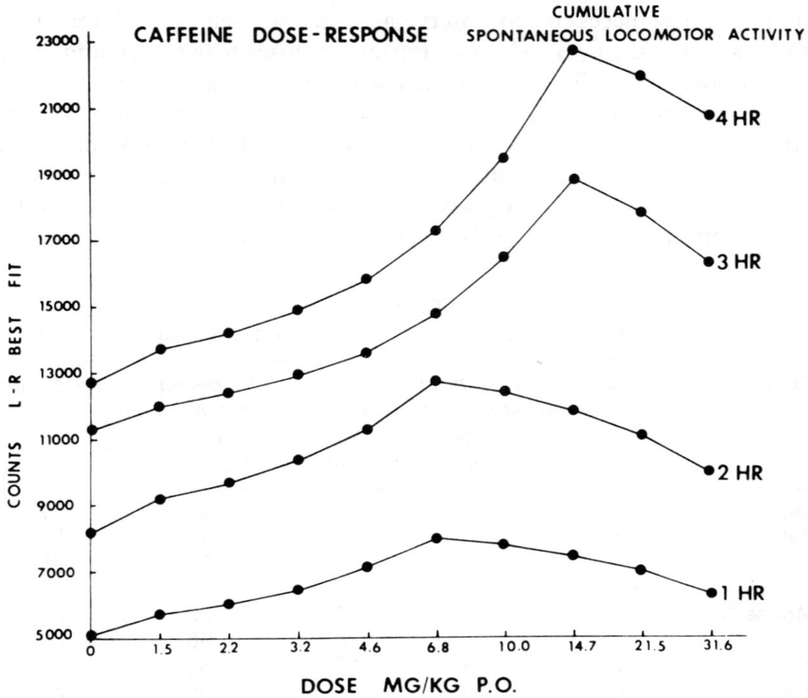

Fig. 1. Caffeine dose response on spontaneous locomotor activity (SLA) in mice. Cumulative activity during the first hour, the first 2 h, the first 3 h and the first 4 h of SLA is shown for each dose level. Two characteristics of caffeine dose response are illustrated. (a) *Dose factor:* SLA increases with dose up to the peak stimulant dose (see Table 1) and then decreases with further increasing doses. (b) *Time factor:* the peak stimulant dose is higher at 1 and 2 h than at 3 and 4 h, showing a delayed increase of SLA at high doses. Reproduced from Hirsh,[3] by courtesy of Alan R. Liss, Inc.

2.2.3.2. Human studies. The effects of caffeine on activity in humans have usually been studied together with the effects on sleep and are reviewed in the next section.

2.2.4. Sleep

Effects of caffeine on sleep are certainly the most often observed and admitted among the population. Many studies have therefore been concerned with these effects.

2.2.4.1. Animal studies. To our knowledge very few studies and no reviews have concerned specific investigations in animals. This is perhaps because recordings are more precise in man and permit a more thorough interpretation. However, it is worth mentioning a study which would not have been possible in man: caffeine was given in the diet of pregnant rats (0·025 to 0·1% in the diet, which represents very high doses) throughout gestation. Sleep and some neurochemical parameters were analysed in the progeny when they were 3 months old. The vigilance states analysis showed that the number of paradoxical sleep (PS) phases was increased, thus increasing total PS in the progeny of the caffeine-treated rats, and that dopamine (DA) levels were lower in the locus coeruleus (LC) whereas norepinephrine remained unchanged. Serotonin and its main metabolite, 5-hydroxyindoleacetic acid, were increased in the dorsal raphé.[37,38] It was then shown that while no DA cells are present in the LC,[39] terminals were found, originating from the ventral tegmental area.[40] However, these results do not allow us to hypothesise that a link exists between paradoxical sleep and DA terminals in the LC. Corroborative results were obtained by Sinton *et al.*[41] in the mouse.

2.2.4.2. Human studies. In humans, caffeine has popularly been believed to disturb sleep by preventing it and by inducing restlessness. Hence, before developing sophisticated experiments involving electrical brain and muscle activity recordings, early investigators used mechanical systems to measure motor activity during sleep.[42] Whereas a heavy meal or a hot day provokes increases of activity during the following night, caffeine has no effect. It was shown subsequently that coffee caused a deterioration in sleep quality as compared with decaffeinated coffee or placebo. No difference was observed between habitual and non-habitual coffee drinkers.[43]

In more recent investigations, electroencephalogram (EEG) recordings have been used to analyse sleep patterns, either alone or together with other activity recordings such as electro-oculograms, electrocardiograms and electromyograms to permit a more complete analysis. All these studies concluded that caffeine delays the onset of sleep, has no effect on the REM (rapid eye movement) sleep phases but reduces the duration of the EEG sleep stages 3 and 4 (sometimes called 'deep' sleep) with an increase of the co-called 'light' sleep (stages 1 and 2). These effects were only observed during the first three or four hours of sleep.[44,45]

In one study, the dose effect of caffeine was examined. The subjects were given 1·1, 2·3 and 4·6 mg kg^{-1} on different days before going to bed. An EEG was recorded during the following night and a questionnaire was completed the next morning. Both analyses provided parallel results: increased sleep latency, and decreased total sleep and quality. Surprisingly, a clear dose effect was observed and the authors proposed caffeine as a model for insomnia.[46]

2.2.5. *Individual Variations*

One very surprising observation concerning the effects of caffeine is their variability between individuals: some will not drink a cup of coffee after midday because they would not sleep at all the following night, whereas others claim coffee even helps them to go to sleep.

Even if some studies in humans have shown that the variability of the direct effects of caffeine is not as pronounced,[47,48] some facts exist which could account for it:

(a) Dose. The amount of ingested caffeine depends on the coffee blend used and on the way the beverage is prepared. For a single cup, this amount usually varies between 60 and 100 mg. However, the effects are dependent on the dose expressed in mg kg^{-1} body weight. If one considers the ranges of caffeine content and of body weights, the bioavailability of caffeine can vary by a factor of 4 between two individuals. As mentioned in the preceding paragraph, such a variation can play an important role because of the dose effect on sleep, for example.

(b) Psychological factors. Although these are more difficult to study, the state and mood of a given individual at the time he or she ingests caffeine certainly influences its effects both qualitatively and quantitatively.

(c) Mechanisms of action. These will be discussed later, For the moment it can be said that when they are studied the doses used are usually far above those corresponding to coffee drinking. However, it is probable that individual variations exist on the sites of action, as proposed by Goldstein *et al.*[47]

(d) An important factor which determines the effects of caffeine is its plasma concentration. This depends on the elimination rate of caffeine and on its rate of metabolism. Indeed, depending on the circumstances, the rate of clearance of caffeine can vary substantially: it has been shown to be slowed down in women taking oral contraceptives[49] or in the last term of pregnancy[50] by a factor of 2, as well as in severe liver failure[51,52] and to be increased by cigarette smoking.[53] Premature babies eliminate caffeine more slowly[54] and the foetal liver is even able to metabolise theophylline to caffeine.[55] A recent study has shown that subjects who experience sleep disturbances after drinking coffee have a longer plasma half-life and a slower plasma clearance for caffeine.[56]

2.3. Biochemical Effects

Before interpreting the kind of results obtained in studies of the biochemical effects in the CNS, and—perhaps more important—before attempting to correlate them with behaviour, especially when a dietary substance is concerned, several points must be raised to avoid overestimation or even fundamental errors.

First, as regards caffeine, the dose used in neuropharmacological or neurochemical studies usually far exceeds quantities ingested in normal life. It should also be taken into consideration that usually, in animals, which are non-coffee-consuming subjects, single administrations are used. In humans, the situation is quite different when considering the effects on a habitual coffee drinker or someone who is suddenly given caffeine. As has been mentioned earlier, tolerance to caffeine exists and it has been hypothesised that heavy coffee drinkers could be low responders to caffeine.

Secondly, when neurotransmitter levels are determined, the significance and interpretation of the results depend on the way the experiments were carried out: because of the high specificity of the different regions of the brain, the absence of effects in whole brain does not mean that there would be no effect in specific regions. Moreover, a change in level should be analysed to determine whether it was due to an effect on synthesis, degradation or turnover.

These remarks should be kept in mind when reading this review on recent literature on the biochemical effects and particularly when trying to extrapolate these to man in coffee-drinking situations.

2.3.1. Monoamines

This group of substances includes three neurotransmitters: serotonin (5-HT), norepinephrine (NE) and dopamine (DA). The effects of caffeine on the levels and dynamics of these neurotransmitters are reported in Table 2. They were also recently extensively reviewed by Fernstrom and Fernstrom,[57] and can be briefly summarised as follows:

2.3.1.1. Serotonin. Caffeine increased 5-HT levels in the rat raphé region, cerebral cortex and cerebellum. However, the raphé region, a serotoninergic structure, is the most sensitive to caffeine (the only one to exhibit an increase after 2·5 mg kg^{-1} p.o.).[65] Release and uptake of 5-HT in

Table 2

Effects of caffeine on norepinephrine (NE), dopamine (DA) and serotonin (5-HT) levels and dynamics (from Hirsh[4])

Animal	*Preparation*	*Dose*[a] $(mg\,kg^{-1})$	*Concentration* (*M*)	*Endogenous level*			*Release*			*Uptake*			*Synthesis*			*Turnover*			*SLA*	*Reference*
				NE	*DA*	*5-HT*	*NE*	*DA*	*5-HT*	*NE*	*DA*	*5-HT*	*NE*	*DA*	*5-HT*	*NE*	*DA*	*5-HT*		
C				0																58
R	CC, BS	50		0			+			0			+			+			–	32
M	WB	50, 100		0			0	–					+	+–[b]		+	+			59
R	WB					+												–		60
R	CS, CC	100	10^{-5}, 10^{-4}							0										61
R	WB	25		0	0	0										0	–	0	0	61
R	WB	50		0+	0+	0+										0	–	–	0	61
R	WB	100		0	0	0										+	–	–	0	61
M	WB	25		0	0														+	62
M	WB	100				+												–	–	63
M	WB	10		0	0														+	64
R	CC, CS, HT, HC, PM, CB, MR	2·5–50		0	+[c]	+[d]														65
R	CC, MR		10^{-7}–10^{-4}						–[e]			+								66
R	WB	100		0	0	+									0			0		67
R	CS		10^{-7}–10^{-4}					0+[e]						0						68

0, no change; +, increase; –, decrease; 0–, slight decrease; 0+, slight increase; C, cat; M, mouse; R, rat; BS, brainstem; CB, cerebellum; CC, cerebral cortex; CS, corpus striatum; DE, diencephalon; HC, hippocampus; HT, hypothalamus; MR, midbrain raphé; PM, pons–medulla; WB, whole brain; SLA, spontaneous locomotor activity.

[a] All doses i.p. except 63, which was p.o.

[b] Increase at 30 min, decreased at 2 h.

[c] In CS.

[d] In MR, CC, CB.

[e] High K^{+}.

the raphé regions were respectively reduced and increased.[66] To complete the picture, turnover has been repeatedly shown to be reduced.[59,61,63]

2.3.1.2. Norepinephrine. The effects of caffeine on NE are not as clear as those on 5-HT. They are often contradictory and not as pronounced. The most positive and concordant effects are an increase in NE synthesis in the dose range 16–100 mg kg^{-1} i.p.[59] and an increase in NE turnover.[61]

2.3.1.3. Dopamine. Roughly similar observations were made for dopamine as for norepinephrine.[59,61]

2.3.2. Hormones

As regards the other neurophysiological manifestations of caffeine, the neuroendocrine system can be affected, depending on the administered dose. At high doses in rats (>20 mg kg^{-1}) the effects resemble those of a stress situation: decreased thyrotropin (TSH) and growth hormone (GH), increased corticosterone and β-endorphin. In humans, whereas low doses corresponding to drinking a few cups of coffee do not provoke any observable changes, high doses stimulate the pituitary–adrenal axis. Evidence for these effects has recently been reviewed by Spindel *et al.*[68] and is summarised below.

2.3.2.1. Animal studies. In rats, the dose needed to produce a half-maximal effect (ED_{50}) on TSH and corticosterone is 30–50 mg kg^{-1} i.p.[70] There appears to be a caffeine plasma level threshold (15 $\mu g\,ml^{-1}$) beneath which no effect is observed on any hormone. Moreover, the pulsatile secretion of GH from the pituitary also has an ED_{50} value of 30–50 mg kg^{-1}.[71] These and other observations, such as the persistence of the decrease of TSH in adrenalectomised animals whereas the increase of corticosterone is suppressed[72] and the probable involvement of somatostatin in the effect of caffeine on GH release,[70] suggest that caffeine has effects comparable to those of stress, except that it does not affect serum prolactin levels.[73] Another important feature of the neuroendocrine effects of caffeine is the development of tolerance within two weeks. The lowest dose to induce tolerance is 25 mg kg^{-1},[74] and serum clearance is not increased.

In conclusion, clear neuroendocrine effects of caffeine are observed at high doses only and are similar to those of a stress, but the mechanisms involved are not fully elucidated. Somatostatin and the hypothalamus could be implicated.

2.3.2.2. Human studies. Taking into account the differences in endocrine response between man and rat, caffeine also provokes stress-like effects, probably through the pituitary–adrenal axis: increased levels of serum cortisol,[75] β-endorphin[74] and ACTH. However, high doses (500 mg

kg^{-1}) are required to observe more consistent results, and tolerance also develops.

2.3.3. *Mechanisms of Action of Caffeine*

A surprising aspect of the central effects of caffeine concerns its mechanisms of action. Both the lack of conclusive evidence explaining all its effects and the history and chronology of the different proposed mechanisms form a good example of serendipity in science. Rall has beautifully developed this theme in a recent review[76] in which the evolution of the mechanisms of action of the methylxanthines serves as an illustration of how discoveries are made in science, how concepts develop and how dogmas are hard to dislodge.

To date, at the cellular level, three types of actions have been proposed for caffeine. They are, as they were chronologically discovered: intracellular calcium mobilisation, phosphodiesterase inhibition, and adenosine receptor blackade. They will now be briefly discussed, and their potential to explain the biochemical and behavioural effects observed both qualitatively and quantitatively will be emphasised.

2.3.3.1. Calcium mobilisation. It was first reported that caffeine (1–2 mM) produced contractions in frog skeletal muscle preparations.[77] The first rapid phase of the biphasic action was slowed in the absence of extracellular calcium, but no change occurred in the second phase. In another series of experiments on calcium accumulation in broken-cell preparations of the sarcoplasmic reticulum,[78] it was noted that caffeine provoked a stimulation of 'leakage' rather an an inhibition of 'pumping' of calcium. The minimum effective concentration was 250 μM.

Although caffeine acts on calcium, the range of concentrations needed far exceeds those necessary to produce CNS and cardiovascular effects (less than 100 μM).

2.3.3.2. Phosphodiesterase inhibition. For a long time this effect has been invoked as the cellular action of caffeine. By inhibiting phosphodiesterase activity, the concentration of *c*-AMP, which plays an important role in adrenergic post-synaptic transmission, is increased.[79] However, here again caffeine concentrations needed to inhibit phosphodiesterase activity are much above those producing caffeine effects.

2.3.3.3. Adenosine receptor blockade. This last hypothesis probably arose as an attempt to interpret the effects on phosphodiesterase inhibition. Today both the sites of action and the concentrations correlate to offer the best if not a unifying explanation for the effects of caffeine.

Briefly, it was first suggested that the methylxanthines were antagonists of adenosine at its receptor sites.[80,81] Adenosine is found ubiquitously in organisms. Its physiological actions are depressant both behaviourally and biochemically: it lowers heart rate and blood pressure, decreases lipolysis and synaptic release of catecholamines, dilates the cerebral vessels and acts as a CNS sedative. Caffeine and the other methylxanthine derivatives were extensively used for studying adenosine receptors and the role of adenosine. Caffeine has been shown to be the only derivative able to block both types of adenosine receptors, A1 (high affinity) and A2 (low affinity). More important, the effects of caffeine occur at low concentrations (*c.* 1 mg kg^{-1} in mice) corresponding to 'normal' caffeine circulating levels.[82]

These observations could lead to the hypothesis that caffeine would not act so much as a stimulant by itself but would rather suppress the depressive action of adenosine.[83–85] This would also explain why caffeine can restore performance but not enhance it.

2.4. Central Respiratory Effects

It has been known since 1913 that caffeine stimulates respiration.[86] More recently, it has been shown that this action is central[87] and may be mediated by the neurotransmitters dopamine and serotonin.[88]

The methylxanthine theophylline is commonly used to treat apnoea in the newborn. Recently caffeine has been proposed as an as effective but less toxic substitute.[89] Although this review is concerned with the constituents of coffee, some potential consequences of this type of action should be mentioned. In two case reports it has been hypothesised that theophylline[90] and caffeine[91] withdrawal could provoke apnoea in neonates. Judging by the similarities of the respiratory effects of both methylxanthines, it is important to bear in mind that, in the case of infants at risk, withdrawal of coffee, and therefore of caffeine, could be deleterious.

2.5. Conclusions

Definitive statements concerning the central effects of caffeine are limited by:

(a) their weakness due to the very small doses ingested in normal coffee consumption;
(b) individual variations;
(c) the fact that, altogether, few systematic studies have examined these effects. Under the keyword 'caffeine', a multitude of references appear but only some of them are really relevant, and it is difficult if not impossible to draw clear-cut conclusions.

Even the notion of 'stimulant' is questionable.[92] From a purely electrophysiological point of view, Jouvet *et al.*[93] showed that caffeine has a double stimulatory effect on the cortex and on the reticular mesencephalic formation. But, already in such specific and basic experiments, the action of caffeine is complex and competitive, and does not resemble any other single psychotonic drug but rather bears characteristics of at least two different classes. Therefore it is obviously almost impossible to specify its effects when the general behaviour of an individual is under investigation.

In summary, it seems well established that caffeine:

(a) delays the onset of sleep and disturbs it, within the limits due to individual variation, during the first 3–4 hours;
(b) could have some restorative effects on some functions (physical endurance, monitoring of data) when they have been diminished by fatigue, etc. Finally, it must be kept in mind that, whatever these effects are, they are very discrete and are susceptible to influence by other factors, physical, nutritional and psychological. Any experiment performed on man with a daily consumed and weakly active compound such as coffee can be easily biased by artefacts, and counter-hypotheses are very difficult to confirm or refute.

3. CARDIOVASCULAR SYSTEM

3.1. Introduction

In subjects not regularly consuming coffee, a slight rise in blood pressure is the most consistent acute effect of coffee ingestion.[94,95] A concomitant rise in plasma epinephrine and/or norepinephrine is generally observed[94] and may account for the pressor effect as well as for the occasionally occurring tachycardia[95] or extrasystoles[96] after heavy coffee drinking. Such effects of coffee may be ascribed to caffeine, which, like other methylxanthines, has been found to trigger a release of norepinephrine from sympathetic nerve endings[97,98] and of epinephrine from the medulloadrenals,[99,100] and to produce a central activation of the adrenergic system.[101] Caffeine appears to be the only cardiovascular-active constituent of coffee, at least at usual drinking levels, since similar cardiac and pressor responses have been obtained after ingestion of coffee or water solutions with equivalent amounts of caffeine.[95,102] In addition, decaffeinated coffee was generally found to exert no effect on blood pressure or the heart,[95,103] except for some disturbance of the heart-beat rhythmicity which may be due to reflex mechanisms starting from the gut.[104] Only acute or short-term effects of caffeine or coffee in animals and man will be reviewed in this chapter, and properties of other closely related methylxanthines will only be mentioned in relation to some experimental points when information about caffeine is lacking.

In isolated heart preparations, the demonstration of caffeine effects requires much higher concentrations (usually more than 0·5 mM) than attained in plasma after habitual coffee drinking (usually less than 0·05 mM). In these preparations, the role of caffeine-induced release of norepinephrine from sympathetic nerve endings is negligible, since prior depletion of tissue norepinephrine with reserpine fails to modify or reduces only slightly the actions of caffeine.[104]

3.1.1. Actions on Effector Cells

Most physiological and electrophysiological effects of caffeine and methylxanthines on the cardiovascular system appear to be due to changes in cell calcium content and/or rates of Ca^{2+} distribution in cell compartments.[104] *c*-AMP plays an important role as a mediator of caffeine's effects on cell calcium. A rise in cell *c*-AMP, as a result of *c*-AMP-phosphodiesterase inhibition by methylxanthines,[105] produces an activation of the influx of Ca^{2+} across the slow Ca^{2+} channel.[105] A *c*-AMP-mediated increase of Ca^{2+} influx may also result from antagonistic competition by methylxanthines with adenosine binding to receptors located on the outer surface of the heart cell membranes (R-type receptors) where most of these receptors are linked to adenylcyclase inhibition (Ri or A1 receptors).[106,107] Caffeine may also interfere with the beat-to-beat control of contractility by the sarcoplasmic reticulum. Caffeine activates the release of Ca^{2+} from the sarcoplasmic reticulum which causes the tension rise[108] and inhibits the subsequent removal of cytosolic Ca^{2+} by the organelle.[108,109] The latter effect masks a *c*-AMP-induced activation of Ca^{2+} uptake by the sarcoplasmic reticulum.[110] Furthermore, caffeine was found to increase the sensitivity of the contractile proteins to Ca^{2+} through a direct effect on troponin.[111]

3.1.2. Interaction with Adenosine and Catecholamines

The fact that caffeine, theophylline or aminophylline, at concentrations known to be high enough to compete efficiently with adenosine binding to R-type receptors, may still be unable to produce detectable heart responses in isolated preparations suggests that endogenous adenosine does not participate in physiological functions of the heart in these preparations, except possibly through binding to P-site receptors, which are methylxanthine-insensitive.[112] In the presence of exogenous adenosine, however, theophylline and aminophylline proved to be effective in modifying heart functions or reinforcing β-adrenergic actions of catecholamines on isolated preparations at very low concentrations.[98,113] One would expect similar effects of caffeine at concentrations lower than 0·1 mM. As seen below, the concentrations of adenosine added to heart-perfusion solutions in these experiments are close to those attained in heart tissue under conditions of decreased O_2 supply (ischaemia, anoxia)[114] or increased O_2 consumption, as induced by sympathetic activation.[115,116] Under these conditions, adenosine is released in larger amounts from myocytes and acts to improve aerobic metabolic conditions. Adenosine contributes to increasing the perfusion of the heart through its coronarodilating action,[117] and acts to decrease the O_2 need of the tissue by both opposing the β-mediated stimulatory effect of catecholamines on heart cells[118] and inhibiting norepinephrine release from sympathetic terminals.[119,120] Since methylxanthines stimulate this release through their action on the sympatho-adrenal system and, in addition, antagonise the effects of adenosine on heart cells and sympathetic terminals, their global action is to enhance the release as well as facilitate and prolong the effect of catecholamines. This action of methylxanthines may be expected to be especially pronounced on ischaemic hearts, which are subject to overstimulation by catecholamines.[121,122] Numerous studies are now designed to define the role of adenosine as a modulator of cardiac responses to catecholamines under physiological and pathological conditions.

3.2. Studies on Animals

Acute effects exerted by caffeine on heart contractility, beat rhythmicity, coronary perfusion and blood pressure will be reviewed. Although simultaneously occurring, *c*-AMP- or Ca^{2+}-induced changes of cell metabolic processes will not be reviewed as they do not account to an appreciable extent for the functional effects of methylxanthines.[123] Most recent information related to functional effects of low concentrations of methylxanthines on isolated heart tissue was obtained in the presence of exogenous adenosine. Although obtained with methylxanthines other than caffeine, data from these studies will be mentioned at the end of each subsection.

3.2.1. Contractility

Both positive and, less frequently, negative inotropic effects of caffeine have been observed in isolated heart preparations.[104] A general effect of caffeine on isometric contractions of mammalian heart preparations perfused *in vitro* is to prolong the time to peak tension and the duration of contraction.[124–126] Usually the rate of heart relaxation is depressed. Whether the peak tension and the rate of force development ($+\mathrm{d}F/\mathrm{d}t$) are increased or decreased depends on several factors, notably the heart rate, the concentration of caffeine, the type of preparation used and the species of the donor. Schematically caffeine effects on heart contractility may be classified as procaine-sensitive and procaine-insensitive effects. Procaine is a local anaesthetic which, in addition to inhibiting the Na^+ current across the sarcolemma, opposes the inhibitory effect of caffeine on Ca^{2+} uptake by the sarcoplasmic reticulum.[109]

3.2.1.1. Procaine-sensitive effects. Through its procaine-sensitive effects (i.e. those resulting from inhibition of Ca^{2+} uptake by the sarcoplasmic reticulum), caffeine tends to maintain cytosolic Ca^{2+} above the mechanical threshold for a prolonged period of time. This mechanism is responsible for the increase of the duration of contraction and time to peak tension, as well as for the slower relaxation.[124] In addition to delaying Ca^{2+} dissociation from troponin C and the onset of relaxation, a sustained elevation of cytosolic Ca^{2+} between pulses tends to activate Ca^{2+} uptake by mitochondria and Ca^{2+} efflux from the cell, which occurs largely through exchange with extracellular Na^+. Preventing this exchange process by perfusing coronary vessels with a Na-free solution readily results in contracture in the presence of high levels of caffeine.[124] Procaine-sensitive effects of caffeine tend to reduce the non-mitochondrial pool of cell calcium and thereby depress the heart contractility.

3.2.1.2. Procaine-insensitive effects. Through its procaine-insensitive effects, caffeine induces a *c*-AMP-mediated increase of Ca^{2+} influx across the sarcolemma during the systole so that the amount of Ca^{2+} releasable from the sarcoplasmic reticulum upon subsequent stimulation is increased. In the presence of procaine, the inotropic positive effect exerted by caffeine on isolated heart preparations is accompanied by shortening of the contraction duration and faster relaxation, i.e. effects similar to those induced by β-adrenergic agonists.[124] Similar effects are produced in the absence of procaine by alkylated xanthines with no perceptible action on the sarcoplasmic reticulum.[126] These effects of caffeine can be ascribed essentially to *c*-AMP-phosphodiesterase inhibition, since most adenosine receptors were presumably blocked throughout the whole range of concentrations used in these studies (more than 0·25 mM) owing to the high affinity of caffeine for adenosine R-type receptors. An increase in the sensitivity of troponin C to Ca^{2+} may also contribute to the positive inotropic effect of high concentrations of methylxanthines.[111]

3.2.1.3. Variability of responses. Through its procaine-sensitive and -insensitive sites of action, caffeine exerts opposite effects on the magnitude of the non-mitochondrial pool of calcium and contractility. The resulting effect is either a positive or a negative inotropism according, among other things, to the development of the sarcoplasmic reticulum and the duration of the action potential, which both vary with species or type of preparation used, to the concentration of caffeine, and to the frequency of stimulation. At low frequency, a positive inotropic effect is obtained in heart atrial, ventricular or papillary muscle preparations from mammals, except the rat, unless excessively high concentrations are used (20 mM).[124,125] In all species, however, the maximal peak tension and dF/dt which can be obtained by pacing the preparations at optimal frequency were depressed by caffeine, which also exerted negative effects at frequencies above the optimal value.[125] Owing to the prolongation of the active state, fusion of contractions may occur at fast heart rates, despite the fact that increasing the rate tends to shorten the action potential duration.[124] In these studies on isolated heart preparations, positive inotropic responses to caffeine or other methylxanthines were mostly direct and not due to norepinephrine release from adrenergic nerve endings, since pretreatment with reserpine, or propranolol addition, failed to depress them or only slightly reduced them.[124,126,127] Caffeine concentrations above 0·25 mM were required to elicit detectable contractile responses. Such concentrations are considerably higher than those achieved *in vivo* after drinking caffeine-containing beverages.

Relatively high concentrations of caffeine were also used for investigations on heart–lung preparations, which again showed variable effects on heart contractile force according to the dose and the animal species used (rat, cat or dog).[104]

In intact animals (dog, cat), methylxanthine effects, when present, were characterised by an increase in the amplitude of contraction and/or stroke volume.[104,128,129] Although inconsistent, such effects could be obtained at caffeine doses as low as 5 mg kg^{-1} i.v., which would be expected to establish concentrations around myocytes below 0·1 mM. At such concentrations, no effect was detected on isolated heart preparations.

3.2.1.4. Interaction with adenosine. In the presence of exogenous adenosine, theophylline was found to enhance the inotropic effect of isoprenaline on rat Langendorf hearts at a concentration (5×10^{-5} M) too low to inhibit *c*-AMP-phosphodiesterase activity appreciably. Theophylline acted to suppress the antagonistic effect of adenosine against the isoprenaline-induced rise of cell *c*-AMP.[113] Both adenosine and isoprenaline were required for theophylline to exert its effect. The fact that theophylline failed to reinforce the inotropic effect of isoprenaline in the absence of exogenous adenosine suggests that endogenous adenosine, presumably released under stimulation by isoprenaline,[116] failed to attenuate the activation of adenylcyclase by the β-agonist. In another study, however, endogenous adenosine was found to prevent full mechanical responsiveness to β-adrenoceptor stimulation while exerting no negative inotropic action by itself.[116]

3.2.2. Heart Rate and Beat Rhythm

3.2.2.1. Heart rate. Methylxanthines exert mostly positive chronotropic effects on isolated heart and heart–lung preparations from mammals.[104,126,130] High concentrations of caffeine similar to those eliciting positive inotropic effects are required. At concentrations above 0·25 mM, caffeine produced a dose-dependent increase of heart rate in rabbit right atria, albeit a decline was noted at a very high concentration (10 mM).[126] These effects were propranolol-insensitive, as a confirmation of previous observations indicating that chronotropic effects of caffeine on isolated heart preparations are not (or only slightly) prevented by propranolol or reserpine pretreatment.[104]

On a series of xanthine derivatives, including caffeine, a striking positive correlation was noted between the potency of these agents to increase the rate of spontaneously beating right atria from rabbits and their potency to inhibit *c*-AMP-phosphodiesterase activity of atrial extracts.[126] Therefore the tachycardia induced by high doses of caffeine probably relies on a *c*-AMP-dependent Ca^{2+} current contributing to diastolic depolarisation of nodal pacemaker cells.

In situ, the heart is somewhat more sensitive to chronotropic effects of caffeine. Relatively low doses tend to produce a bradycardia, while larger doses increase the heart rate.[104] The bradycardia was prevented by vagotomy or atropine administration, i.e. mediated by a parasympathetic activation.[131] After i.v. administration of 6–18 mg kg^{-1} caffeine, tachycardia was generally observed in dogs and was paralleled by an increased contractile force.[104] Positive chronotropic effects of caffeine were still noted after heart denervation and ligation of adrenal vessels,[132] and thus appeared to be largely direct.

3.2.2.2. Heart-beat rhythm. Very high doses of caffeine (above 20 mg kg^{-1}) were required to produce arrhythmias in dogs.[133,134] Extrasystoles induced in anaesthetised dogs by i.v. injection of 25 mg kg^{-1} caffeine did not occur in pithed dogs.[134] Pretreatment with reserpine also exerted protection.[134] A release of catecholamines, secondary to a central action of caffeine, was presumably involved. The arrhythmogenic effect of catecholamines is ascribed to the occurrence of after-contractions and triggered automaticity, due to a rise of *c*-AMP and cell Ca^{2+} overload.[118,135] It is probable that catecholamines also mediated, at least partly, the decrease of the ventricular fibrillation threshold observed in anaesthetised dogs after i.v. administration of 12·5 mg kg^{-1} caffeine.[136] The decrease of the ventricular fibrillation threshold, which reflects a lack of synchrony of cell repolarisation in ventricles,[137] was prevented by propranolol.[136] This finding does not exclude a direct contribution of caffeine to the phenomenon, because propranolol may exert anti-arrhythmic effects independently on its β-adrenoceptor-blocking action.[138] Atenolol, a more selective β-blocking agent, failed to prevent the reduction of the ventricular fibrillation threshold produced by a high concentration of theophylline in isolated perfused rat hearts.[139] The prolongation of the vulnerable period and decrease of the ventricular fibrillation threshold induced by either epinephrine or dibutyryl *c*-AMP on the same preparation were accentuated in the presence of 1 mM theophylline.[122] At concentrations high enough to inhibit *c*-AMP-phosphodiesterase activity or interfere with the function of the sarcoplasmic reticulum, caffeine was found to produce an increase in excitability and induce various electrophysiological modifications in different types of heart-tissue preparations *in vitro*. These include an increase in impulse conduction velocity,[128] a prolongation of the action potential in guinea-pig left atria[127,140] and dog Purkinje fibres,[141] and a shortening of action potential duration and refractory period in rabbit and cat papillary muscles, respectively.[127,142] Such changes are likely to predispose the heart to spontaneous arrhythmias and to ventricular tachycardia or fibrillation triggered by electric stimuli. Curiously, an inversion from a decrease to an increase of the ventricular tachycardia threshold was observed in rats following daily administrations of high doses of caffeine (90 mg kg^{-1}) for 11 weeks.[143]

3.2.2.3. Interaction with adenosine. All the data mentioned above indicate that high doses of caffeine, unrelated to the low amounts ingested with usual caffeine-containing beverages, are required to produce a tachycardia and still higher doses to trigger arrhythmias in either isolated heart preparations or intact animals. In the presence of adenosine, however, methylxanthines may produce detectable modifications of electrophysiological processes in isolated heart preparations or heart cells at much lower concentrations. Adenosine exerts negative chronotropic[127] and dromotropic effects,[98,127] decreases the automaticity of sino-atrial,[144] atrio-ventricular junctional[98,145] and ventricular pacemakers,[144,146] and opposes isoprenaline effects on membrane potentials.[118,135] At low concentrations (10^{-5} to 10^{-4}M), theophylline was found to antagonise the chronotropic negative effect of adenosine analogues[150] and inhibit the depression of rat heart ventricular automaticity induced by adenosine, even in the presence of propranolol.[146] The adenosine-induced slowing of conduction in the atrio-ventricular node of guinea-pig hearts was prevented by aminophylline at a concentration (10^{-5}M) too low to exert a detectable inhibition of *c*-AMP-phosphodiesterase activity.[98,145] The effect of aminophylline was propranolol insensitive.

3.2.2.4. Effects on hypoxic hearts. Although aminophylline (10^{-5}M) had no effect by itself on the conduction time of oxygenated guinea-pig hearts in the experiments just mentioned, it was effective in decreasing the prolongation of conduction occurring during heart hypoxia,[145] which is a strong stimulus of adenosine formation and release by myocytes.[114] For this reason, methylxanthines were proposed to be of therapeutic value in combating atrio-ventricular blocks produced by endogenous adenosine under hypoxia.[145] On the other hand, adenosine was found to protect isolated working rat hearts from the damaging action of a high concentration of epinephrine (10^{-6}M).[147] Failure of catecholamines to produce high elevations of cell *c*-AMP and Ca-overload in cardiac slices under anoxia may be partly explained by adenosine release.[148] On this basis, methylxanthines may be expected to sensitise the heart to the action of catecholamines and increase the tendency for ventricular tachycardia or fibrillation to complicate myocardial ischaemia.

Partial cell depolarisation secondary to K^+ release from ischaemic tissue and accumulation in the extracellular fluid, and the associated shortening of action potential duration, may be important factors in the genesis of early arrhythmias occurring after coronary occlusion.[122,149] Either *c*-AMP-phosphodiesterase inhibition or antagonism to adenosine action might account for the sensitisation exerted by theophylline at a high concentration (1 mM) to K^+-induced premature ventricular systoles or sustained ventricular tachycardia and fibrillation in isolated rat working hearts with coronary ligation.[122]

3.2.3. Coronary Flow

Most investigations on isolated heart preparations or anaesthetised animals showed a coronarodilating effect of caffeine and other methylxanthines at relatively high doses.[151] *c*-AMP-phosphodiesterase inhibition was likely to account for the coronarodilation, since a rise of the *c*-AMP content of vascular smooth muscle cells is generally associated with vasodilation.[152] A hyperaemia secondary to positive inotropic and chronotropic effects of the xanthine could participate in the response. As seen below, metabolites other than adenosine, the vasodilating effect of which was certainly blocked at high xanthine concentrations, may contribute to reactive hyperaemia. In non-anaesthetised animals, this response may be modified by norepinephrine released under the action of methylxanthines. An increase in coronary resistance to flow may result from coronary adrenergic receptor occupancy due to the prevalence of α-mediated vasoconstriction, and obscure the hyperaemic response to positive inotropic and chronotropic effects of the catecholamine and/or methylxanthine. This mechanism was claimed to be responsible for the reduction of coronary flow noted in conscious dogs after aminophylline administration.[153] In hearts from conscious dogs with experimental coronary ligation, however, the heart transmural blood flow was found to be improved by administration of aminophylline (9–15 mg kg^{-1} i.v.).[154] A concomitant inhibition of the ischaemia-induced increase of the left ventricular end-diastolic pressure was proposed as a factor responsible for the improvement. On the contrary, caffeine (20 mg kg^{-1} i.v.) was found to reinforce the ischaemia-induced increase of left ventricular end-diastolic pressure in propranolol-inhibited hearts from dogs with coronary stenosis and pacing-induced tachycardia.[155] The effect of caffeine was ascribed to an inhibition of Ca^{2+} removal from the cytosol by sarcoplasmic reticulum. These data suggest that methylxanthines, at concentrations low enough not to interfere directly with the function of sarcoplasmic reticulum, may reduce the ischaemia-induced stiffness of the ventricular walls and thereby improve diastolic blood flow distribution to the heart. Either haemodynamic actions resulting in pre-load reduction or antagonistic actions to adenosine binding to myocyte A1 receptors resulting in improved relaxation might be involved. A rise in myocyte *c*-AMP, which is known to activate Ca^{2+} removal by the sarcoplasmic reticulum, would account for the methylxanthine-induced improvement of diastolic relaxation.[110] It is thus possible that, under ischaemia, beneficial effects of methylxanthines on coronary perfusion, secondary to binding to myocyte adenosine receptors, may overcome the adverse effects expected to result from binding to coronary adenosine receptors.

3.2.3.1. Interaction with adenosine on coronary vessels. Aminophylline (10 mg kg^{-1} i.v.) was found to prevent the increase of coronary flow induced by exogenous adenosine in open-chest dogs with coronary ligation.[156] Although endogenous adenosine did not seem to participate in coronary flow regulation in non-ischaemic and unstressed dog or pig hearts,[157–159] the hyperaemic response to coronary occlusion in dog hearts was partially depressed after intracoronary infusion of either adenosine deaminase or theophylline,[160,161] i.e. adenosine takes part in the response together with other vasodilating agents. In addition, aminophylline was found to depress the hyperaemic response to a pacing-induced increase of heart rate, while vasodilating metabolites other than adenosine appeared mainly to mediate the hyperaemia reactive to an increase in heart stroke work.[162]

3.2.4. Blood Pressure

In animals the effects of caffeine on blood pressure vary greatly, in particular according to the dose and the route of administration. A typical response to an intravenous injection of caffeine is an early decrease in blood pressure, which starts within seconds after injection and corresponds in time to the peak arterial concentration before caffeine is distributed in the whole body water.[163–165] This response is likely to result from *c*-AMP-phosphodiesterase inhibition, since *c*-AMP acts to relax vascular smooth muscle by preventing myosin light-chain phosphorylation.[166] A rise in blood pressure usually ensues unless very high doses of caffeine are administered. A secondary rise in blood pressure was clearly observed at caffeine doses even lower than 5 mg kg^{-1} i.v. in rabbits, cats and rats.[163–165] In rats, pretreatment with reserpine was found to prevent it whereas prior adrenalectomy was ineffective.[164] These data suggest that the pressor effect of caffeine, whether due to peripheral arteriolar constriction or a cardiotonic action, involves a release of norepinephrine from sympathetic terminals. A central activation of the adrenergic system is presumably involved, since relatively high levels of methylxanthines are required to produce a detectable release of norepinephrine from sympathetic terminals in isolated hearts.[97,98]

3.2.4.1. Interaction with adenosine. A very slight increase in the arterial plasma level of adenosine produced in the rat by an intravenous infusion of adenosine was accompanied by a decrease in blood pressure.[167] The hypotensive response was not prevented by pretreatment with reserpine and appeared to be due to a direct vasorelaxant effect of adenosine. Recent investigations in hamsters and dogs indicate that endogenous adenosine may exert a control on the blood perfusion of skeletal muscle (during exercise)[168] and adipose tissue,[169] through its vasorelaxant effect. An antagonism to adenosine action on peripheral arterioles probably contributed to the inhibition exerted by caffeine (15 mg kg^{-1} intra-arterially 5 min before adenosine) on the adenosine-induced hypotension.[167] Theophylline was also found to inhibit the hypotensive action of adenosine.[170]

3.2.4.2. Short- and long-term effects of caffeine on blood pressure. Caffeine consumption by rats (0·1% in drinking water) for 3 weeks produced no change in the basal mean arterial pressure but appeared to increase the blood-pressure response to arterial plasma adenosine.[167] On the other hand, the long-term rise in blood pressure in mice induced by a 'psycho-social' stress appeared to be reinforced by adding a very small amount of caffeine (0·002%) to daily drinking water.[171] Within 3–5 months a slight increase in blood pressure (less than 10 mm Hg) was induced by caffeine in mice submitted to a low-intensity stress. The effect vanished at greater stress intensity.[171]

3.3. Studies in Man

3.3.1. Blood Pressure, Heart Rate and Contractility

A regular effect produced in subjects unused to coffee by ingestion of 150 to 970 mg caffeine was a slight rise in blood pressure.[94,95,102,172–176] Diastolic and systolic pressures attained maximal values 15–90 min after ingestion and came back to pre-ingestion values after 3–4 h. The pressor effect was generally associated with a change in heart rate. At low doses (150–280 mg), a decrease in heart rate was noted in most studies,[102,175,177,178] but a rise in heart rate may also occur.[95,179] At high doses (380–970 mg), tachycardia predominates.[103,172] In the presence of bradycardia, which may represent a vagally mediated response to elevation of blood pressure, significant increases of heart left ventricular pre-ejection or ejection times were noted.[102,178] Such changes are those expected to result from an increase in afterload and point to a greater peripheral resistance to flow as primarily responsible for the rise in blood pressure. Because of the longer diastolic filling time, the cardiac output did not decrease despite the bradycardia. However, an increase in cardiac output was observed in other studies, concomitantly with an increase in heart rate.[95,172,179] These data indicate that cardiotonic effects of caffeine may occur at relatively low doses and contribute to the rise in blood pressure.

3.3.1.1. Role of catecholamines and adenosine. In subjects unused to coffee, a common observation after peroral caffeine intake is a rise of plasma norepinephrine and/or epinephrine,[173,180] which may indirectly account for positive chronotropic and inotropic effects of caffeine as well as for peripheral vasoconstriction. Because the pressor effect of caffeine is not always paralleled by a detectable increase in plasma catecholamines,[94] an antagonism to a hypotensive action of endogenous adenosine has been proposed to be involved. However, it is known that a central activation of the sympatho-adrenal system, as produced by a psychological stress, is not consistently reflected in a rise of plasma catecholamines, even though sufficient to induce a marked elevation in blood pressure.[181]

A hypotensive action of adenosine was verified in man,[182] and exogenous adenosine was found to oppose the inotropic positive effect of isoprenaline on a human heart ventricular preparation.[183] However, the role of endogenous adenosine in the control of cardiovascular functions in man is not established. It has been suggested that the role of adenosine in this control might be greater in those subjects who present a tachycardia and other signs of hypersensitivity to caffeine after coffee ingestion.[184]

3.3.1.2. Tolerance. A rapid tolerance to the acute cardiac and pressor effects of caffeine has been observed in several studies.[94,95,174,177] The fact that a tolerance to the caffeine-induced release of catecholamines occurs simultaneously gives support to the contention that catecholamines are responsible for these effects.[94] At withdrawal the sensitivity to caffeine resumes rapidly with clearance of plasma caffeine.[94,102,176] Under conditions simulating regular and heavy coffee consumption (170–250 mg caffeine three times a day for 1–4 weeks), acute effects of caffeine on blood pressure and plasma catecholamines disappeared within a few days in both normotensive and hypertensive subjects, and did not reappear until withdrawal.[94,174] On this basis a role of coffee consumption as a risk factor in the development of hypertension seems unlikely. Information from epidemiological studies about the relationship between blood pressure and coffee consumption is contradictory (see Chapter 3).

3.3.2. Heart-beat Rhythm

Heavy coffee consumers (nine or more cups per day) were found to present a higher incidence of premature ventricular beats during a 2-min ECG monitoring.[96] With plasmatic half-lives generally above 3 h[102,185] the plasma caffeine level was presumably rising to excessively high values under such conditions. Electrophysiological changes may be produced in the human heart by relatively low doses of caffeine. A decrease in the effective refractory period in the atrio-ventricular node as well as in the right atrium and ventricle, and an increase in the left atrium refractory period, were noted after acute intake of 200 mg caffeine.[186,187] Despite these potentially arrhythmogenic changes, which were tentatively attributed to catecholamines released under the action of caffeine, short-term studies in healthy subjects failed to detect a higher incidence of ectopic beats or sustained tachyarrhythmias after caffeine intake.[94,185] Whereas caffeine ingestion (1–3·5 mg kg^{-1} at 3- to 9-h intervals) during a 24-h ECG monitoring was not found to increase the number of atrial or ventricular ectopic beats in healthy subjects, similar measurements in patients presenting spontaneously an elevated incidence of ventricular ectopic beats showed a higher incidence of ectopic beats after caffeine ingestion.[185] In addition, the duration of atrial tachyarrhythmias provoked by electrical stimulation of the heart was prolonged after caffeine ingestion (200 mg), especially in arrhythmia-prone patients.[187] The clinical importance of these observations, notably those obtained under extrastimulation of the heart, is questionable.[185,188] Altogether these data reinforce the old impression that some people are sensitive to dysrhythmic actions of moderate amounts of caffeine. Whether a tolerance may develop with regular coffee ingestion (which would be expected if catecholamines are involved) cannot be established, because the subjects under study were generally asked not to drink coffee for 2–3 days before investigations.

3.3.3. Effects on Ischaemic Hearts

The belief, essentially based on animal studies, that methylxanthines increase the coronary blood flow has led to their use in patients with angina.[189] A therapeutic use of aminophylline in the management of patients with left ventricular failure associated with myocardial ischaemia has been proposed,[154] because the slight inotropic and chronotropic positive effects of the xanthine noted in dogs with coronary occlusion were accompanied by a favourable influence on the perfusion of the heart. On the other hand, aminophylline was found to produce a coronary constriction in healthy subjects and patients with heart disease,[190] whereas adenosine is thought to take part in the reactive hyperaemia after coronary occlusion[160,161] and exert a protection against mechanical overstimulation by catecholamines in ischaemic hearts.[116] Methylxanthines might thus produce undesirable effects through binding to adenosine receptors. According to the site of occlusion, locally produced adenosine is suspected to either favour an evolution to atrio-ventricular block[191] or to protect the heart against ventricular arrhythmias triggered by excessive catecholamine stimulation.[192] Both acute and long-term effects of coffee consumption on the incidence and severity of post-ischaemic arrhythmias are poorly documented. In a recent study on chronic coffee drinkers with coronary heart disease,[193] coffee drinking did not appear to increase the incidence of atrial and ventricular premature contractions induced by exercise or to shorten exercise duration until an ST-segment depression was detected on the ECG. The duration of exercise until angina was even prolonged. However, some studies mentioned above suggest that caffeine may exert untoward effects in subjects unused or intolerant to coffee.

4. COFFEE AND THE GASTROINTESTINAL TRACT

4.1. Introduction

Most physiological effects of coffee depend entirely on its caffeine content; their study requires the use of experimental methods which are very often performed on animals at cellular or subcellular levels, and aimed at the exploration of fundamental biological mechanisms. On the other hand, caffeine explains only partially the digestive effects of coffee, which are principally investigated in humans by way of clinical methods and, mostly, with practical objectives.

Coffee, which is frequently consumed during or after meals, is usually granted a positive image as regards the digestion. However, patients suffering from various digestive diseases feel, or are advised, that it could aggravate their illness. Other subjects, referred to as coffee 'sensitive' or 'intolerant', associate the ingestion of coffee with the appearance of subjective gastrointestinal symptoms which are mainly located in the upper abdomen; they complain frequently of heartburn, nausea, belching, 'heavy' stomach and discomfort. No clear clinical picture emerges from these symptoms or from their association, since they are very common, unspecific and, sometimes, related to an underlying digestive disease. Because of its complexity and its still incompletely known chemical composition, the coffee brew can easily be presented as carrying substances able to 'irritate' the gastrointestinal mucosa. Although no such irritant has been identified up to now, several technological treatments of the unroasted coffee beans have been developed in order to remove or to neutralise them or their precursors. Exposure of the beans to hot water vapour or to liquid carbon dioxide is currently employed for producing 'mild', 'wholesome' or 'free of irritants' coffees aimed at coffee 'intolerant' subjects.

The research on the digestive effects of coffee suffers from many drawbacks which hinder a purely experimental approach. The evaluation of the subjective symptoms needs the participation of specially trained physicians. The measurement of digestive functions such as motility or secretion is often performed within a medical ward by way of unpleasant intubations of the gastrointestinal tract. Part of the digestive effects of coffee are probably linked to its usual oral intake and can be modified if another route of administration (such as gastric instillation) is chosen. The cup of coffee is by no means a simple pharmacological concept; it cannot be viewed as a mere water solution of a known amount of pure drug, be it caffeine or some hypothetical irritant, when coffee is drunk together with a meal the components of which often display a direct effect on the digestive system, as also do the various ingredients (sugar, cream...) which are commonly added to the beverage. The proportions of the many components of the coffee brew depend on a multiplicity of factors such as the species of the beans, the roasting process or the mode of preparation. Decaffeination and treatments for the removal of 'irritating' substances are achieved by way of various procedures, some of which are shared by both technologies, and whose effects on the final composition of the brew are still not fully known (see Volume 2, Chapter 2). Finally, the experimental manipulation of the infusion is rather restricted; for instance, due to reasons of taste and of gastric response to sheer distension, concentrations and volumes have to be kept within rather narrow limits.

However, a predominantly pharmacological and experimental approach to coffee 'intolerance' has been tried in the past. The first coherent presentation[194] of the hypothesis on the presence of 'irritant' substances other than caffeine within the coffee brew, and its testing by the way of an animal model, dates back to 1940. The authors employed the emetic effects of coffee on the dog, and in addition on the pigeon, as a criterion for the presence of 'irritants' (so-called roasting products). They reported a positive relation between acidity of the brews and ematogenic effects, and found that pretreatment of the beans by hot-water steaming, but not decaffeination, lessens the ematogenic effects; they also started to investigate various coffee fractions. Various coffees and coffee components were later systematically tested on digestive functions of rats and mice.[195] This pharmaceutical approach has been neglected nowadays and the relevance of its first results for human physiology has been questioned.

The great majority of the digestive complaints associated with the consumption of coffee point towards, or are interpreted as, troubles of motility and of secretion centred around the stomach. The pancreas is implicated in discussions about the carcinogenicity of coffee, and the liver plays a role in the metabolic pathways, such as the ones of caffeine and of cholesterol; these organs are mentioned elsewhere in this book. Recent extensive and critical reviews whose scope is larger than that of the present section are available.[196,197]

The definition of a 'standard cup of coffee' is still widely debated. For practical purposes we will assume during the following discussion that a 'cup of coffee' is approximately equivalent to 2·2 g instant or 11 g ground coffee, providing 80 mg caffeine dissolved or extracted in 150 ml water. Coffee, without any further specification, refers to regular instant or ground coffee, as opposed to 'decaffeinated' or 'treated' coffees (treatment means a technological pretreatment of the beans, other than decaffeination). In fact, physiological probes have been used for comparing regular coffees with decaffeinated or treated coffees, but not for comparing different types of ground, instant or decaffeinated coffees resulting from various decaffeination processes.

4.2. Effects of Coffee

4.2.1. Lower Oesophageal Sphincter (LES)

Heartburn, a very frequent complaint voiced by coffee 'intolerant' subjects, describes a wave-like retrosternal burning sensation and is often associated with acidic refluxes from the stomach into the oesophagus. Such refluxes are usually prevented by the lower oesophageal sphincter (LES). When this sphincter is incompetent, the acidic content of the stomach may enter the lower oesophagus and irritate its unprotected mucosa, causing eventually a chronic oesophagitis. Diagnosis of LES incompetence relies essentially on manometric monitoring of intra-oesophageal pressure and on detecting acidity of the luminal contents by way of a glass electrode; endoscopy and histological examinations of oesophageal mucosa biopsies are also useful complementary diagnostic procedures.

Intragastric installation of caffeine solutions (40 to 500 mg in 100 or 150 ml) failed to induce any significant changes in LES pressure,[198,199] but a similar volume of solution containing amounts of coffee corresponding to that found in half a cup or in one, two or four cups of coffee, or of decaffeinated coffee, caused a significant dose-dependent increase in this pressure.[199] On the other hand, intragastric administration of caffeine (250 mg in 300 ml water) was reported to increase LES pressure; one cup of coffee, but not a cup of decaffeinated coffee, elicited the same response as caffeine.[200]

Coffee was also tested in conjunction with a test meal. Healthy volunteers and patients suffering from reflux oesophagitis drank either a plain cup of coffee or the same cup to which potassium hydroxide was added in order to neutralise its acidity.[201] A decrease in LES pressure was seen in both groups of subjects after both coffee solutions, whether drunk under fasting conditions or after a test meal; the drop in pressure was, however, more important with the plain acidic coffee. These results were not confirmed: in healthy volunteers the ingestion of a cup of coffee, with or without milk, showed no effect on the LES pressure; a standard Lundh meal caused a significant decrease in this pressure, which was unaffected by a subsequent intake of coffee.[202]

Coffee 'intolerant' subjects were found to display a lower LES pressure than 'tolerant' subjects; in contrast to the latter group, the former did not respond to increasing doses of coffee (the equivalent of half a cup or of one, two or four cups in 150 ml) by an increase in the LES pressure.[203] It was also reported that coffee 'intolerant' subjects, but not coffee 'tolerant' ones, responded to the ingestion of a test meal with coffee by a significant drop in LES pressure; however, in both 'intolerant' and 'tolerant' subjects, LES pressure does not change significantly whenever coffee is replaced by decaffeinated coffee, tea or water.[204]

4.2.2. Stomach Motility

Few experiments have been performed in the past on the effects of coffee on stomach muscles; inhibition as well as stimulation of smooth muscle motility by methylxanthines have both been reported.[205] *In vitro*, coffee induces a contraction followed by a relaxation of rat stomach muscle strips, but these effects are due essentially to unspecific factors such as acidity and potassium or choline content.[206] In rats the blood levels of ethanol or of paracetamol were depressed when the drug was given orally together with or shortly after caffeine; this dose-dependent effect of caffeine is probably related to gastric muscle relaxation and consequent delayed gastric emptying.[207,208]

In humans, gastric-emptying studies rely mostly on roentgenological observations of the progression of radio-opaque barium meals along the gastrointestinal tract. However, such methods provide only qualitative or at best semiquantitative results. Caffeine has been reported to accelerate gastric emptying.[209] When comparing coffee and decaffeinated coffee, a faster emptying rate was found for the former,[210] or similar rates were observed for both coffees.[211] In coffee 'intolerant' subjects, but not in controls, the addition of cream to coffee elicited the prolongation of the gastric emptying time.[212] Stomach intubation permits the sequential collection of liquids marked by means of unabsorbable dyes and provides quantitative data for the analysis of gastric emptying rates. This technique has frequently been used in conjunction with the study of gastric secretion. Thus it was seen that coffee 'intolerant' subjects displayed a slower emptying rate for water or for decaffeinated coffee than coffee 'tolerant' subjects.[213]

4.2.3. Gastric Secretion

The stimulation of gastric secretion is certainly the best known and the most striking digestive property of caffeine. For many years the caffeine probe has been used as a standard clinical test by gastroenterologists.[214,215] Although caffeine is nowadays replaced by more potent and more consistent stimulants (namely histamine, histamine analogues and pentagastrin), the basic philosophy of the probe is still commonly accepted. Briefly stated, the probe aims to provoke a maximal secretory response from the gastric mucosa, in order to classify the patients as normo-, hypo- or hyperacidic responders (which is of importance for diagnostic purposes). The action of caffeine on gastric secretion is, however, species-dependent: the drug is ineffective in dogs[216] and in rats.[217]

Caffeine is effective at a dose (200 mg) corresponding to the content of two or three cups of coffee. It is thus not surprising that this property of caffeine is also shared by coffee, but it was somewhat unexpected to find that it was also shared by decaffeinated coffee.[199,218] The mode of action of caffeine differs from that of the unknown coffee factor which remains after decaffeination. For instance, the plasma level of gastrin (a hormone secreted by the gastric mucosa which promotes acid secretion) is unaffected by caffeine but enhanced by coffee and decaffeinated coffee.[200,219–221] Decaffeinated coffee is an even more potent stimulant of gastric acid secretion and of gastrin release than a protein test meal.[222] However, cimethidine, a drug antagonising H2-receptor, is able to block altogether the acid output stimulation due to caffeine[223] or to coffee.[224] If one assumes that an 'excessive' acid secretion could irritate the gastric mucosa or favour gastro-oesophageal refluxes in 'sensitive' persons, then the removal or the neutralisation of coffee components able to stimulate acid output would generate a 'better tolerated' or 'wholesome' coffee. By reversing this line of thinking, measurement of gastric secretion stimulation could provide a means for discriminating regular and 'wholesome' coffees. Several clinical trials were performed according to this philosophy, and proceeded to first assays based only on clinical analysis of subjective complaints voiced by the participants. However, in order to avoid the rather cumbersome classical methodology for measuring gastric secretion (which requires gastric intubation and an elaborate analysis of the results), substitute methods were generally preferred. These last methods, based on the urine analysis of a dye whose elimination after oral administration is a function of stomach acidity, or on the monitoring of stomach acidity by the way of a telemetric device, are convenient and non-invasive but less informative and less precise than the classical intubation method.

One regular and two 'wholesome' coffees (hot water vapour steaming or 'controlled' roasting procedures) were given to patients suffering from heart and circulatory or various gastrointestinal diseases; the treated coffees were reportedly better tolerated than the regular one.[225] A further purely clinical study showed that the assessment of 'wholesomeness' has to be extracted from a complex network of individual histories. Two coffees, a regular and a treated one (hot water vapour steaming), were compared during a double-blind trial performed on 572 volunteers under conditions very near to those of actual coffee consumption.[226] The two coffees could not be discriminated on the basis of the analysis of subjective symptoms: about 25% of the subjects experienced unpleasant feelings at least once after drinking one or the other of the two tested coffees (more than half of these complaints were related to the digestive tract).

Two groups of authors have reported reduced gastric acid output after treated coffees in comparison with that which occurs after regular coffee ingestion. The first[227] tested a coffee treated by a combination of hot water vapour steaming and of liquid carbon dioxide washing on convalescent patients, and evaluated their gastric acid output by using a urinary dye elimination method. The second[228] administered a coffee treated by hot water vapour steaming to patients suffering from various digestive diseases, and monitored the intragastric acidity by way of a telemetric device. This device also served to demonstrate in patients with peptic ulcers, but not in patients suffering from gastritis, a lesser decrease in intragastric acidity after ingestion of chemically neutralised coffee than after ingestion of the original unaltered coffee.[229]

More recently the accent has switched from a comparison between coffees to one between coffee 'tolerant' and 'intolerant' subjects. Gastric acid production was also measured by way of classic intubation methods. A lesser acid secretion after coffee intake[203] and a greater acid output after ingestion of decaffeinated coffee combined with a test meal[204] were reported in coffee 'intolerant' subjects. However, a similar dose-dependent response of acid output after decaffeinated coffee was found in both 'intolerant' and 'tolerant' subjects.[213]

4.2.4. Ulcerogenic Effects on the Gastric Mucosa

In the rat, high doses of caffeine cause erosions and ulcerations of the gastric mucosa[230] and prevent the healing of chronic gastritis.[231] The primary effect of caffeine could be a reduction of the gastric tissue blood flow, resulting in an inhibition of motility and secretion which favours the formation of ulcers.[232] On the other hand, in dogs caffeine is unlikely to injure the gastric mucosa since it did not alter significantly net ionic fluxes and transmural potentials.[233]

These experiments suggest that the question of caffeine representing a potential irritant for the human gastric mucosa may be raised, especially in the presence of gastritis, but the problem has not been investigated. The peptic ulcer constitutes a pathological entity very different from the above-mentioned ulcerations. Caffeine has not been incriminated in its formation, and an epidemiological study has concluded that it is not correlated with the prevalence of peptic ulcers.[234]

4.2.5. Intestine

The effects of caffeine and coffee on the intestine have been investigated only cursorily. Among the complaints of coffee 'intolerant' subjects, those pointing towards the lower abdomen (such as diarrhoea) are the least frequent. The implication of coffee in the irritable bowel syndrome, which is essentially due to abnormal intestinal motility resulting from emotional or physical stress, can be hypothesised *a priori* but has not yet been supported by clinical observations.

In animals, caffeine has been reported to stimulate or to inhibit intestinal motility, according to species, dose or methodology.[196] In man, X-ray examinations detected a stimulation of intestinal motility by coffee, and in a lesser way by decaffeinated coffee;[211] however, coffee failed to alter the mouth-to-caecum transit time as measured by the breath hydrogen test.[235]

During intestinal perfusions in humans, caffeine promoted net fluid movements[236,237] but did not affect intestinal motility;[237] on the other hand, coffee did not mimic the action of caffeine on fluid movements.[237]

Some effects of caffeine on intestinal absorption have also been reported. In rats, caffeine stimulated the absorption of tritiated water, urea, antipyrine and salicylic acid,[238] as well as that of ergotamine.[239] In man it enhanced persorption, a mechanism by which large molecules cross the intestinal mucosa through intercellular pathways.[240] Coffee, but not caffeine, increased the absorption of ethanol when ingested before its administration.[241]

4.2.6. Gall Bladder

Gallstones appear frequently within the gall bladder and are often unnoticed. Symptoms arise when contractions force calculi into the cystic duct; the patient experiences a steady pain in the right upper abdomen. It was postulated that coffee could induce contractions of the gall bladder and that this effect could be prevented by the technological removal of the active agent. However, it should be emphasised that gall-bladder contractions are usually not painful and permit the physiological emptying of this organ.

Clinical studies have demonstrated the effect of two cups of coffee on the gall bladder. X-Ray examination showed that coffee and also treated coffee (hot water vapour steaming), and to a lesser extent decaffeinated coffee, were able to induce gall-bladder contractions;[210] this stimulation was also observed when coffee or, even more, decaffeinated coffee was given in conjunction with a test meal.[211] The use of echography, a more precise method, permitted confirmation of the stimulation of gall-bladder contraction by a water vapour treated coffee.[242]

4.3. Conclusions

The picture of the relationship between coffee and the digestive tract has not been fundamentally altered recently. However, its investigation from a pharmacological point of view, which prevailed up to the 1960s, has increasingly given way to a more empirical and clinical one. Certainly coffee acts on several gastrointestinal functions, but many of these effects appear of little practical importance under usual conditions of coffee consumption. Some of them can be traced to the presence of caffeine, but others appear as proper effects of coffee itself and cannot be related to any specific component of physicochemical property of the brew.

The stimulation of gastric acid secretion by both coffee and decaffeinated coffee still remains their most outstanding digestive effect. Further physiological actions, on other organs or on other functions—mainly on motility, are often secondary to the secretory effect and are not presently regarded as important in practice. Generally speaking, a promotion of gastric acid secretion can reasonably be judged as a positive contribution to digestion. However, in the presence of diseases of the gastrointestinal tract, especially when symptoms are dependent on an excessive acid output (peptic ulcers or some form of gastritis, for instance) or even when they appear to be related in some way to the intake of coffee, safe management of these patients requires a restriction on coffee consumption even if a positive relation of causality cannot be established.

The existence of a digestive syndrome caused by coffee is still questionable. Older reports substantiating this hypothesis or claiming beneficial effects of various technological treatments of the coffee beans have not been confirmed. More recent investigations failed to link coffee ingestion and incompetency of the lower oesophageal sphincter, a condition which is frequently at the origin of heartburn. The occurrence of some kind of 'intolerance' to coffee is taken for granted by many authors, especially in the United States and in Germany; it is recognised in subjects complaining from essentially upper-abdominal symptoms following coffee ingestion. Minor differences between such subjects and coffee 'tolerant' ones were observed during examinations of lower oesophagael pressure or of gastric secretion and motility, but these findings are still preliminary.

Further progress in methodology, more extensive research on the digestive effects of various food items and closer clinical investigations on the background of subjective complaints will undoubtedly improve in the very near future the understanding of the implications of coffee in the complex and intricate physiopathological processes of digestion.

Methodology could benefit from the adoption of more elaborate experimental designs based on well founded hypothesis.[243] Responses to coffee are rarely studied as functions of dose and volume, even within rather narrow physiological limits, nor as functions of time. The usual coffee-intake patterns of the subjects, the origin of the coffee and its mode of preparation are seldom controlled.

Improvement on these common weaknesses, as well as a return to more fundamental approaches, would certainly help the advance of future research.

Measurements of digestive functions frequently require intra-enteral intubations, whose procedures are not always fully standardised and which could display effects of their own, such as being a factor of stress[244] and affecting directly gastric and intestinal motility.[245] Developments of classical methods provide opportunities for more precise and more relevant observations. For instance, intragastric titration permits the *in situ* monitoring of acid output in the presence of a liquid meal.[246] A family of non-invasive methods, which avoid intra-enteral intubation, employ radioactive tracers in order to probe digestive function, such as gastric secretion[247] or rates of emptying of solid foods.[248] Furthermore, a combination of classical and new non-invasive methods offers the opportunity to quantify simultaneously gastric secretion and the emptying of liquids and solids after a mixed meal.[249]

The digestive effects of coffee have not been systematically compared with those of other foods. Indeed some of the observations made after coffee ingestion are also common to other drinks. For instance, the occurrence of heartburn is not uncommon after ingestion of orange juice and is not yet fully understood.[250] Besides coffee and decaffeinated coffee, various drinks were found to stimulate gastric acid output (caffeinated and non-caffeinated soft drinks, tea, beer, milk), but this property could not be predicted from their physicochemical composition.[251]

The digestive effects of coffee should also be evaluated against a more general physiological and clinical background. The gastric acid output response to coffee is certainly important; it is nevertheless of the same magnitude as that induced by other food items, and smaller than those provoked by histamine or pentagastrin (the so-called 'maximal acid output'); much smaller also than those observed during hyperacidic pathological conditions. Other weaker digestive effects of coffee can also be regarded as mere physiological variations rather than as pathological deviations.

Digestive symptoms are frequent and often remain unexplained, a situation which has led to the concept of 'functional gastrointestinal disorders'.[252] This clinical entity refers to patients whose symptoms cannot be related to any organic disease, but where psychosocial and psychological factors are often involved. Such patients do happen to incriminate food in the genesis of their disease. They may also display hypersensitivity to distension of the gastrointestinal tract, such as that provoked by clinical examination procedures.

REFERENCES

1. Rall, T. W., in *The Pharmacological Basis of Therapeutics*, Eds A. Goodman Gilman, L. S. Goodman and A. Gilman, Macmillan, New York, 1980, 592.
2. Eichler, O., *Kaffee und Coffein*, Springer-Verlag, New York, 1976.
3. Dews, P. B., *Caffeine*, Springer-Verlag, New York, 1984.
4. Hirsh, K., in *The Methylxanthine Beverages and Foods: Chemistry, Consumption and Health and Effects*, Ed. G. A. Spiller, Alan M. Liss, New York, 1984, 235–301.
5. Bättig, K., in *Coffee: Botany, Biochemistry and Production of Beans and Beverages*, Eds M. N. Clifford and K. C. Willson, AVI, Westport, Conn., 1985, 394–439.
6. Sivetz, M., in *Coffee Technology*, Eds M. Sivetz and N. Desrosier, AVI, Westport, Conn., 1979, 575–597.
7. Gilbert, R. M., Marshmann, J. A., Schwieder, M. and Berg, R., *Can. Med. Assoc. J.*, 1976, **114**, 205–208.
8. Rozin, P. and Cines, B. M., *Ecol. Food Nutr.*, 1984, **12**, 79–88.
9. Cines, B. M. and Rozin, P., *Appetite J. Int. Res.*, 1982, **3**, 23–34.
10. Goldstein, A. and Kaizer, S., *Clin. Pharmacol. Ther.*, 1969, **10**, 477–488.
11. Rahmann, H., *Pflügers Arch.*, 1963, **276**, 384–397.
12. Hoffmeister, F. and Wüttke, W., *J. Pharmacol. Exp. Ther.*, 1973, **186**, 266–275.
13. Griffiths, R. R., Brady, J. V. and Bradford, L. D., in *Advances in Behavioural Pharmacology, Vol. 2*, Eds T. Thomas and P. B. Dews, Academic Press, London, 1979, 163–208.
14. Alpern, H. P. and Jackson, S. J., in *Psychopharmacology: A Generation of*

Progress, Eds M. A. Lipton, A. DiMascio and K. F. Killam, Raven Press, New York, 1978, 663–675.
15. Cooper, B. R., Potts, W. J., Morse, D. L. and Black, W. C., *Psychonomic Sci.*, 1969, **14**, 225–226.
16. Burov, Y. V. and Borisenko, S. A., *Bull. Biol. Med.*, 1975, **79**, 140–143.
17. Weiss, B. and Laties, V. G., *Pharmacol. Rev.*, 1962, **14**, 1–36.
18. Foltz, C. D., Ivy, A. C. and Barborka, C. J., *Am. J. Physiol.*, 1942, **136**, 79–86.
19. Carpenter, J. A., *J. Comp. Physiol. Psychol.*, 1959, **52**, 491–496.
20. Hull, C. L., *J. Gen. Psychol.*, 1935, **13**, 249–274.
21. Payne, R. B. and Hauty, G. T., *J. Exp. Psychol.*, 1954, **47**, 267–273.
22. Hauty, G. T. and Payne, R. B., *J. Exp. Psychol.*, 1955, **49**, 60–67.
23. Switzer, S. A., *J. Comp. Psychol.*, 1935, **19**, 155–175.
24. Switzer, S. A., *J. Gen. Psychol.*, 1935, **12**, 78–90.
25. Graf, O., in *German Aviation Medicine, World War II, Vol. 2*, US Government Printing Office, Washington, DC., 1950, 1080–1103.
26. Hirsh, K. R., Pinzone, M. G. and Forde, J. M., *Fed. Proc.*, 1974, **33**, 466.
27. Waldeck, B., *Acta Pharmacol. Toxicol.*, 1975, **36** (Suppl. IV), 1–23.
28. Scott, C. C., Anderson, R. C. and Chen, K. K., *J. Pharmacol. Exp. Ther.*, 1946, **86**, 113–119.
29. Strom, S., *Acta Pharma Suec.*, 1964, **1** (Suppl. 1), 22–31.
30. Boissier, J. R. and Simon, P., *Arch. Int. Pharmacodyn.*, 1965, **158**, 212–221.
31. Herz, A., Neteler, B. and Teschemacher, H. J., *Naunyn Schmiedebergs Arch. Pharmakol. Exp. Pathol.*, 1968, **261**, 486–502.
32. Berkowitz, B. A., Tarver, J. H. and Spector, S., *Eur. J. Pharmacol.*, 1970, **10**, 64–71.
33. Natsuno, Y. and Inada, S., *Jpn J. Vet. Sci.*, 1972, **34**, 243–253.
34. Thithapandha, A., Maling, H. M. and Gillette, J. R., *Proc. Soc. Exp. Biol. Med.*, 1972, **139**, 582–586.
35. Snyder, S. M., Katims, J. J., Annau, Z. and Bruns, R. F., *Proc. Natl Acad. Sci.*, 1981, **78**, 3260–3264.
36. Kurtbat, N. M. and Korablev, M. V., *Farmakol. Toksikol.*, 1981, **44**, 559–561.
37. Würzner, H. P., Enslen, M. and Milon, H., *Proc. 10th Coll. ASIC*, 1982, 309–317.
38. Enslen, M., Milon, H. and Würzner, H. P., *Experientia*, 1980, **36**, 1105–1106.
39. Milon, H. and McRae-Degueurce, A., *Neurosci. Lett.*, 1982, **30**, 297–301.
40. McRae-Degueurce, A. and Milon, H., *Brain Res.*, 1983, **263**, 344–347.
41. Sinton, Ch., Valatx, J. L. and Jouvet, M., *Neurosci. Lett.*, 1981, **24**, 169–174.
42. Giddings, G., *J. Am. Med. Assoc.*, 1934, **102**, 525–529.
43. Stradomsky, N., *Med. Klin.*, 1970, **65**, 1372–1376.
44. Müller-Limmroth, W., *Proc. 5th Coll. ASIC*, 1971, 375–382.
45. Brezinova, V., *Br. J. Clin. Pharmacol.*, 1974, **1**, 203–208.
46. Karacan, I., Thornby, J. I., Anch, A. N., Booth, G. M., Williams, R. L. and Salis, P. J., *Clin. Pharmacol. Ther.*, 1976, **20**, 682–689.
47. Goldstein, A., Warrer, R. and Kaiser, S., *J. Pharmacol. Exp. Ther.*, 1965, **149**, 156–159.
48. Dews, P. B., *Ann. Rev. Nutr.*, 1982, **2**, 323–341.
49. Callahan, M. M., Robertson, R. S., Arnaud, M. J., Branfman, A. R., McComish, M. F. and Yesair, D. W., *Drug Metab. Dispos.*, 1983, **11**, 211–217.

50. Aldridge, A., Bailey, J. and Neims, A. H., *Sem. Perinatol.*, 1981, **5**, 310–314.
51. Statland, B. E. and Demas, T. J., *Am. J. Clin. Pathol.*, 1980, **73**, 390–393.
52. Desmond, P. V., Patwardham, R. V., Johnson, R. F. and Schenker, S., *Digest. Dis. Sci.*, 1980, **25**, 193.
53. Parsons, W. D. and Neims, A. H., *Clin. Pharmacol. Ther.*, 1978, **24**, 40–45.
54. Aranda, J. W., Cook, C. E., Gorman, W., Collinge, J. M., Loughnan, P. M., Outerbridge, E. W., Aldridge, A. and Neims, A. H., *J. Pediatr.*, 1979, **94**, 663.
55. Aranda, J. W., Louridas, T., Vittulo, B. B., Thom, B., Aldridge, A. and Haber, R., *Science*, 1979, **206**, 1319–1321.
56. Levy, M. L. and Zylber-Katz, E., *Clin. Pharmacol. Ther.*, 1983, **33**, 770–775.
57. Fernstrom, J. D. and Fernstrom, M. H., in *Caffeine*, Ed. P. B. Dews, Springer-Verlag, New York, 1984, 107–117.
58. Vogt, M., in *Metabolism of the Nervous System*, Ed. D. Richter, Pergamon Press, London, 1957, 553–564.
59. Waldeck, B., *J. Pharm. Pharmacol.*, 1971, **23**, 824–830.
60. Berkowitz, B. A. and Spector, S., *Eur. J. Pharmacol.*, 1971, **16**, 322–325.
61. Corrodi, H., Fuxe, K. and Jonsson, G., *J. Pharm. Pharmacol.*, 1972, **24**, 155–158.
62. Strömberg, U. and Waldeck, B., *J. Pharm. Pharmacol.*, 1973, **25**, 302–308.
63. Valzelli, L. and Bernasconi, S., *Pharmacol. Biochem. Behav.*, 1973, **1**, 251–254.
64. Waldeck, B., *Acta Pharmacol. Toxicol.*, 1975, **36**, 172–180.
65. Chou, D. T., Cuzzone, H., Springstead, J., Ali, R. and Hirsh, K., *Neurosci. Abstr.*, 1979, **5**, 551.
66. Cuzzone, H., Hirsh, K., Ali, R. and Chou, D. T., *Fed. Proc.*, 1981, **40**, 266.
67. Schlosberg, A. J., Fernstrom, J. D., Kopczynski, M. C., Cusak, B. M. and Gillis, M. A., *Life Sci.*, 1981, **29**, 173–183.
68. Chou, D. T., Cuzzone, H., Springstead, J. and Hirsh, K., *Fed. Proc.*, 1982, **41**, 1062.
69. Spindel, E. R. and Wurtman, R. J., in *Caffeine*, Ed. P. B. Dews, Springer-Verlag, New York, 1984, 119–128.
70. Spindel, E. R., Arnold, M. A., Cusack, B. and Wurtman, R. J., *J. Pharmacol. Exp. Ther.*, 1980, **214**, 58–62.
71. Aldridge, A., Parsons, W. D. and Neims, A. M., *Life Sci.*, 1977, **21**, 967–974.
72. Spindel, E. R., Griffith, L. and Wurtman, R. J., *J. Pharmacol. Exp. Ther.*, 1983, **225**, 346–350.
73. Arnold, M. A., Carr, D. B., Togosaki, D. M., Pian, M. C. and Martin, J. B., *Life Sci.*, 1982, **32**, 1017–1024.
74. Spindel, E. R., Wurtman, R. J., McCall, A., Carr, D. B., Conlay, L., Griffith, L. and Arnold, M. A., *Clin. Pharmacol. Ther.*, 1984, **36**, 402–407.
75. Arogaro, P., Capri, C., Pais, M. and Cazzoleto, G., *Israel J. Med. Sci.*, 1973, **9**, 114–119.
76. Rall, T. W., *Pharmacologist*, 1982, **24**, 277–287.
77. Bianchi, C. P., in *Cellular Pharmacology of Excitable Tissues*, Ed. T. Narahashi, C. C. Thomas, Springfield, IL, 1975, 485–519.
78. Katz, A. M., Repke, D. I. and Hasselbach, W., *J. Biol. Chem.*, 1977, **252**, 1938–1949.
79. Berthet, J., Sutherland, E. W. and Rall, T. W., *J. Biol. Chem.*, 1957, **229**, 351–361.
80. Sattin, A. and Rall, T. W., *Mol. Pharmacol.*, 1970, **6**, 13–23.

81. Huang, M. and Daly, J. M., *Life Sci.*, 1974, **14**, 489–503.
82. Snyder, S. H., in *Caffeine*, Ed. P. B. Dews, Springer-Verlag, New York, 1984, 129–141.
83. Phillis, J. W. and Kostopoulos, G. K., *Life Sci.*, 1975, **17**, 1085–1094.
84. Phillis, J. W., Edstrom, J. P., Kostopoulos, G. K. and Kirkpatrick, J. R., *Can. J. Physiol. Pharmacol.*, 1979, **57**, 1289–1312.
85. Perkins, M. N. and Stone, T. W., *Arch. Int. Pharmacodyn. Ther.*, 1980, **246**, 205–214.
86. Cushny, A. R., *J. Pharmacol. Exp. Ther.*, 1913, **4**, 363.
87. Trippenbach, T., Zinman, R. and Milic-Emili, J., *Respir. Physiol.*, 1980, **40**, 211.
88. Mueller, R. A., Lundberg, D. B. and Breese, G. R., *J. Pharmacol. Exp. Ther.*, 1981, **218**, 593.
89. Turmen, T., Davis, J. and Aranda, J. V., *Semin. Perinatol.*, 1981, **5**, 326.
90. Horowitz, D. A., Jablonski, W. and Mehta, K. A., *Am. J. Dis. Childh.*, 1982, **136**, 73.
91. Khanna, N. K. and Somani, S. M., *Clin. Toxicol.*, 1984, **22**, 473–483.
92. Dews, P. B., in *Caffeine*, Ed. P. B. Dews, Springer-Verlag, New York, 1984, 86–103.
93. Jouvet, M., Benoit, O., Marsallon, A. and Courjon, J., *C.R. Soc. Biol. Lyon*, 1957, **151**, 1542–1545.
94. Robertson, D., Wade, D., Workman, R., Woosley, R. L. and Oates, J. A., *J. Clin. Invest.*, 1981, **67**, 1111–1117.
95. Buhr, G., *Arzneumittel-Forsch.*, 1958, **12**, 584–590.
96. Prineas, R. F., Jacobs, D. R., Crow, R. S. and Blackburn, H., *J. Chron. Dis.*, 1980, **33**, 62–72.
97. Westfall, T. C. and Brasted, M., *J. Pharmacol. Exp. Ther.*, 1974, **189**, 659–665.
98. Belardinelli, L., Fenton, R. A., West, A., Linden, J., Althaus, J. S. and Berne, R. M., *Circ. Res.*, 1982, **51**, 569–579.
99. Berkowitz, B. A. and Spector, S., *Eur. J. Pharmacol.*, 1971, **13**, 193–196.
100. Poisner, A. M., *Biochem. Pharmacol.*, 1973, **22**, 469–476.
101. Bondaryov, M. G., *Progr. Brain Res.*, 1967, **20**, 171–186.
102. Whitsett, T. L., Manion, C. V. and Christensen, H. D., *Am. J. Cardiol.*, 1984, **53**, 918–922.
103. Dalla Torre, L. and Boldrini, R., *Minerva Med.*, 1952, **1**, 853–856.
104. Eichler, O., in *Kaffee und Coffein*, Ed. O. Eichler, Springer-Verlag, Berlin, 1976, 281–286.
105. Tsien, R. W., in *Advances in Cyclic Nucleotide Research*, Eds A. P. Greengard and G. A. Robison, Raven Press, New York, 1977, 363–420.
106. Wolff, J., Londos, C. and Cooper, D. M. F., *Adv. Cycl. Nucl. Res.*, 1981, **14**, 199–214.
107. LaMonica, D. A., Frohloff, N. and Dobson, J. G., *Am. J. Physiol.*, 1985, **248**, H737–H744.
108. Blaney, L., Thomas, H., Daries, P. S. and Opie, L. H., *J. Clin. Invest.*, 1978, **61**, 1260–1269.
109. Weber, A. and Herz, R., *J. Gen. Physiol.*, 1968, **52**, 750–759.
110. Katz, A. M., *Adv. Cycl. Nucl. Res.*, 1979, **11**, 303–343.
111. Fabiato, A. and Fabiato, F., *Ann. Rev. Physiol.*, 1979, **41**, 473–484.

112. Londos, C., Wolff, J. and Cooper, D. M. F., in *Regulatory Function of Adenosine*, Eds R. M. Berne, T. W. Rall and R. Rubio, Martinus/Nijhoff, Boston, Mass., 1983, 17–32.
113. Dobson, J. G. Jr, *Circ. Res.*, 1983, **52**, 151–160.
114. Rubio, R., Wiedmeyer, V. T. and Berne, R. M., *J. Mol. Cell. Cardiol.*, 1974, **6**, 561–566.
115. Wiedmeyer, V. T. and Spell, L. H., *Circ. Res.*, 1977, **41**, 503–508.
116. Dobson, J. G., Ordway, R. W. and Fenton, R. A., *Am. J. Physiol.*, 1986, **251**, H455–H462.
117. Schnarr, R. L. and Sparks, H. V., *Am. J. Physiol.*, 1972, **223**, 223–228.
118. Belardinelli, L. and Isenberg, G., *Circ. Res.*, 1983, **53**, 287–297.
119. Hedquist, P. and Fredholm, B. B., *Naunyn Schmiedebergs Arch. Pharmacol.*, 1976, **293**, 217–223.
120. Verhaeghe, R. H. and Vanhoutte, P. M., *Circ. Res.*, 1977, **40**, 208–215.
121. Strange, R. C., Vetter, N., Rowe, M. J. and Oliver, M. F., *Eur. J. Clin. Invest.*, 1974, **4**, 115–119.
122. Opie, L. H., *Am. J. Cardiol.*, 1975, **36**, 938–953.
123. Williamson, J. R., *Pharmacol. Rev.*, 1966, **18**, 205–210.
124. Blinks, J. R., Olson, C. B., Jewell, B. R. and Braveny, P., *Circ. Res.*, 1972, **30**, 367–392.
125. Henderson, A. H., Brutsaert, D. L., Forman, R. and Sonnenblick, E. H., *Cardiovasc. Res.*, 1974, **8**, 162–172.
126. Mushlin, P., Boerth, R. C. and Wells, J. N., *Mol. Pharmacol.*, 1981, **20**, 179–189.
127. DeGubareff, T. and Sleater, W. Jr, *J. Pharmacol. Exp. Ther.*, 1965, **148**, 202–214.
128. Schaff, G., Duchène-Marullaz, P., Lautier, F. and Lavaranne, J., *C.R. Acad. Sci. sér. D (Paris)*, 1967, **161**, 1415–1419.
129. Kulkarni, D. R. and Gambhir, S. S., *Ind. J. Physiol. Pharmacol.*, 1966, **9**, 75–80.
130. Hughes, M. J. and Coret, I. A., *J. Mol. Cell. Cardiol.*, 1975, **7**, 613–624.
131. Block, J. and Buchholtz, J., *Arch. Exp. Pharmacol.*, 1980, **88**, 192–215.
132. Schaff, G. and Faucon, G., *Arch. Int. Pharmacodyn.*, 1963, **145**, 213–232.
133. Sollman, T. and Pilcher, J. D., *J. Pharmacol.*, 1912, **3**, 66–93.
134. Ueda, I., Loehning, R. W. and Ueyama, H., *Anesthesiol.*, 1961, **22**, 926–932.
135. Isenberg, G. and Belardinelli, L., *Circ. Res.*, 1984, **55**, 309–325.
136. Bellet, S., Horstmann, E., Roman, L. R., deGuzman, N. T. and Kostis, J. B., *Am. Heart J.*, 1972, **84**, 215–227.
137. Zipes, D. P., *Circulation*, 1975, **51/52** (Suppl. 111), 120–130.
138. Kupersmith, J., Shiang, H. and Litwak, R. S., *Circ. Res.*, 1976, **38**, 302–307.
139. Lubbe, W. F., Podzuweit, T., Daries, P. S. and Opie, L. H., *J. Clin. Invest.*, 1978, **61**, 1260–1269.
140. Kimoto, Y., *Jpn J. Physiol.*, 1972, **22**, 225–238.
141. DiGennaro, M. and Vassalle, M., *Am. J. Physiol.*, 1985, **249**, H520–H533.
142. Kuschinsky, G. and Muscholl, E., *Arch. Exp. Path. Pharmacol.*, 1956, **229**, 348–359.
143. Rottenberg, F. A., deFeo, J. J. and Swonger, A. K., *Lancet*, 1976, **11**, 140–141.
144. Szentmiklosi, A. J., Nemeth, M., Szegi, J., Papp, J. G. and Skekeres, L., *Naunyn Schmiedebergs Arch. Pharmacol.*, 1980, **311**, 147–149.
145. Belardinelli, L., Belloni, F. L., Rubio, R. and Berne, R. M., *Circ. Res.*, 1980, **47**, 684–691.

146. Heller, L. J. and Olsson, R. A., *Am. J. Physiol.*, 1985, **248**, H907–H913.
147. Opie, L. H., Walpoth, B. and Barsacchi, R., *J. Mol. Cell. Cardiol.*, 1985, **17** (Suppl. 2), 21–34.
148. Dobson, J. G. Jr, *Circ. Res.*, 1978, **43**, 785–792.
149. Fozzard, H. A. and Makielski, J. C., *Ann. Rev. Med.*, 1985, **36**, 275–284.
150. Evans, D. B., Schenden, J. A. and Bristol, J. A., *Life Sci.*, 1982, **31**, 2425–2432.
151. Iwai, M. and Sassa, K., *Arch. Exp. Path. Pharmacol.*, 1973, **99**, 215–225.
152. Kukovetz, W. R., Pöch, G. and Holzmann, S., in *Vasodilation*, Raven Press, New York, 1981, 339–353.
153. Rutherford, J. D., Vatner, S. F. and Braunwald, E., *Circulation*, 1981, **63**, 378–387.
154. Rutherford, J. D., Vatner, S. F. and Braunwald, E., *Am. J. Cardiol.*, 1981, **48**, 1071–1076.
155. Paulus, W. J., Serizawa, T. and Grossman, W., *Circ. Res.*, 1982, **50**, 218–227.
156. Saito, D., Hyodo, T., Takeda, K., Abe, Y., Tani, H., Yamada, N., Ueeda, M. and Nakatsu, T., *Am. J. Physiol.*, 1985, **248**, H812–H817.
157. Kroll, K. and Feigl, E. O., *Am. J. Physiol.*, 1985, **249**, H1176–H1187.
158. Gewirtz, H., Olsson, R. A., Brautigan, D. L., Brown, P. R. and Most, A. S., *Am. J. Physiol.*, 1986, **250**, H1030–H1036.
159. Hanley, F. L., Grattan, M. T., Stevens, M. B. and Hoffman, J. I. E., *Am. J. Physiol.*, 1986, **250**, H558–H566.
160. Dole, W. P., Yamada, N., Bishop, V. S. and Olsson, R. A., *Circ. Res.*, 1985, **56**, 517–524.
161. Hori, M., Inoue, M., Kitakaze, M., Koretsune, Y., Iwai, K., Tamai, J., Ito, H., Kitabatake, A., Sato, T. and Kamada, T., *Am. J. Physiol.*, 1986, **250**, H509–H518.
162. Randall, J. R. and Jones, C. E., *Am. J. Physiol.*, 1985, **248**, H1–H7.
163. Hahn, F., *Arch. Exp. Path. Pharmacol.*, 1941, **198**, 491–508.
164. Strubelt, O. and Siegers, C. P., *Arzneimittel-Forsch.*, 1968, **18**, 1278–1281.
165. Raff, W. K., Scholtholt, J. and Lochner, W., *Arzneimittel-Forsch.*, 1971, **21**, 1177–1179.
166. Conti, M. A. and Adelstein, R. S., *Fed. Proc.*, 1980, **39**, 1569–1573.
167. Von Borstel, R. W., Wurtman, R. J. and Conlay, L. A., *Life Sci.*, 1983, **32**, 1151–1158.
168. Proctor, K. G., *Am. J. Physiol.*, 1984, **247**, H195–H205.
169. Martin, S. E. and Bockman, E. L., *Am. J. Physiol.*, 1986, **250**, H1127–H1135.
170. Schaumann, W., Juhran, W. and Dietmann, K., *Arzneimittel-Forsch.*, 1970, **20**, 372–377.
171. Henry, J. P. and Stephens, P. M., *Pharmacol. Biochem. Behav.*, 1980, **13**, 719–727.
172. Grollmann, A., *J. Pharmacol. Exp. Ther.*, 1930, **39**, 313–327.
173. Robertson, D., Fröhlich, J. C., Carr, R. K., Watson, J. T., Hollifield, J. W., Shand, D. G. and Oates, J. A., *N. Engl. J. Med.*, 1978, **298**, 181–186.
174. Ammon, H. P. T., Bieck, P. R., Mandalaz, D. and Verspohl, E. J., *Br. J. Clin. Pharmacol.*, 1983, **15**, 701–706.
175. Smits, P., Hoffmann, H., Thien, T., Houben, H. and van't Laar, A., *Clin. Pharmacol. Ther.*, 1983, **34**, 153–158.
176. Smits, P., Thien, T. and van't Laar, A., *Am. J. Cardiol.*, 1985, **56**, 958–963.

177. Colton, T., Gosselin, R. E. and Smith, R. P., *Clin. Pharmacol. Ther.*, 1967, **9**, 31–39.
178. Pincomb, G. A., Lavallo, W. R., Passey, R. B., Whitsett, T. L., Silverstein, S. M. and Wilson, M. F., *Am. J. Cardiol.*, 1985, **56**, 119–122.
179. Gould, L., Venkataraman, K., Goswami, M. and Gomprecht, R. F., *Angiol.*, 1973, **24**, 455–463.
180. Levi, L., *Acta Med. Scand.*, 1967, **181**, 431–438.
181. Floras, J., Jones, J. V., Hassan, M. O., Osikowska, B. A., Sever, P. S. and Sleight, P., *Hypertension*, 1986, **8**, 641–649.
182. Sollevi, A., Lagerkranser, M., Irestedt, L., Gordon, E. and Lindquist, C., *Anesthesiol.*, 1984, **61**, 400–405.
183. Böhm, M., Meyer, W., Mügge, A., Schmitz, W. and Scholz, H., *Eur. J. Pharmacol.*, 1985, **116**, 323–326.
184. Fredholm, B. B., in *The Methylxanthine Beverages and Foods: Chemistry, Consumption and Health Effects*, Ed. G. A. Spiller, Alan R. Liss, New York, 1984, 303–330.
185. Sutherland, D. J., McPherson, D. D., Renton, K. W., Spencer, C. A. and Montague, T. J., *Chest*, 1985, **87**, 319–324.
186. Gould, L., Reddy, C. V. R., Oh, K. C., Kim, S. G. and Becker, W., *J. Clin. Pharmacol.*, 1979, **19**, 46–55.
187. Dobomeyer, D. J., Stine, R. A., Leier, C. V., Greenberg, R. and Schaal, S. F., *N. Engl. J. Med.*, 1983, **308**, 814–816.
188. Graboys, T. B. and Lown, B., *N. Engl. J. Med.*, 1983, **308**, 835–837.
189. Russek, H. I., *Am. J. Med. Sci.*, 1960, **239**, 87–93.
190. Maxwell, G. M., Crumpton, C. W., Rowe, G. G., White, D. H. and Castillo, C. A., *J. Lab. Clin. Med.*, 1959, **54**, 88–95.
191. Berne, R. M., DiMarco, J. P. and Belardinelli, L., *Circulation*, 1984, **69**, 1195–1197.
192. Lerman, B. B., Belardinelli, L. and DiMarco, J. P., *J. Am. Coll. Cardiol.*, 1986, **7**, 155.
193. Piters, K. M., Colombo, A., Olson, H. G. and Butman, S. M., *Am. J. Cardiol.*, 1985, **55**, 277–280.
194. Behrens, B. and Malorny, G., *Arch. F. Exp. Path. Pharmakol.*, 1940, **194**, 369–388.
195. Czok, G., *Z. Ernährungswiss.*, 1966 (Suppl. 5), 1–96.
196. Schmid, E., in *Kaffee und Coffein*, Ed. O. Eichler, Springer-Verlag, New York, 1976, 297–321.
197. Fredholm, B. B., in *The Methylxanthine Beverages and Food*, Ed. G. A. Spiller, Alan R. Liss, New York, 1984, 331–354.
198. Dennish, G. W. and Castell, D. O., *Dig. Dis.*, 1972, **17**, 993–996.
199. Cohen, S. and Booth, G. H., *N. Engl. J. Med.*, 1975, **293**, 897–899.
200. Westermann, H., Muller-Wieland, K. and Spielman, M., *Med. Klin.*, 1977, **72**, 2201–2203.
201. Thomas, F. B., Steinbaugh, J. T., Fromkes, J. J., Mekhjian, H. S. and Caldwell, J. H., *Gastroenterology*, 1980, **79**, 1262–1266.
202. Salmon, P. R., Fedail, S. S., Wurzner, H. P., Harvey, R. F. and Read, A. E., *Digestion*, 1981, **21**, 69–73.
203. Cohen, S., *N. Engl. J. Med.*, 1980, **303**, 122–124.

204. Cole, S. G., Hogan, D. L., Sandersfeld, M. and Isenberg, J. I., *Gastroenterology*, 1984, **86**, 1051.
205. Mitznegg, P., in *Kaffee und Coffein*, Ed. O. Eichler, Springer-Verlag, New York, 1976, 125–131.
206. Atkinson, J., *Z. Ernährungswiss.*, 1976, **15**, 156–163.
207. Siegers, C. P., Strubelt, O. and Back, G., *Eur. J. Pharmacol.*, 1972, **20**, 181–187.
208. Siegers, C. P., *Pharmacology*, 1973, **10**, 19–27.
209. Fabian, G., *Arzneimittel-Forsch.*, 1954, **8/I**, 145–150.
210. Apel, G. and Schenk, V., *Angew. Med.*, 1961, **15/I**, 18–32.
211. Glatzel, H. and Hackenberg, K., *Med. Klin.*, 1967, **62**, 625–628.
212. Brummer, P. and Sysimetsa, E., *Acta Med. Scand.*, 1957, **CLIX/II**, 141–145.
213. Dubois, A., Shea-Donohue, P. T., Max, F. and Castell; D. O., *Gastroenterology*, 1984, **84**, 1142.
214. Roth, J.-L. and Bockus, H. L., *Gastroenterology*, 1951, **18**, 546–565.
215. Roth, J.-L., *Gastroenterology*, 1951, **19**, 199–215.
216. Roth, J. A. and Ivy, A. C., *Am. J. Physiol.*, 1944, **141**, 454–461.
217. Antonsen, S., *Acta Pharmacol. Toxicol.*, 1965, **23**, 165–178.
218. Roth, J. A. and Ivy, A. C., *Am. J. Physiol.*, 1944, **142**, 107–113.
219. Berkowitz, J. M., Shields, J. E., Andersen, I. H. and Praisman, M., *Gastroenterology*, 1974, **66**, 666.
220. Borger, H. W., Schafmeyer, A., Arnold, R., Becker, H. D. and Creutzfeld, W., *Deutsche Med. Wschr.*, 1976, **101**, 455–457.
221. Wright, L. F., Gibson, R. G. and Hirschowitz, R. I., *Proc. Soc. Exp. Biol.*, 1977, **154**, 538–539.
222. Feldman, E. J., Isenberg, J. I. and Grossman, M. I., *J. Am. Med. Assoc.*, 1981, **246**, 248–250.
223. Cano, R., Isenberg, J. I. and Grossman, M. I., *Gastroenterology*, 1976, **70**, 1055–1057.
224. Salmon, P. R. and Barton, T., *Dig. Dis. Sci.*, 1986, **31**, Suppl. Feb., 558–628.
225. Geisbusch, P., *Med. Klin.*, 1961, **56**, 1918–1925.
226. Sziegoleit, W. and Forster, W., *Pharmazie*, 1972, **27**, 768–772.
227. Rosner, P., Keiner, F. and Kuhn, U., *Med. Clin.*, 1971, **66**, 238–242.
228. Fintelmann, V. and Haase, W., *Z. All. Med.*, 1977, **53**, 1888–1895.
229. Schusdziarra, V., Mathiessen, A. and Heisig, N., *Z. Gastroenterol.*, 1977, **15**, 448–456.
230. Pfeiffer, C. J. and Gass, G. H., *Can. J. Biochem. Physiol.*, 1962, **40**, 1473–1476.
231. Okabe, S., *Gastroenterology*, 1981, **80**, 1241.
232. Yano, Sh., Isobe, Y. and Harada, M., *J. Pharm. Dyn.*, 1982, **5**, 485–494.
233. Chvasta, T. E. and Cooke, A. R., *J. Lab. Clin. Med.*, 1972, **79**, 302–315.
234. Frjedman, G. D., Siegelaub, A. B. and Seltzer, C. C., *N. Engl. J. Med.*, 1974, **290**, 469–473.
235. Brettholz, E. M. and Meshkinpour, H., *Gastroenterology*, 1985, **88**, 1335.
236. Wald, A., Back, C. and Bayless, T. M., *Gastroenterology*, 1976, **71**, 738–742.
237. Wagner, S. W., Nekhjian, H. S., Caldwell, J. H. and Thomas, F. B., *Gastroenterology*, 1978, **73**, 379–381.
238. Beubler, E. and Lembeck, F., *Naunyn Schmiedebergs Arch. Pharmacol.*, 1976, **292**, 73–77.

239. Anderson, J. R., Drehsen, G. and Pitman, I. H., *J. Pharm. Sci.*, 1981, **70**, 651–657.
240. Volheimer, G., Schulz, F. H., Hofer, E. and Schicht, J., *Nutr. Dieta*, 1969, **11**, 13–22.
241. Strubelt, O., Bohme, K., Siegers, C. P. and Bruhn, P., *Z. Ernährungswiss.*, 1976, **15**, 125–131.
242. Matkies, F. and Perisoara, A., *Fortschrift. Med.*, 1985, **103**, 713–714.
243. Grossman, E. M., *Food Chem. Toxicol.*, 1984, **22**, 245–249.
244. Erckenbrecht, J. F., Rehm, S., Schope-Stiller, A., Lubke, H. J. and Wienbeck, M., *Gastroenterology*, 1985, **88**, 1375.
245. Read, N. W., Al Janabi, M. N., Bates, T. E. and Barber, D. C., *Gastroenterology*, 1983, **84**, 1568–1572.
246. Richardson, C. T., Walsh, J. H., Hicks, M. I. and Fordtran, J. S., *J. Clin. Invest.*, 1976, **58**, 623–631.
247. Stephensen, C. B., Leon Barua, R., Sack, R. B. and Sack, D. A., *Gastroenterology*, 1985, **88**, 1599.
248. Weiner, K., Graham, S. L., Reedy, T., Elashoff, J. and Meyer, J. H., *Gastroenterology*, 1981, **81**, 257–266.
249. Malagelada, J. R., *Gastroenterology*, 1977, **72**, 1264–1267.
250. Lloyd, D. A. and Borda, I. T., *Gastroenterology*, 1981, **80**, 740–741.
251. McArthur, K., Hogan, D. and Isenberg, J. I., *Gastroenterology*, 1982, **83**, 199–203.
252. Lennars-Jones, J. E., *N. Engl. J. Med.*, 1983, **308**, 431–435.

Chapter 5

Nutritional Factors

R. MACRAE
Department of Food Science and Technology, University of Reading, UK

1. INTRODUCTION

The desirable characteristics of coffee brews are associated with their sensory and stimulatory effects rather than their nutritional ones. However, there are a number of components present in coffee brews which can have a positive effect on the nutritional status of the consumer. For most people with a varied and well-balanced diet the contribution of these nutrients from coffee will probably be insignificant when compared with other sources in the diet. In some instances, where sections of the population are on marginal diets, the contribution from coffee may, however, be vital for prevention of deficiency. An example of this is to be found in regions where the staple foods are based on maize (Central America and Central Africa). Here pellagra is quite common and it has been shown that the symptoms can be alleviated by the consumption of coffee brew, which contains sufficient levels of nicotinic acid. Nicotinic acid is clearly the most important micronutrient in coffee but a wide range of other vitamins have also been reported, though at levels where their nutritional significance is very limited. Coffee brews also contain a wide range of mineral salts, and it is conceivable that some of these may be nutritionally significant for people on extreme diets.

Roasted coffee, and indeed subsequent coffee brews, contain significant amounts of both unchanged and degraded polysaccharides and melanoidins, which are mainly indigestible in the gut. It is possible that these components from coffee brews will make a contribution to total dietary fibre intake, and they may also become associated with mineral components and may therefore affect the absorption of the latter from the gut.

The nutritional significance of coffee comes more from its role as a vehicle for nutritious additives, in particular milk and sugar, both in terms of energy and other nutrients.

2. VITAMINS

Green coffee beans contain a considerable amount of lipid material and, like other beans or seeds of similar composition, contain fat-soluble as well as water-soluble vitamins. The levels of these vitamins have been extensively studied in green coffee and to a lesser extent in roast coffee, where they are found at considerably lower levels, except for nicotinic acid, which is actually produced during the roasting process.

2.1. Fat-soluble Vitamins

The major fat-soluble vitamin in coffee oil appears to be vitamin E. Indeed there are no reports of the other fat-soluble vitamins (vitamins A and D) having been detected. The greatest collection of data has come from Folstar and his group,[1,2] who have used gas–liquid chromatography for their analyses. This is not able to separate β- and γ-tocopherol, and so values for combined β- and γ-forms as well as values for α-tocopherol are reported.

Typical results for a fresh Colombian arabica green sample are 140 ppm α-tocopherol and 465 ppm $(\beta+\gamma)$-tocopherol, and for an Ivory Coast robusta green sample (1–2 years old) a somewhat higher value for α-tocopherol of 191 ppm and lower for $(\beta+\gamma)$-tocopherol of 252 ppm (results as fraction of coffee oil). The exact levels found will depend on the age of the sample and its processing (wet or dry). A more recent application of high-performance liquid chromatography (HPLC) to the analysis of coffee oil from a green arabica coffee showed that there is in fact little β-tocopherol present, so the figures quoted by Folstar for $(\beta+\gamma)$-tocopherol probably correspond reasonably well to the true levels of γ-tocopherol.[3] The HPLC technique also showed the presence of low levels of α-tocotrienol.

The interest in the presence of tocopherols in coffee oil is not so much from the point of view of their nutritional role but rather from that of their antioxidant properties. Green coffee, provided it is well stored at a reasonable moisture content, is quite stable, and rancidity of the lipid fraction is rarely encountered. This is important as green coffee is often stored for long periods prior to roasting. Tocopherols are stable to heat but in the presence of oxygen they are rapidly oxidised, especially in the presence of traces of catalytic metals such as copper or iron. It would therefore be expected that tocopherols in green coffee would be partially destroyed during roasting and that the lipid fraction in the resulting roasted coffee would be more susceptible to oxidation. This appears to be the case, with oxidative rancidity being a contributing factor to advanced coffee staling. However, roasted coffee is rarely held for long periods unless special packaging is used to exclude oxygen, e.g. vacuum packing.

The tocopherols, being fat-soluble, are not soluble in water and so the residual amounts do not pass from the roasted coffee to the brew.

2.2. Water-soluble Vitamins

The full spectrum of water-soluble vitamins have been detected in green coffee, although there are comparatively little quantitative data. A recent compilation of these numerical data has been given by Maier[4] based on the earlier work of Teply and Prier[5] and Souci *et al.*,[6] reproduced here in modified form in Table 1. The levels of most of the B-group vitamins are extremely low in green coffee and, as they are thermally unstable, their levels are drastically reduced on roasting, the actual percentage loss depending on the roast conditions. It is therefore extremely unlikely that the trace amounts of these compounds that would be expected in coffee brews have any nutritional significance at all when compared with other sources in the diet. The one clear exception among the water-soluble vitamins is niacin (nicotinic acid), which is produced during the roasting

Table 1
Water-soluble vitamins in coffee ($mg\,kg^{-1}$)

Vitamin	*Green coffee*	*Roast coffee*
Thiamin (B_1)	2·1	0–0·7
Riboflavin (B_2)	2·3	0·5–3·0
Niacin	22·0	93–436
Pyridoxine (B_6)	1·4	0·01–0·1
Pantothenic acid	10·0	2·3
Folic acid	0·12	0·03–0·22
Cobalamin (B_{12})	0·001 1	0·000 6
(Choline)	(590)	(840)
Ascorbic acid (C)	460–610[a]	—

[a] Data from Vasudeva and Gopal.[8]

procedure and is found at considerable levels in coffee brews (see Section 2.2.1). Choline has been placed in parentheses in Table 1 since, although it has been shown to be essential for certain organisms, its essentiality as a vitamin for humans has not yet been established.[7] There do not appear to be any data on the levels of vitamin C in roast coffee comparable to those reported for green coffees by Vasudeva and Gopal.[8] However, ascorbic acid is thermally unstable and susceptible to oxidation, and would react with many of the nucleophilic components present in green coffee during roasting, so that none would be expected to be present in the final coffee brew.

2.2.1 Niacin

The dietary requirement of humans for niacin (nicotinic acid) is satisfied by two sources: niacin directly available from foodstuffs consumed, and the conversion of dietary tryptophan into niacin in the liver. The exact daily requirement for actual niacin therefore depends also on the intake of tryptophan. The recommended daily intake for niacin (16 mg for men and 12 mg for women) includes the contribution from tryptophan. Various factors have been calculated for the equivalence between niacin and tryptophan, with a value of 1:60 being most widely accepted.[9] A well-balanced mixed-protein diet (60 g protein per day) may well contain about 600 mg of tryptophan, which would be equivalent to 10 mg of niacin, or 63% of the recommended daily intake for men and about 83% for women. Clearly, if the protein source is deficient in tryptophan, as for example with maize, then the requirement from other sources increases in order to avoid deficiency symptoms. It is under these circumstances that the contributions from other food sources such as coffee become important.

The level of nicotinic acid in green coffee depends on species and other factors (e.g. processing conditions) and a wide range has been quoted by Hughes and Smith[10] of between 1·6 and 4·4 mg per 100 g, which encompasses the figure quoted by Maier (Table 1). On roasting this value is increased considerably as a result of the thermal demethylation of trigonelline which is present at much higher levels of about 1% db.[11]

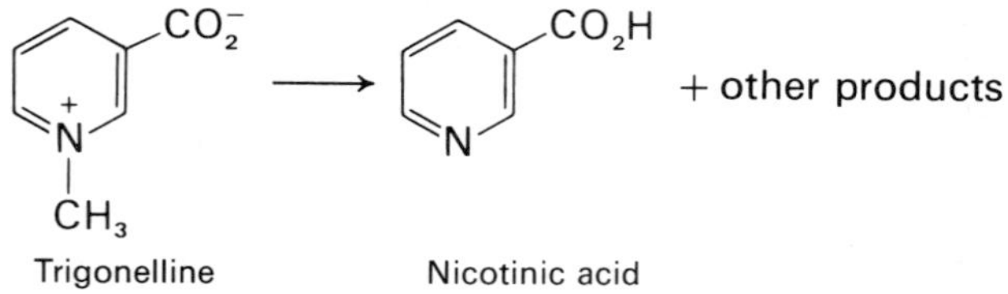

The final level of nicotinic acid in the roast coffee depends more on the degree of roast than on the original level of nicotinic acid present. Adrian[12] has shown that the maximum levels of about 40 mg per 100 g for nicotinic acid occur at total roast losses of about 20% db and that above this value (i.e. with extremely severe roasting) the nicotinic acid formed is further degraded. A more realistic mass loss on roasting for a commercial sample would be 5–10% db, which would result in a final nicotinic acid content of some 15–20 mg per 100 g. Nicotinic acid is highly water-soluble so that it is readily extracted on brewing into the resulting beverage. Thus a cup of coffee prepared from 10 g of roast and ground coffee containing 20 mg per 100 g nicotinic acid would result in a beverage containing some 2 mg nicotinic acid which is 12% of the recommended daily intake for men. It is quite clear, then, that a few cups of coffee contribute significantly to the dietary intake of niacin. In addition, there will be a small contribution from the tryptophan which is also present in the roast and ground coffee.[13]

Under the processing conditions used for the industrial extraction of ground coffee nicotinic acid is thermally quite stable in solution and will remain largely intact, therefore appearing in instant coffees in proportionate amounts. A recent survey of 13 commercial instant coffees, using both HPLC and microbiological techniques to determine nicotinic acid, showed results in the range 21–47 mg per 100 g.[14] A comparison of the results from the two methods, and also further results with and without preliminary hydrolysis, was used to show that the nicotinic acid is present mainly in the free form.[14] Studies on both rats and human subjects have also confirmed that the nicotinic acid is present in a biologically available form.[5,15]

Coffee, instant or freshly brewed from roast and ground beans, does provide a significant source of niacin which can be important for those people on poorly balanced diets. The effects of niacin deficiency (pellagra) are unpleasant, including symptoms such as body lesions, diarrhoea, anorexia, oral lesions and reddening of the tongue. Neurological symptoms, such as depression, apathy, headaches, dizziness, irritability and tremors, may also arise.

3. MINERAL CONTENT

The total mineral content of green coffees is normally expressed as oxide ash content, with a mean value of about 4% db being reported for both arabica and robusta coffees. The available data have been collated by Clarke in a recent publication.[16] The major metallic component of green coffee is potassium, which is present at levels of 1·2–2·1% K db.[17] Narrower ranges have been reported in individual studies, e.g. 1·6–2·0% K db.[18] These figures represent about 40% of the total oxide ash content. Other metals which are present in considerable amounts include calcium, 0·07–0·35% db, and magnesium, 0·16–0·31% db. In addition, some 33 other elements have been reported, many of which are only present at levels below 1 mg kg^{-1}.[17] It would be anticipated that the level of minerals found in green coffee would be highly dependent on the levels present in the soil. There appears to have been very little work carried out on this aspect, with the exception of studies on the effect of nitrogen, phosphorus and potassium fertilisers on the content of minerals in the resulting green coffee. Surprisingly it was shown that addition of potassium to the soil did not result in increased levels in the beans although some differences were observed in the combined fruit wall and mucilage.[19] This may be due to the plant's inability to 'fix' in the coffee beans the increasing amounts of potassium absorbed through the root. The same may not be true for elements present at much lower levels, and it would be interesting to see if indeed there is any correlation between trace element levels in the soil and in the green coffee beans, although this would be difficult to investigate on account of commercial green coffee blending.

The amounts of the metallic elements present will not change on roasting, although their form of chemical combination with other elements may alter slightly. An apparent increase in their percentage composition will be observed as a result of the loss of organic material, the size of this increase depending on the degree of roast and hence dry matter loss.

Most of the metallic elements are present in coffee beans (green or roast) as soluble salts, so that they are readily extracted during brewing and to an even greater extent during instant coffee manufacture. For example, as much as 99% of the potassium content of roast coffee is reported to be extracted in commercial instant coffee production. This finding led to the suggestion that the potassium content of instant coffee could be used to determine the processing conditions employed during its manufacture. However, in practice the wide variability of potassium content in green coffees reduced the diagnostic value of this method.[18] The solubility of the other mineral components is more variable. In general, however, the majority of each element present in the roast coffee appears in proportionate amount in the subsequent brew, or instant coffee, the actual percentage composition depending on the degree of extraction of other components as well. Maier[17] has produced a compilation of the levels of minerals and trace elements in dry coffee extracts, taken from a number of publications.[6,18,20–24] The ranges quoted for many elements are very wide indeed—for example calcium, 100–2350 mg kg^{-1} db; sodium, 22–11 670 mg kg^{-1} db; and lead, 0·1–4·0 mg kg^{-1} db—and are in many cases much wider than the ranges observed in individual studies, even when a wide range of instant coffees have been studied. Thus Angelucci *et al.*,[25] in their study of Brazilian coffees, reported narrower ranges—for example calcium, 1620–3010 mg kg^{-1} db, and sodium, 24–7595 mg kg^{-1} db. Obviously some of these differences may be associated with the differences in coffees used and their different trace element contents, but a much larger source of variation is likely to be found in the trace elements present in the processing water used. If an extract of 20% w/v solubles concentration is prepared at the extractors, then approximately five times the concentration of ions present in the water will appear on a dry weight basis in the final product. In particular, this explains the very high values quoted in some instant coffees for calcium or sodium, arising from hard water used in processing or from hard water that has been softened. Similarly, any trace metals derived from process equipment, e.g. extractors, pipework, etc., will be concentrated in instant coffee powders. This may explain the elevated levels of iron and copper reported in some cases. Man has a nutritional requirement for a wide range of elements, ranging from large quantities of carbon, hydrogen, nitrogen and oxygen, as carbohydrates, lipids and proteins, to smaller amounts of minerals (sodium, magnesium, potassium and calcium) and to even smaller amounts of trace elements (manganese, iron, cobalt, nickel, copper and zinc). Other elements, such as phosphorus, sulphur, chlorine, iodine and possibly also fluorine, are also required for the maintenance of health. The small amounts of coffee consumed per day, in terms of roast and ground beans or their equivalent, means that the contribution they make as a macronutrient is very small indeed, especially when much of the coffee solubles are probably not assimilable from the human gut. The amounts of sodium, magnesium, potassium and calcium are also likely to be insignificant when considered in relation to other sources in the diet. Even instant coffees which have accumulated high levels of sodium from processing water are unlikely to contribute, to any appreciable extent, to the daily intake of sodium. With the present trend to reduce daily intake of sodium, the consumption of such coffee powders with extremely high levels is likely to be discouraged. The significance of the trace elements in coffee is even less certain. The amounts present are unlikely to have any significance in a well-balanced diet containing these elements from a variety of sources. Furthermore, it is far from certain that these trace elements are present in a biologically available form.

4. DIETARY FIBRE

The term 'dietary fibre' has been variously defined over the years as the fraction of the diet 'indigestible in the human gut' or 'inaccessible to the normal digestive enzymes of man'.[26] A more suitable definition for analytical purposes may be found from a consideration of the chemical structures of the compounds encompassed by the above definitions. These consist mainly of polysaccharides, excluding starch which is digested by pancreatic amylase. Lignin is also included as a component of dietary fibre but this is not present in significant amounts in foods used for human consumption, and only small amounts have been reported in green coffee (*c.* 2%) and this will not be solubilised by roasting and subsequent extraction to form coffee beverage. Cellulose (glucan) is present in green coffee, and hence in roasted coffee in small amounts, where it is largely unchanged, but being insoluble in water it will not be found to any appreciable extent in the subsequent coffee brew or in instant coffee. The other coffee polysaccharides are less stable during roasting, with arabinose-containing species being particularly unstable as determined by the amount of arabinose recoverable after acid hydrolysis of the roasted product (see Volume 1, Chapter 3). None the less, polysaccharides, possibly now in a modified/condensed form, are major components of roast coffee and are extractable to a considerable extent during domestic brewing and more so during industrial extraction to prepare instant coffee. The amount of carbohydrate extracted depends on the roast and extraction conditions as well as the species in question. However, some 15–20% of the roasted bean would appear to be extractable as carbohydrates at extraction yields of some 50%, i.e. leading to instant coffee powders with carbohydrate contents of 30–40%.[27] The composition of this polysaccharide depends on the extraction yield. It consists mainly of galactan and mannan, the latter increasing in amount at higher extraction rates. By analogy with other polysaccharides of plant material of similar structures (e.g. guar gum, a galactomannan with a backbone of $(1 \rightarrow 4)$-β-D-mannopyranosyl units[28]) it would be expected that those found in coffee brews would be indigestible in the human gut and would therefore act as dietary fibre. On this basis a cup of coffee, particularly of instant coffee which contains a higher proportion of polysaccharides because of the higher extraction rates realised, then becomes a source of dietary fibre in the diet. Each cup prepared from instant coffee (2–4 g powder) would then contain in the region of 0·5–1·5 g of indigestible polysaccharide, or dietary fibre. When this is compared with the various recommended daily intakes of fibre of between 20 and 40 g, the role of coffee may become significant. Considering the analogy with guar gum further it may be postulated that the polysaccharides from coffee, if taken in sufficient quantities, could have an effect on the absorption of low molecular weight nutrients from the gut. Thus guar gum, for example, significantly reduces the absorption of glucose from the gut, which has implications for diabetic patients. No similar effects appear to have been reported for coffee polysaccharides.

Coffee is known to reduce the availability of minerals in the diet. For example, a cup of coffee was shown to reduce the iron absorption from a hamburger meal by some 39%.[29] The major binding components are thought to be phenolic compounds and some of these may well be associated with the degraded polysaccharide fraction. Instant coffees are also known to contain low levels of phytate which will reduce the availability of minerals, such as iron and zinc, but this is not considered to be significant in the overall diet[30] (see Chapter 2 in this volume).

5. ADDITIVES

5.1. Introduction

The discussion so far concerning the nutritional factors in coffee has been restricted to the components present in brewed coffee itself, either from roast and ground beans or from instant coffee. However, in terms of the contribution which coffee beverages actually make to the total diet it is more relevant to consider the beverage as consumed. This then includes contributions from the various additives used. The range of such additives is considerable, reflecting both national traditions and individual tastes. In the case of speciality coffees, for example the French cordial with liqueur, café Borgia with chocolate or even Hawaiian coffee with coconut,[31] these are only consumed in relatively small amounts and should be considered for the pleasure they may give rather than their nutritional value. However, this is certainly not the case with the more common additives, in particular milk and sugar, and here it is the nutrient content of the additives which is significant rather than the coffee itself. This then brings into question a number of factors, namely the *per capita* consumption of coffee, the proportion of this coffee consumed with or without additives and the nutritional composition of these additives. These points will be considered in turn, with the discussion restricted to the most important additives, milk and sugar.

Table 2
Derived *per capita* disappearance[a] of coffee in importing member countries January–December 1981 to 1986, kg
(data from International Coffee Organisation[32])

Importing members	*January–December*					
	1981	*1982*	*1983*	*1984*	*1985*	*1986*
Total	4·40	4·37	4·42	4·31	4·33	4·34[b]
USA[c]	4·80	4·77	4·63	4·71	4·65	4·41
EEC	4·88	4·89	5·03	4·74	4·88	4·93
Belgium/Luxembourg	8·58	7·15	8·84	7·25	7·60	7·14
Denmark	11·79	11·46	11·15	11·05	11·04	11·00
France	6·05	5·91	5·94	5·39	5·47	5·49
Greece	2·63	2·65	2·81	3·00	2·96	2·18
Ireland	1·10	1·00	1·16	1·47	1·59	1·81
Italy	3·98	4·33	4·34	3·89	4·93	4·37
The Netherlands	9·09	8·97	9·58	9·46	9·41	9·65
Portugal	1·48	1·42	1·90	1·96	2·19	1·64
Spain	2·75	2·76	3·19	2·92	2·74	3·44
United Kingdom	2·55	2·43	2·41	2·51	2·44	2·42
West Germany	7·06	7·34	7·29	7·03	6·84	7·38
Other importing members	3·27	3·18	3·32	3·26	3·17	3·44
Australia	2·51	2·62	2·31	2·44	2·11	2·24
Austria	6·55	7·92	8·53	7·73	7·34	7·75
Canada	4·79	4·33	4·25	4·27	4·41	4·15
Cyprus	2·95	2·63	6·55	[d]	2·87	1·34
Fiji	0·37	0·18	0·18	0·17	0·09	0·09
Finland	13·52	12·78	12·93	14·59	10·09	12·09
Japan	1·68	1·85	1·94	2·01	2·14	2·23
New Zealand	2·09	1·96	2·25	2·01	1·94	1·88
Norway	10·26	10·51	11·36	10·39	10·47	10·09
Singapore	[d]	[d]	[d]	2·36	[d]	[d]
Sweden	12·91	11·73	12·14	11·29	11·55	11·64
Switzerland	6·57	5·58	6·00	6·04	6·17	6·59
Yugoslavia	2·01	1·05	1·51	0·89	0·58	2·31

[a] Derived *per capita* disappearance is expressed in kilograms of green coffee equivalent and based on total population (i.e. all age groups).
[b] Preliminary.
[c] Based on estimates of civilian population by the United States Department of Commerce.
[d] Re-exports exceeded imports in these years.

5.2. Coffee Consumption

It is extremely difficult to obtain reliable data on the *per capita* consumption of coffee. One approach is to carry out statistical surveys of representative samples of the population and extrapolate these data to the entire population. Clearly this approach is liable to be subject to considerable sampling error and may not be able to take into account all the possible variables, for example differences due to regions or social classes. A more global approach may be attempted by considering disappearance figures for a particular country, which are based on the data of net imports of all forms of coffee adjusted for changes in inventories. The most recent data available are reproduced in Table 2, which shows quite remarkable differences in consumption between countries throughout the world and, perhaps surprisingly, even large differences between European countries. These figures are for all types of coffee and do not distinguish between roast and ground or instant coffee. It is therefore not possible to equate the data to cups of beverage consumed. The majority of instant coffee is manufactured in the country where it is finally consumed, but instant coffee powder is exported from Brazil and other countries to Europe. The average consumption of over 4 kg in the EEC and USA does, however, show the considerable potential of coffee as a vehicle for the consumption of other food components as additives.

5.3. Coffee Consumption Patterns

The proportion of coffee consumed black without any additives varies considerably from country to country, and a comprehensive set of data is not readily available. A number of national surveys have been carried out, and it may be of interest to study the results of some of these as representative examples. Here again there is no differentiation between instant coffee and roast and ground, and the proportion of beverages prepared from these products will vary enormously from country to country. It has been estimated that in 1980 some 19% of the coffee consumed worldwide went into instant coffee production whereas in the UK nearer 90% of the coffee consumed derived from instant products.[33] Table 3 shows how additives are used with coffee in Great Britain. It is clear from the data that the most common way to take coffee is with milk or some other whitener (non-dairy creamer or cream) and furthermore approximately half of those people drinking white coffee also take sugar. A significant proportion of coffee drinkers use non-nutritive sweeteners (12%). The fact that most coffee drinkers in Great Britain drink white instant coffee needs to be considered in more detail. Approximately half of

Table 3
Use of additives in coffee in Great Britain
(data from Mil Research[34])

	All coffee drinkers	*Coffee drinkers according to sex and social class*					
		Men	*Women*	*AB*	*C1*	*C2*	*DE*
Bases	1 106	513	593	209	256	355	286
Drank at least one cup with (%)							
Milk	88	90	87	83	89	89	90
Cream	3	2	4	6	2	3	3
Non-dairy creamer	6	5	7	10	6	6	5
Sugar	54	69	40	44	53	57	58
Artificial sweetener	12	10	13	16	10	12	11
None of these (i.e. black, without sweetener)	5	4	6	8	7	5	3

these people prepare their coffee with water and add a small amount (say 10–20%) of milk, a quarter prepare coffee 'half and half' (50% hot milk), while the remaining consumers prepare instant coffee with milk only. Particularly in the final category any nutritional value from the instant coffee powder *per se* is insignificant when compared with that from the milk, with the possible exception of nicotinic acid (see Section 2.2).

Tables 4 and 5 show comparable data for the USA and West Germany. Unfortunately the survey questions are not directly equivalent to those used in Table 3 (Great Britain). For example, the type of whitener is not differentiated, but the higher proportion of coffee that is consumed black is

Table 4
Use of additives in coffee in USA
(percentage drinking[a]; data from International Coffee Organisation[35])

	1984		*1985*	
	All persons	*Drinkers only*	*All persons*	*Drinkers only*
All coffee	57·3	100	55·2	100
Black	24·4	43	23·6	43
Sweetener only	5·3	9	4·9	9
Creaming agent only	12·0	21	13·5	24
Sweetener and creaming agent	17·4	30	14·9	27

[a] The sum of the components may be more than the total because of multiple responses.

Table 5
Use of additives in coffee in Germany
(percentage drinking[a]; data from International Coffee Organisation[36])

	1980	*1981*	*1982*	*1983*	*1984*
All coffee	90·9	89·5	89·5	88·0	88·4
Black	19·4	20·6	19·8	18·6	18·5
Sweetener only	9·0	8·4	8·9	7·9	8·2
Creaming agent only	28·1	28·1	27·0	28·3	27·8
Sweetener and creaming agent	43·3	41·9	42·7	42·0	43·0

[a] The sum of the components may be more than the total because of multiple responses.

quite evident. There will also be a marked reduction in the proportion of instant coffee consumed but this may not be significant in terms of the amount of milk consumed. Coffee brewed from roast and ground beans, when taken white, is often consumed with up to 50–60% warm milk and so is similar to the beverage prepared 'half and half' from instant coffee with respect to milk content. The data in these tables do not allow the significance of sugar to be interpreted since artificial sweeteners (non-nutritive) are not differentiated from sugar, both being grouped as sweeteners. It is probably reasonable to assume that the proportion of consumers using artificial sweeteners is similar to that in Table 3, although this may be slightly higher in the USA.

These data do not reveal other variables in coffee consumption, such as the time of day. In many countries coffee is consumed at breakfast as part of a nutritious meal, and in this case will be consumed with a very high proportion of hot milk (and possibly sugar). The Italian capuccino would be an example of such a drink. In the evening the same consumers may turn to strong black coffee after a meal as a supposed aid to digestion.

5.4. Nutritional Composition of Additives

The nutritional significance of sugar (sucrose) consumed in coffee is restricted mainly to its role as an energy source (1680 kJ per 100 g),[37] so that a heavily sweetened cup of coffee containing, say, 5–10 g sucrose may make a significant contribution to the consumer's daily energy intake. Indeed it may be that for many people this additional energy intake is undesirable, and elimination of sugar from beverages is often achieved by the use of artificial (non-nutritive) sweeteners. There is also a trend to reduce sucrose consumption in beverages on account of its cariogenic properties.

Milk, and its derived products, make a considerable contribution to the nutritional content of many people's diet. Milk is consumed on its own as a drink and in many other forms, including its use as an important additive to coffee. In addition to providing a source of energy (272 kJ per 100 g liquid), milk also provides a significant source of protein (3·3 g per 100 g liquid), carbohydrate, mainly lactose (4·7 g per 100 g liquid), and fat (3·8 g per 100 g liquid).[37] Thus, for those people consuming 'milky' coffee this provides a significant source of energy and protein. For example, a person drinking three or four cups of 'milky' coffee a day could well be consuming 300 g of milk from this source alone, i.e. *c.* 10 g of protein and *c.* 800 kJ. Other nutrients in milk should also be considered, in particular minerals (calcium, 120 mg per 100 g liquid) and vitamins (retinol, 35; carotene, 22; riboflavin, 190; vitamin B_{12}, 0·3 μg per 100 g liquid).[37] Milk provides an especially valuable source of calcium (recommended daily allowance (RDA), 400–500 mg),[38] riboflavin (RDA, 1–2 mg or 130 $\mu g\, MJ^{-1}$)[38] and vitamin B_{12} (RDA, 2 μg).[38] Whether the contribution from milk in coffee for these nutrients is nutritionally important can only be judged in relation to other sources in the total diet. However, from the above figures it can be seen that the RDA for some of these nutrients could be achieved by drinking 'milky' coffee.

6. CONCLUSION

Coffee is a pleasant and stimulating drink, but only for those people on marginal diets can it be considered as truly nutritious and here it is probably only the nicotinic acid content that is significant. However, coffee does provide a vehicle for the consumption of more nutritious additives, and for many people 'milky' coffee can provide an important source of energy, protein and certain micronutrients.

REFERENCES

1. Folstar, P., van der Plas, H. C., Pilnik, W. and de Heus, J. G., *J. Agric. Food Chem.*, 1977, **25**, 283–285.
2. Folstar, P., in *Coffee. Vol. 1: Chemistry*, Eds R. J. Clarke and R. Macrae, Elsevier Applied Science Publishers, London, 1985, 215.
3. Macrae, R. and Tsimidou, M., unpublished data, 1984.
4. Maier, H. G., *Kaffee*, Paul Parey, Berlin, 1981, 64.
5. Teply, L. J. and Prier, R. F., *J. Agric. Food Chem.*, 1957, **5**, 375–377.
6. Souci, S. W., Fachmann, W., Kraut, H. and Bosch, H., *Die Zusammensetzung der Lebensmittel Nährwerttabellen*, Wiss. Verl.-Ges., Stuttgart, 1979.
7. Chan, M. M., in *Handbook of Vitamins*, Ed. L. J. Machlin, Marcel Dekker, New York, 1984, 549–570.

8. Vasudeva, N. and Gopal, N. H., *J. Coffee Res.*, 1974, **4**, 25–28.
9. Hankes, L. V., in *Handbook of Vitamins*, Ed. L. J. Machlin, Marcel Dekker, New York, 1984, 358.
10. Hughes, E. B. and Smith, R. F., *J. Soc. Chem. Ind.*, 1946, **65**, 284–286.
11. Macrae, R., in *Coffee. Vol. 1: Chemistry*, Eds R. J. Clarke and R. Macrae, Elsevier Applied Science Publishers, London, 1985, 127–137.
12. Adrian, J., *Proc. 1st Coll. ASIC*, 1963, 141–149.
13. Macrae, R. and MacDonald, I. A., *Proc. 11th Coll. ASIC*, 1985, 333–340.
14. Trugo, L. C., Macrae, R. and Trugo, N. M. F., *J. Micronutr. Anal.*, 1985, **1**, 55–64.
15. Goldsmith, G. A. and Miller, O. N., in *The Vitamins*, 2nd edn, Eds P. György and D. Pearson, Academic Press, New York, 1967, 137–143.
16. Clarke, R. J., in *Coffee. Vol. 1: Chemistry*, Eds R. J. Clarke and R. Macrae, Elsevier Applied Science Publishers, London, 1985, 42–82.
17. Maier, H. G., *Kaffee*, Paul Parey, Berlin, 1981, 34.
18. Clarke, R. J. and Walker, L. J., *J. Sci. Food Agric.*, 1974, **25**, 1389–1404.
19. Verlière, G., *Café, Cacao, Thé*, 1973, **17**, 269–280.
20. Ferreira, L. A. B., Fragoso, M. A. C., Peralta, M. F., Silva, M. C. C. and Rebelo, M. C., *Proc. 5th Coll. ASIC*, 1971, 51–62.
21. Quijano Rico, M. and Spettel, yB., *Proc. 6th Coll. ASIC*, 1973, 102–105.
22. Quijano Rico, M. and Spettel, yB., *Proc. 7th Coll. ASIC*, 1975, 165–173.
23. Roffi, J., Corte Dos Santos, A., Mexia, J. T., Busson, F. and Maigrot, M., *Proc. 5th Coll. ASIC*, 1971, 179–200.
24. Thorold, C. A., *Ann. Appl. Biol.*, 1946, **33**, 177–178.
25. Angelucci, E., Yokomizo, Y., de Moraes, R. M., de Campos, R. B., Miya, E. E. and Figueiredo, I. B., *Proc. 6th Coll. ASIC*, 1973, 178–183.
26. Englyst, H., in *Dietary Fibre: Progress Towards the Future*, Eds I. McLean Baird and M. H. Ornstein, Kellogg Co., Manchester, 1981, 5–11.
27. Thaler, H., *Food Chem.*, 1979, **4**, 13–22.
28. Whistler, R. L. and Daniel, J. R., in *Food Chemistry*, 2nd edn, Ed. O. R. Fennema, Marcel Dekker, New York, 1985, 126.
29. Morck, T. A., Lynch, S. R. and Cook, J. D., *Am. J. Clin. Nutr.*, 1983, **37**, 416–420.
30. McKenzie, J. M., *Nutr. Rep. Int.*, 1984, **29**, 387–395.
31. Roden, C., *Coffee*, Faber & Faber, London, 1977, 95–100.
32. International Coffee Organisation, *Quarterly Statistical Bulletin*, No. 40, October–December 1986, London, 1987.
33. Smith, A. W., in *Coffee. Vol. 1: Chemistry*, Eds R. J. Clarke and R. Macrae, Elsevier Applied Science Publishers, London, 1985, 27.
34. Mil Research Ltd, London, *Research on Usage and Attitudes Towards Coffee in Great Britain*, Report prepared for International Coffee Organisation, August 1982.
35. International Coffee Organisation, *USA Coffee Drinking Study*, Winter 1985 (PC-418/85).
36. International Coffee Organisation, *Federal Republic of Germany Summary of National Coffee Drinking Study*, 1984 (PC-402/85).
37. Paul, A. A. and Southgate, D. A. T., *McCance and Widdowson's The Composition of Foods*, 4th edn, HMSO, London, 1978.
38. Davidson, S., Passmore, R., Brock, J. F. and Truswell, A. S., *Human Nutrition and Dietetics*, Churchill Livingstone, London, 1979, 152–157.

Chapter 6

The Physiology of Flavour: Taste and Aroma Perception

H. B. HEATH
Consultant, Shrewsbury, UK

1. SENSORY PERCEPTION

1.1. Distinctions between Odour, Taste and Flavour

To the majority the flavour of what we eat and drink is something that is taken for granted, and at best is judged only on the basis of a pleasurable or distasteful response. Quality judgements are generally committed to memory in terms of the recognition and recall of these purely hedonic responses, which, once formed, are difficult to eradicate. The flavour of food and drink may be defined as the simultaneous appreciation of stimuli on the receptors of taste on the tongue and smell in the upper nasal cavity and of general pain, tactile and temperature receptors located throughout the mouth and throat. The totality of these stimuli, however they may be perceived, is what is called the flavour profile of a product. Unfortunately, in colloquial terms flavour is often referred to as taste, and this can lead to misunderstandings. In a scientific context, it is desirable to adhere to the following usage published by the International Standards Organisation (ISO/TC34-5492/1) and the British Standards Institution (BS 5098:1975):

Odour—Describes the sensation perceived via the olfactory organ in smelling or sniffing volatile substances. It is also applied to the quality of this particular sensation due to these substances.

Aroma—Often used to describe an odour perceived by the olfactory organ during consumption of a product. It may also be used to describe an odour having a pleasant connotation.

Taste—Describes the sensation perceived via the taste buds on the tongue as a result of the presence of certain water-soluble substances. There are four primary taste sensations: acid (sour), bitter, salty and sweet, although there is evidence for other taste categories.[1] The word is also used to describe the quality of this particular sensation due to these substances.

Flavour—A complex sensation describing the total perception derived from taste and odour, influenced by sensations of pain, heat and cold, as well as tactile sensations.

1.2. Sensory Perceptions that Influence Flavour

Most of the foods and drinks we ingest have highly complex chemical compositions, comprising both volatile and non-volatile substances, the sensory response to which may be significantly affected by their relative concentrations and the temperature of the product as consumed. Whereas volatile constituents are directly responsible for the perceived odour of the product, it is the non-volatile components that may either stimulate the taste buds or activate the network of free sensory nerve endings and organised nerve terminations within the soft tissues of the mouth and throat, producing sensations such as astringency, pungency and cooling as well as tactile, kinaesthetic and pain responses in the relevant sensory areas. Although odour and taste responses are paramount in the appreciation of food and drink, these other sensations, coupled with visual and auditory responses, directly influence the perceived quality and preference for what we consume.

Most of the mucous linings of the nose, mouth and throat are particularly sensitive to irritants. In the nose such stimuli may give rise to sneezing and in the mouth to a sharp sense of physical pain, often associated with watering of the eyes and even a choking sensation. These responses are detected via the receptors of the free nerve endings served by the trigeminal nerve (V of the cranial nerves). In most foods and drinks the production of pain during ingestion, particularly that produced by extremes of temperature, either very cold or very hot, is a negative factor in its acceptance. Indeed, in the case of hot beverages such as coffee, temperature plays a significant part in the enjoyment of the drink.

Some level of pungency, induced by specific components in spices such as capsicum, pepper, ginger and mustard, is a positive attribute in the flavour profile. But excessive astringency, characterised by a dry feeling in the mouth or a puckering of the linings of the oral cavity, such as occurs in overstrong tea, detracts from the refreshing qualities of this beverage.

Astringency is generally recognised as being due to the presence of polymerised phenols having a high molecular weight (500–3000), at least two 1,2-dihydroxyphenyl residues and an ability to precipitate salivary proteins. The degree of astringency is determined by the concentration of these phenols and by their degree of polymerisation. Such substances are collectively known as tannins. Unlike tea, in coffee there are no tannins or polymerised phenols. Clifford[2] concludes that some of the chlorogenic acids (i.e. the dicaffeoyl quinic acids) are the main astringent compounds of a coffee brew. Fortunately, milk, which is usually added to tea and coffee, has the property of reducing any astringency factor, so making the drink more generally smooth and acceptable.

1.3. Sensory Stimuli Associated with Odour and Taste

Stimuli associated with the sensory perception of odour and taste are recognised as having three distinct effects which may differ in magnitude depending not only on the source of the stimulation and the sensory receptors involved but also on the overall receptivity of the subject. These characteristic effects are: (a) arousal—a non-specific means of maintaining a state of readiness and a level of awareness in the central nervous system; (b) discriminative—enabling the subject to react to and distinguish differences between various stimuli (this informative aspect of stimulation recognition may be trained); (c) reactive—the basis of hedonic response involving degrees of likes and dislikes as well as other behavioural reactions.

These three aspects of stimulation may be variously separated but generally they occur simultaneously and at great speed. The discriminative element can be developed to a high level of sensitivity and, with training, can totally override any reactive response. However, in flavour evaluation situations this possibly subconscious bias must be recognised and eliminated.

Responses to odour and taste stimuli may appear to be relatively simple but in fact are complex, particularly when they arise from multiple sources such as foods and drinks. Without doubt specific quality attributes can be recognised and selected from such complex stimulations, but they do not necessarily correlate with equivalent isolated standards or established levels of specific stimulants in the source material.

1.4. Nature of Coffee Aroma and Flavour

Aroma is a term especially used in connection with the sensory evaluation of roasted coffee and its products (ground, brews, extracts and instant coffees). There are, however, several distinct usages of the term. First, there is the head-space aroma arising from the dry roasted product (e.g. roasted and ground coffee or instant coffee) at room temperature and perceived by sniffing only through the front nasal orifices. The term is also used to describe the odorous sensation from the gases evolved during the roasting of coffee, though the aroma is not quite the same in composition or type. Secondly, there are the volatile components constituting the aroma (and therefore part of the flavour) arising from a prepared aqueous brew from roasted coffee at drinking temperature and perceived essentially from within the mouth by their passage to the olfactory region via the posterior nares (see page 155), though also by simultaneous passage through the front of the nose. The volatile components of the aroma are necessarily different in amount (both total and proportionately) and character from those of the first. So-called brew aroma may also be that which is merely perceived by the nose, without drinking, from a prepared cup or during brewing.

The distinctive flavour of brewed coffee is certainly the main reason for its wide popularity and almost universal appeal as a refreshing beverage. This flavour is not, however, a single distinguishable entity; actual flavour depends on the originating green coffee, the degree of roasting to which it has been subjected and, to a lesser extent, the conditions of brewing (e.g. coffee/water ratio, temperature, time and type of brewing equipment). The chemistry of the pyrolytic changes which take place on roasting under different conditions is fully discussed in Volume 1. The different substances, in different amounts according to the degree of roasting—both volatile and non-volatile—will be responsible for the various sensory effects to be described in the following sections. Differences in flavour arising as a result of brewing the same roasted coffee by different methods, and therefore concentrations (and even proportions) of these substances taken into solution, are especially discussed in Volume 2, Chapter 7.

2. TASTE: PHYSIOLOGICAL MECHANISMS OF THE TONGUE

2.1. Taste Receptors—the Taste Buds

For over 200 years the tongue has been recognised as the organ of taste and, as the various forms of papillae are the only visible specialised structures on the tongue, it was soon discovered that these were the site of the sensory receptors—the taste buds. The taste sensation is initiated by the contact of an aqueous solution of certain non-volatile, water-soluble substances with some or all of the 9000 to 10 000 taste buds on the human tongue, but also possibly with a matrix of free nerve endings throughout the mouth tissues. Obviously the salivary glands play an important role in the taste sensation by providing the aqueous medium in which tasteful materials dissolve or are diluted prior to exposure to the receptor cells. Saliva, which is a complex solution containing enzymes as well as potassium and thiocyanate ions, may act as a sensitiser to the taste receptor cells, variously affecting their sensitivity to sweet and bitter substances. In addition, the saliva acts as a buffering agent, maintaining an acceptable level of pH around the taste buds. It also to some extent helps in temperature control.

There are four types of papillae on the human tongue (see Fig. 1):

(a) Filiform papillae—widely distributed over the front section of the tongue. These are devoid of taste buds and so are not receptor areas.
(b) Fungiform papillae—these well defined, round, mushroom-shaped papillae number between 150 and 400, distributed over the front two-thirds of the tongue but mainly concentrated at the tip and sides.
(c) Foliate papillae—not well developed but are located in folds on the sides of the posterior third of the tongue. They play only a small part in taste perception.
(d) Circumvallate papillae—6 to 15 of these form a clearly defined V shape on the back of the tongue. They are mound-shaped and surrounded by a groove.

The total number of papillae begins to decrease from about the age of 45 years, and as an individual grows older the number of taste buds also diminishes.

With the exception of the fungiform papillae, adult taste buds, containing the receptor cells, are located mainly in depressions of the papillae. In children they may also occur on the sides and roof of the mouth but these are gradually lost with age. Each taste bud looks not unlike a brandy glass having a narrow opening at the top. Each bud contains between 5 and 18 long, slender, sensory cells together with other supplementary supporting cells grouped together. These are modified epithelial cells which are developed at the edges of the bud and move into its centre where they appear to be constantly changed and removed by phagocytic action. They have a life cycle of only 5 to 7 days. Structurally these sensory cells carry numerous microvilli, about 2 μm long and 0·12 μm wide, extending into the taste pore. It is thought that the microvilli could expedite rapid absorption of the tasteful substance. Work by Murray,[3]

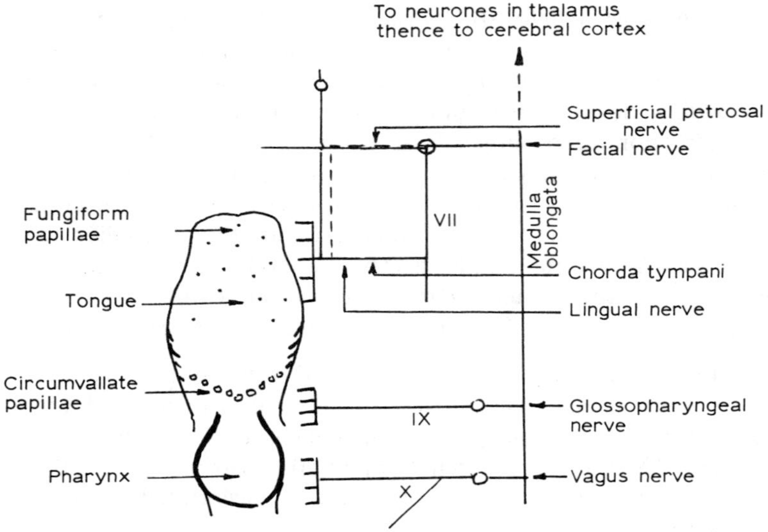

Fig. 1. Neural pathways serving the mouth and pharynx (diagrammatic).

Shimamura *et al.*[4] and Takeda and Hoshino[5] has shown the presence of two types of terminal structure, the microvilli and bulbous endings, which may be related to specific receptor responses.

Three sensory ganglia innervate the taste buds, each serving a particular zone (see Fig. 1). The fungiform papillae and the anterior soft palate are innervated by sensory neurones in the lingual nerve, the taste fibres from which become part of the chorda tympani nerve which passes through the middle ear and joins the facial nerve (VII) of the cranial nerve en route to the brainstem. The foliate and circumvallate papillae, as well as the posterior palate, are innervated by neurones in the glossopharyngeal nerve (IX) whereas taste buds located in the epiglottis and larynx are served by neurones in branches of the vagus nerve (X). The taste fibres from all the sensory nerves from the mouth come together in a single tract with its nucleus in the medulla oblongata. Individual nerve fibres branch profusely within the taste buds, each making contact with several taste cells.

Any damage to, or severance of, the facial nerve or chorda tympani results in a loss of taste sensation from the front two-thirds of the tongue accompanied by a degeneration of the taste buds served by this nerve. Sensation is slowly restored when the nerves regenerate.

The free nerve endings found on surfaces within the mouth and pharynx provide a sensory matrix which may be associated with taste but is of greater importance in the perception of touch, temperature and pain. These sensory sites are provided by neurones of the trigeminal ganglion, the severance of which does not appear to affect the perception of taste.

2.2. Taste Perception

Taste sensations result from stimulation of the excitatory portion of the taste receptor cells, the microvilli, by a combination of diffusion and of hydrostatic forces induced by the strong muscular movements of the tongue during mastication in the case of food. The aqueous tasteful materials are only weakly adsorbed on to the surface of the sensory tissues, but this is sufficient to cause a temporary breakdown in the cell membrane which allows the transfer of potassium ions. This in turn alters the electrical potential across the cell membrane, giving rise to a microelectrical impulse which is transmitted to the brain in the associated nerve fibre. The resting neural activity is typified by a steady series of discrete pulses which display a spontaneous spike discharge pattern. If the sensory source is stimulated above this spontaneous level, the neurone is excited and an appropriate response is perceived. The frequency of these induced electrical impulses, which are measured in milliseconds, is directly related to the magnitude of the perceived taste stimulus. Over the past two decades, numerous studies have been made into the nature of these stimuli and their related electrical activity in the determination of taste quality. The present state of knowledge of gustatory neural coding has been discussed by Scott and Chang.[6] Theoretically it might be expected that the taste buds, being anatomically different in structure, innervation and distribution, might be specific to one taste stimulus. Studies on receptor potentials indicate that different receptor types do exist, and some taste cells appear to respond only to salts and acids, others to salts, acids and sugars. However, the complex innervation of the taste buds makes differential taste cell responses difficult to detect, and the pattern of receptor potentials and spike potentials that characterise specific taste responses is still in doubt. It is generally accepted that it is the total pattern of nerve activity coming from numerous taste cells which determines taste quality. Boudreau[7] describes four separate chemoresponsive neurone groups which respond to different types of compound, but as yet there is not complete agreement on the part this plays in taste perception.

The middle top surface of the tongue is insensitive to all taste stimuli. Beidler[8] described the spatial patterns of taste activity and noted that the sweet sensation is perceived primarily on the tip of the tongue, bitter at the base, sour (acid) at the sides and saline on the tip and sides of the tongue, but this distribution is very imprecise and is of little practical concern in the evaluation of a complex food or drink.

2.3. Theories of Taste Perception

The mechanisms of human taste perception have been critically reviewed by Meiselmann.[9,10] Current ongoing research into specific aspects of this topic embraces the relationship between molecular structure and perceived taste quality and strength, as well as the nature of the electrophysical response to neural signals and their relationship with the autonomic nervous system.[11]

Any theory to explain taste perception must embrace certain clearly demonstrable conditions, including:

- (a) the non-specificity of the taste receptors;
- (b) the magnitude of response relative to the concentration of the sapid material (material having power to affect taste receptors);
- (c) the pattern of neural discharge associated with specific taste sensations and intensities;
- (d) the occurrence of 'taste blindness' to specific compounds, e.g. phenylthiocarbamide.

To date there is little agreement on the precise physicochemical reactions involved. Numerous theories have been proposed to explain the neurophysiology of the two sensory systems in the oral cavity—the taste buds and the free nerve endings—but anatomical and neurophysical studies have so far yielded little to an understanding of the mechanisms of taste perception.

Traditionally it is widely accepted that there are four primary taste sensations: sweet, sour (acidic), salty and bitter, but Moncrieff[12] suggested that there are two further well defined taste responses: alkaline and metallic, which fall outside the original classification. Teranishi[1] proposed a listing of eleven specific taste sensations, viz.: sweet, sour, sharp, salty, bitter, fatty, insipid, astringent, viscous, aqueous and nauseating, but some of these are debatable as being true taste sensations. What is of greater interest is the way in which certain substances (e.g. monosodium glutamate, a group of L-amino acids and the 5′-ribonucleotides and their derivatives) are able to act as flavour enhancers and taste modifiers. What is called the 'umami' sensation has been described by Yamaguchi.[13]

Boudreau[7] has tabulated many of the above sensations in terms of the nerves involved in their perception, but the complex neurophysiology of the taste receptors makes it very difficult to attribute specific taste sensations to specific neural activity. What does seem to be clear is that associated taste sensations such as astringency, warmth, cooling and burning are detected by the trigeminal nerve (V); true taste sensations including metallic and umami are recognised by nerves VII (facial) and IX (glossopharyngeal) and irritations, such as pungency, as well as some 'aftertaste' sensations, are perceived by the vagus (X) nerve. The taste cells and associated nerve fibres which carry neural responses are heterogeneous, and most simultaneously respond to two of the four prime taste sensations; response to all four taste sensations is not uncommon.

Taste is largely influenced by the presence of specific functional groups in the molecule, and is more directly related to its molecular make-up than to its physicochemical properties.

2.4. Molecular Structure and Taste

2.4.1. Sweetness

Among the four prime taste sensations, sweetness is paramount. It is the first taste which the newly born experience and appreciate. The other taste sensations develop only gradually. The sweet sensation is produced by several different kinds of substance, most of which are non-ionic. These include:

(a) sugars—monosaccharides (e.g. glucose, fructose) and disaccharides (e.g. sucrose, lactose, maltose);
(b) polyhydric alcohols—glycols and glycerol;
(c) α-amino acids;
(d) synthetic sweeteners—e.g. acesulfame K, aspartame, saccharin and thaumatin.

These groups of non-volatile compounds display wide variations in molecular structure, and attempts to explain their characteristic sweetness in terms of functional groups or molecular architecture have met with little success. It is generally accepted that even small changes in molecular configuration can result in the total removal of sweetness from a particular molecule (e.g. dulcin is 250 times sweeter than sucrose, but the structurally related *p*-ethoxyphenyl thiourea is not sweet but bitter). Sucrose is usually taken as a standard for determining comparative sweetening power of other compounds.

2.4.1.1. Sugars. The most widely occurring natural sweeteners are, of course, the crystalline polyhydric carbohydrate sugars, either monosaccharides with five (pentoses) or six (hexoses) carbon atoms or disaccharides with 12 carbon atoms. The geometric structure of sugar molecules is well established;[14] most exist in what is called the $^4C_1$ conformation, but the sweetest of the synthetic sugars, β-D-fructopyranose, has the $^1C_4$ conformation. Structural relationships to observable sweetness have been studied by Shallenberger and Acree.[15]

Monosaccharides are sweeter than their disaccharide analogues (e.g. fructose plus glucose vs. sucrose); and in an homologous series of oligosaccharides, such as the maltodextrins (based on glucose), the higher the molecular weight of the sugar, the less sweetness it displays. Shallenberger and Acree[15] proposed that a common constellation of structural features in sweet-tasting molecules was a weak proton-donating group, AH, and an electronegative atom, B, located about 2·5 to 4 Å apart. Kier[16] suggests that this AH–B moiety, the so-called glucophore, interacts with a pair of opposite groups on the receptor. Lee and Birch[17] considered that the ability of molecular patterns to elicit sweetness is related to hydrophilic and hydrophobic bonding activity associated with ring size and shape, and that the sweetness/bitterness relationship can be modified by altering these properties.

Birch[18] suggests that large oligosaccharide molecules may be sterically excluded from the taste receptor. Available evidence suggests that only one sugar residue in an oligosaccharide actually binds with the taste-bud receptor. Individual sugar residues appear to be polarised so that one end of the molecule elicits sweetness while the other end may elicit bitterness; indeed these studies indicate that, when sweetness is diminished, bitterness is increased.

Sweetness is not generally associated with the taste of black coffee brews. Green coffee beans contain between 3–8% of sucrose and up to 0·5% of reducing sugars (see Volume 1, Chapter 3), but the roasting converts these into caramelised products or combines them with proteins and free amino acids in Maillard-type reactions. Some reducing sugars will, however, be present in roasted coffee so that the water-soluble part may constitute some 3–5% of monosaccharide sugars, especially in instant coffee, and may therefore contribute to the overall flavour of the beverage. Sweeteners, especially of course sucrose, are commonly used as additives to coffee brews by consumers, and are therefore important flavour modifiers.

2.4.1.2. Synthetic sweeteners. Many substances have an intensely sweet taste, and many are now legally permitted for use in foods in the UK. The structure/activity relationships of such materials as the nitroanilides, sulphamates, oximes, isocoumarins, dipeptides, ureas, saccharins, tryptophans, chlorosugars and acesulfame derivatives in relation to their AH–B moieties have been studied by Heijden *et al.*[19] Studies to clarify the role of sugar moieties in the generation of the bitter taste of flavanone glycosides and the sweet taste of dihydrochalcone glycosides have been made by Esaki *et al.*[20]

2.4.2. Bitterness

The human palate is very sensitive to bitterness, but strangely sweetness and bitterness are often associated and may be produced by the same compound in different concentrations. There is no obvious similarity between the structures of different molecules having a bitter taste, but three classes of substances can be recognised, namely:

(a) alkaloids—bases with a nitrogen-containing ring structure (e.g. caffeine);
(b) glycosides—readily hydrolysed into two fractions, a sugar moiety and an aglycone, the reactive part of which is a hydroxyl group;
(c) peptides—protein hydrolysates.

In coffee brews, the particular components responsible for bitterness are not known. Caffeine is present to the extent of 1% or more in roasted arabica coffee and 2% or more in roasted robusta, and is readily water-soluble; it is present in complexed form. Voilley *et al.*,[21] in their sensory studies of coffee brews, have shown that caffeine contributes less than 10% of the perceived bitterness. Particular peptides are now the prime candidates, with some contribution from the chlorogenic acids, though some Maillard products will also be bitter.

2.4.3. Sourness or Acidity

Sourness and acidity are terms used synonymously in scientific taste terminology, but among coffee tasters sourness has a particular connotation, generally unfavourable, whereas acidity is a favourable characteristic. Pure acidity arises from the presence of hydrogen ions from the ionisation of constituent acids (both strong inorganic and weak organic) in aqueous solution. Organic acids, present in foods and drink, are usually weakly ionised and their undissociated molecules are often volatile and may separately contribute to overall flavour through their aroma in an amount determinable by the degree of dissociation (see Volume 1, Chapter 8). Acidity perceived in sensory evaluation of foods and drinks is not generally correlated with hydrogen ion concentration (pH), though opinions differ.

Green coffee contains up to 12% of different acids, but roasting changes the amount and proportion of acidic compounds present and introduces new acids, though many occur at low levels. The nature of these acids has been fully discussed in Volume 1, Chapter 8. A medium-roasted arabica coffee will have some 1·5% aliphatic acids, 1% quinic acid and 2·5% unchanged chlorogenic acids on a dry roasted coffee basis. All of these are highly water-soluble and will enter coffee brews at nearly 100% extraction. Nevertheless, it is not clear which of these acids are really important for coffee brew acidity. Maier, reported by Clifford,[2] considers that the phosphoric acid is important. There is general agreement that total aliphatic acids go through a maximum at about 15–16% roasting loss (as is), corresponding to a medium roast level, which is of interest in that such roast levels are often stated to provide optimum quality in respect of 'fine' acidity. Whether this is true of total acids is not so clear, since the amount of the unchanged chlorogenic acids continuously decreases whereas that of the generated quinic acid increases. Acidity in coffee brews is generally simply assessed instrumentally by pH values: a medium-roast arabica coffee gives brews between 4·9 and 5·2 (a generally preferred range); robustas tend to be less acid (5·0 to 5·5) whereas more darkly roasted of either species will be less acid, and the more lightly roasted the more acid. Instant coffees, in solution, tend to be slightly more acid than their corresponding roast coffee brews. Poorly processed instant coffees may have a pH as low as 4·7 at the usual aqueous strength, at which point added milk may be curdled; similarly, brewed coffee on holding at elevated temperatures will develop excess acidity. Coffee brews with a pH as high as 6·0 are possible, but these are regarded as flat and uninteresting, as are instant coffees with added alkali. In sensory studies by Voilley *et al.*,[22] the pH of coffee brews was found to be poorly correlated statistically with perceived acidity, though, on the other hand, titratable acidity, to a pH of 8·0, was found to be quite well correlated for reasons which are not clear.

2.4.4. Salty Taste

The salty taste is stimulated by most soluble salts but typically by those of low molecular weight. Cardello and Murphy[23] suggest that ionic imbalance between aqueous solutions of salt and the saliva constituents results in a gradual change from sweet to sour through salty to bitter as the concentration increases. The mechanisms involved are still open to conjecture.

Saltiness is not a normal feature of coffee brews but may occur by use of zeolite-softened waters. It is generally regarded as an undesirable feature, but household hints in culinary articles often suggest the favourable taste advantage of adding a 'pinch of salt' to coffee brews (see next section).

2.5. Taste Interrelationships

Most of what we eat and drink is complex and produces simultaneous multistimulation of both the taste and odour receptors. The nature of the effect of the reactions of such complex stimulation on perceived taste has been studied by Fabian and Blum.[24] Although considerable variation in responses was reported, the following general observations are of interest. In the majority of subjects, salt at subthreshold levels reduces the sharpness of acids and increases the apparent sweetness of sucrose. Acids at subthreshold concentrations intensify the saltiness of sodium chloride, but sugar at subthreshold levels has the reverse effect. Pilgrim[25] carried out a similar series of tests at suprathreshold levels but found even wider variations. Again the following observations have been reported: salt and acids result in a decrease in the apparent sweetness of sucrose; sucrose produces a decrease in the apparent saltiness of salt; and acids produce an increase in the apparent sweetness of sucrose and bitterness of caffeine. It is a common observation that sugar reduces the sourness of acids and the bitterness of caffeine. These findings may well depend on the relative concentrations involved, and a desired effect may only be achieved by trial and error.

3. ODOUR: PHYSIOLOGICAL MECHANISMS OF THE NASAL CAVITY

3.1. Odour Receptors in Man

Smell is the most primitive of our senses, although its relative importance in the assessment of one's environment has diminished significantly as compared with this sense in many of the lower animals. The sense of smell still, however, plays a major part in recognition and in establishing the acceptance of foods and drinks. There is often a marked, rapid, hedonic response to what we smell as well as to the vapours arising from what we eat and drink. It would be wrong to consider that odour plays a dominant role in food flavour, but it is, nevertheless, a very important member of a highly sophisticated group of sensory systems which together provide the basis for judgement.

In mammals the sense of smell is located in the upper part of the nose; in man there are two nasal cavities separated by a thin, smooth septum (see Fig. 2). The lateral walls of each cavity are covered with vascular tissue in a series of two to six horizontal folds. The lower folds extend the whole length of the cavity whereas the upper ones, being shorter, form three well-defined

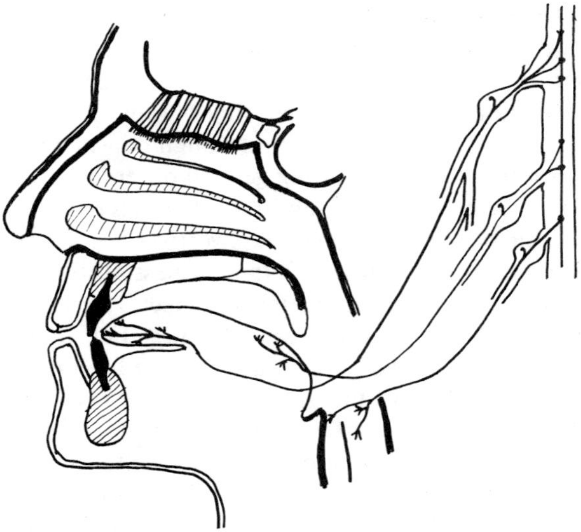

Fig. 2. Cross-section through nasal cavity and mouth.

channels known respectively as the inferior, median and superior meatus. These combine in a series of interconnected passages to form a common meatus opening directly into the olfactory cleft, which is also open to the pharynx. It is estimated that 5 to 10% of inspired air passes directly over the olfactory slit, although probably only as little as 2% reaches the olfactory sensory region. This proportion may be significantly increased by sniffing, which causes a rapid movement of air through the upper nasal passages, greatly enhancing the perceived odour effect. When consuming liquids, the action of slurping, i.e. drawing in air at the same time as the liquid, induces the vigorous movement of odour-filled vapours into the interior nares, which connect the pharynx with the olfactory region in the nose. This action also considerably enhances the perceived aroma impact of liquids taken into the mouth, and is of significance in coffee tasting.

The olfactory receptors are located (see Figs 3 and 4) in a patch of tissue called the olfactory mucosa sited in the dorsal (or back) nasal cavities. This comprises tens of millions of yellow to yellowish brown receptor cells together with support cells and closely associated mucous and serous secretory alveolo-tubular glands. The sensory cells are bipolar neurones with oval cell bodies. The olfactory rod or dendrite extends from their peripheral part and ends in an enlarged, rounded, olfactory vesicle from the surface of which some 9–16 fine, hairlike projections extend into the

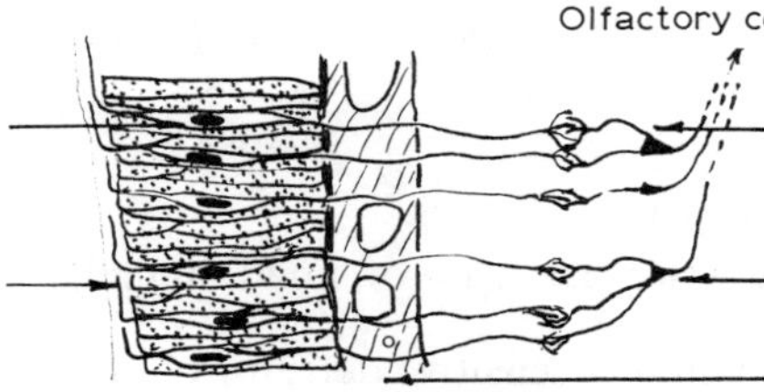

Fig. 3. Structure of olfactory sensory region.

covering layer of mucus. The proximal pole of each sensory cell extends backwards to form an unmyelinated axon, which continues as an olfactory nerve fibre, I. The millions of separate axons pass through minute holes in the cribriform plate to synapse directly with cells in the olfactory bulb. This is the only sense in which neural responses pass straight from the sensor to the brain, the neurone acting as both receiver and conductor. The complexity of carrying out neurophysiological research has limited our understanding of the working of this sensory system.

The process of olfaction consists of the carrying of molecules of volatile odorous substances in a stream of moist air into the nose and over the sensitive receptor areas. Here the odorous molecules, having first diffused through the overlying coating of mucus, interact with the receptor cells causing an electrical imbalance, so generating action potentials which are conveyed directly to the brain.

There is another sensory system located in the nasal cavity, the trigeminal system (V), with a matrix of free nerve endings. The response of the trigeminal system is little understood, but Tucker[26] has demonstrated that this system does respond to odour stimuli in a way very much like that in

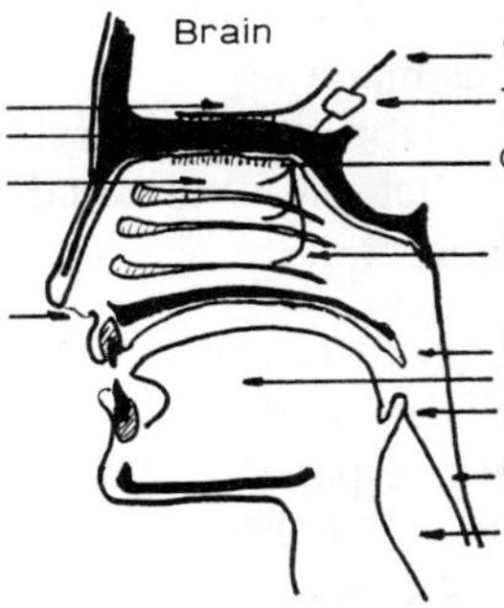

Fig. 4. Structures associated with smelling.

which the olfactory system responds and with an even lower threshold of response to certain odorants.

3.2. Odour Perception

Five things characterise the sense of small, namely:

(a) the nose is a highly sensitive detector of odorous molecules having an extremely low threshold of perception;
(b) perception is very fast and versatile;
(c) it is highly specific and can be trained;
(d) it can induce fast psychological responses dependent on the odour and on the subject (i.e. it has a hedonic content); and
(e) it demonstrates accommodation and fatigue with continuous stimulation.

Many theories of olfaction have been proposed over the past 50 years.[27] All have been subjected to criticism, as none can fully reconcile and explain all of the following criteria:

(a) the poor relationships between the physicochemical characteristics of odorous substances and their observed aromatic profiles;
(b) the nature of the interactions which take place between the odorous molecules and the receptor, particularly the speed and sensitivity of perception;
(c) the way in which different odorous molecules produce different, specific and recognisable response patterns depending on concentration;
(d) the pattern and frequency of action potentials in the olfactory neurones;
(e) the factors which determine odour thresholds, quality and intensity of response to odorous materials;
(f) the condition of anosmia in certain individuals;
(g) consistent but unpredictable individual differences in odour sensitivity and discrimination.

In addition, any valid theory of olfaction should be able to provide a basis for the systematic classification of odorous substances, and even lead to the prediction of olfactory response from a knowledge of molecular structure. To date little progress has been made in this area.

Obviously, whatever the precise mechanism for odour perception, it must involve two quite separate factors: the odorous molecule of almost infinite chemical variety, and the living tissue of the receptor site. Both must be in intimate and active contact at the molecular level to initiate an olfactory response.

Various theories of olfaction have been suggested, and the following are most widely accepted as partially explaining observable facts:

(a) *The vibrational theory:*[28,29] Odour-related attributes are said to be determined by certain far-infrared or Raman frequencies, which are a direct measure of the intramolecular vibrations of a substance; perception then results from equivalent resonance frequencies in the sensory receptors.

(b) *The stereochemical theory:*[30] This is based on the size and three-dimensional shape of molecules of odorous substances and their observed profiles. Functional groupings, electrophilic/nucleophilic characteristics and rotational moments are also taken into account to give seven 'primary odours' which act like keys fitting locks. The seven odour types described are: camphoraceous, musky, floral, pepperminty, ethereal, pungent and putrid. The main objections to this theory are the need for specific sensory receptors, molecular orientation, anosmia and odour fatigue.

(c) *The interfacial adsorption (or penetration) theory:*[31] This proposes that olfaction is stimulated by interactions occurring at hydrophobic//hydrophilic interfaces. Response is thought to be characterised by molecular cross-sectional area and free energy of adsorption at the interface leading to the penetration of the thin-walled site and the interchange of Na^+ and K^+ ions to produce an electrical imbalance.

(d) *The profile functional group theory:*[32] This theory is based on peripheral functional groupings, molecular size and shape. Odour response is attributed to the nature and reactivity of functional groups on the molecule, which determine its interreactivity with the receptor, and also the reactivity of peripheral functional groups, which determine the orientation and hence the profile of the molecule presented to the receptor. It is this orientation which is thought to determine odour profile.

(e) *The electron donor theory:*[33] This proposes that the human odour spectrum can be explained in terms of a two-dimensional continuum with the quality and impact of an odour molecule being determined by its directional dipole and electron-donor ability. This concept is still being studied.

The mechanism of stimulation of olfactory neurones, particularly the concept of the part played by receptor proteins in which specific odorants are considered to interact with specific receptor proteins, has been critically reviewed by Price.[34] It has been demonstrated that there is differential response of sensory units to different groups of substances; recent work on some animal species has yielded evidence of specific receptors or sites sensitive to specific substances, but this may be a case of inherent defence mechanisms. Such evidence cannot really be directly translated in human terms. However, this concept may ultimately result in yet another theory of olfaction.

All of these theories attempt to explain how one is able to recognise and discriminate between a vast number of different odours. All of them acknowledge the importance of the size and shape of molecules of odorous substances and the modifying effects of the nature and position of functional groups, electronic donor-acceptance characteristics and the sorptive capacity of odorants on receptor sites. Only the vibrational theory considers the far-infrared and Raman spectral characteristics of odorants. The other theories centre on molecular structure. Wright[35] postulates that interaction and stimulation probably do not involve the whole molecule but may be limited to a direct link of two atoms or radicals on the surface of a small but very sensitive sensor site. He has also applied his vibrational theory to odour blends.[36] McGill and Kowalski[33] carried out extensive statistical studies of known data about olfactory response. They concluded that no one property of a molecule is sufficient to explain its odour characteristics and that at least two molecular properties must be taken into account. Work by Moulton[37] indicated that spatial organisation may well play an important part in encoding olfactory stimuli. Current research in this field is mainly into molecular geometry, internal molecular energies and vibrational patterns, electronic structure, membrane potentials, and enzymatic and other biochemical processes.

On the anatomical side, studies into the olfactory sensory system and the closely related nasal trigeminal neural matrix have not as yet yielded much understanding of their mechanisms. Neurophysiological studies[38] have shown that different odorants induce different spatiotemporal patterns of excitation which may have two components, the first depending on retentivity of the odorant on the powerfully sorptive mucous sheet, and the second inherent and depending on the aggregation of receptors with similar peak odour specificities. Different odour types may excite different regions of the mucosa and these may overlap. The possibility of a third non-spatial component is postulated.

3.3. Molecular Structure and Odour

The components present in naturally occurring odorous materials embrace the whole spectrum of organic compounds. In spite of many studies having been carried out, there is very little correlation between chemical structure and observed sensory attributes, there being literally thousands of different and readily distinguishable odour qualities which defy more than a superficial classification on either a chemical or physicochemical basis.

The relationship between molecular structure and odour quality can be studied in two different ways, namely: (a) by examining specific groups of organic substances (e.g. alcohols, carbonyls, lactones) and observing the changes in odour that occur with increasing molecular weight, increasing complexity of functional groups and specific location of functional groups; and (b) by association of similarities and differences in structure and/or odour qualities. McLeod[39] has reviewed this topic and concludes that the most important factors affecting these relationships appear to be overall shape and size of the molecule, its detailed stereochemistry and certain chemical (e.g. polarity, functional groups) and physical (e.g. volatility, solubility) properties.

3.4. Coffee Aroma

The chemistry of roasted coffee products (ground, brews, extracts and instant coffee) in relation to their aroma has been extensively covered in Volume 1, and only the sensory aspects will be covered in this section.

It has now been established, primarily from gas chromatographic evidence, that there are over 700 different, more or less volatile compounds present in roasted coffee, and the number being detected is still rising. In roasted coffee, their total quantity, apart from volatile acids such as acetic acid, is about 1000 ppm, but many are present in quantities of less than 1 ppm and some at levels much less than this. Some are located in the carbohydrate matrix of the roasted bean whereas others (especially the non-polar ones) are located in the coffee oil distributed within the bean. They comprise a wide range of organic substances, e.g. aldehydes, mercaptans, and ring compounds, furans, pyrazines, thiazoles and oxazoles, with nitrogen- and sulphur-containing compounds being especially important. Not all of these compounds will contribute to coffee aroma from the whole or ground, dry roasted product, being present in insufficient quantities and with insufficient volatility from their bean matrix to the surrounding air that is sniffed, i.e. they are below their odour thresholds. Nevertheless, a number of compounds with especially low threshold values and others in relatively large amounts are present, especially the mercaptans and disulphides. Differences are evident with the degree of roast and with species, whether arabica or robusta. It is generally true to say that differences reside not so much in the presence or absence of particular compounds but rather their relative proportion; for example, robustas tend to have higher levels of sulphur compounds. Many of these compounds have quite objectionable odours when sniffed at high concentrations, but not when present at a level of a few parts per million; indeed, several are reminiscent of coffee aroma. The combination of aromatic compounds essential for roasted coffee aroma (i.e. proportion and quantity) is not clearly known. Several commercial instant coffees have coffee oil (derived from mechanical expression of roasted coffee beans) in quantities less than about 0·5%, sprayed upon them before packing. This provides an excellent level of 'smell' in the jar but is insufficient to influence the flavour of the subsequently prepared brew over identical products without the oil.

Many of the aromatic components of roasted coffee, and particularly of coffee oil itself, are extremely susceptible to deterioration by the action of moisture and oxygen. In commercial practice, therefore, roasted coffee has to be packed in an atmosphere of low oxygen content, either by CO_2 gas-flushing or vacuum-packing, as fully described in Volume 2, Chapter 7. This is also true of instant coffees with a high volatile compound content. This deterioration is generally known as 'staling', though it initially arises through the attack on the volatile compounds and only in the extreme later stages by any oxidation of the coffee oil. Temperature is also a determining factor, so that refrigerated storage is often recommended.

The aroma of coffee brews results from the volatilisation, into the air space, of volatile components which have been extracted from the roasted coffee into aqueous solution. This extraction is not necessarily complete (e.g. it is often only about 70% in home brewing) and is differential according to the polarity of the particular molecules (as discussed in Volume 2, Chapters 5 and 7). Volatilisation is governed by air–liquid partition coefficients for each component, corresponding to relative volatilities at infinite dilution in water, or more strictly in very weak solutions of coffee solubles. These values are known for many of the components that enter coffee extracts or brews. Much gas chromatographic work has been conducted on roast coffee itself by exhaustive extraction methods; but very little on normal strength coffee brews at drinking temperatures, primarily because of the problem of obtaining sufficient amounts of volatile components for detection and quantification from manageable quantities of brews. Nevertheless, attention has been directed to those same substances and groups of substances known to be present in the originating roasted coffee. As relative volatilities are dependent upon temperature to an important extent, the temperature of drinking a coffee brew will also determine concentration of aromatics in the air-space or air-bundle that passes to the olfactory region for evaluation, either directly to the nose or via the pharynx.

4. COFFEE FLAVOUR

4.1. Flavour Integration

Total perceived flavour of a coffee brew comprises not only the taste sensations (Section 2) and aromatic sensations (Section 3), reflecting the aromatics perceived on drinking, but frequently involves other senses. Among these are the senses of sight, sound, touch, pain and temperature. Colour is obviously of great importance, and to some extent conditions consumer expectancy of flavour. The perceived intensity of flavour is often subconsciously affected by the depth of colour when in fact there is little or no actual relationship. Additional sensations, arising from the stimulation of overall physicochemical senses, may also influence discrimination and preference. Substances that affect the latter include compounds giving rise to pungency, cooling/warming effects and astringency, all of which are perceived via the trigeminal nerve and together produce what is known as 'mouthfeel'. Closely related is the term 'body', which has some relationship to viscosity, whether real or imaginary, of the liquid that is consumed. The actual viscosity of clear coffee brews, though containing macromolecular substances, is very little different, in centipoise values, from that of water at the same temperature. Some brews, however, contain colloidal particles (dispersed oil, etc.) which may contribute to this effect. The term 'body', as used by coffee tasters, may have a closer relationship to the overall strength of the aromatics.

A special factor in coffee brews, though not related to flavour attributes in any sense, is the delayed stimulation provided by the relatively mild alkaloid, caffeine present, so that coffee brew is also described as a stimulating drink, with hedonic implications. Taste and aromatic sensations are fairly readily distinguished in coffee brews by tasting or sensory analysis including assessment and naming of separate items within these groups. Similarly, the other senses may be described, though with some greater difficulty, e.g. 'body' and 'mouthfeel'. Astringency is an attribute often confused with or by bitterness in inexperienced tasters.

Although coffee brews are normally prepared hot, i.e. at 90–100°C, cooling is necessary to around 55°C before drinking and the probable temperature within the mouth will be about 38°C (i.e. close to body temperature) for comfort. Attempts to drink at higher temperatures will cause pain and may temporarily damage the sensitivity of the taste buds to taste stimuli, which may take some time to recover. Iced coffee is a separate type of brew and will necessarily have some different characteristics from normal coffee brew.

4.2. Basic Flavour of Coffee Brews

Three separate kinds of coffee flavour are generally distinguished, which, for a medium roast colour in a conventional brew at about 1·2% solubles concentration, extracted at around 20% from roasted coffee, are described as follows:

Wet-processed arabicas (milds), providing a 'fine acidity' with aromatics and a medium body.

Dry-processed arabicas (Brazils), providing lower acidity and lower aromatic level, but generally good body and having good 'blending' qualities with the foregoing.

Robustas (wet- or dry-processed), providing much less acidity and being more neutral, with a tendency to 'harshness', and again having blending qualities.

Although it is possible to characterise the individual aromatic attributes in much greater detail by various flavour terms, coffee buyers are much more concerned with the assessment of expectancy of norms for the main types and with the examination of any off-flavours to give reason for immediate downgrading (often by numerical standards). These off-flavours are specifically characterised by such terms as 'oniony', 'woody', 'grassy', 'sour', etc., in wet-processed arabicas. A fermented note sometimes occurs in milds from some Central American countries. Coffees from Kenya, Colombia and Costa Rica are generally rated very highly at their norm, with some differences (with price differentials) in different shipments, according to precise origin and crop. Brazils have an official grading system, in the scale 'strictly soft' to 'hard', reflecting degrees of freedom from a so-called Rio flavour ('medicinal' or 'iodoform'), which is not desirable but occurs in coffees from certain areas under certain conditions. Robustas are expected to be 'clean tasting', and again certain characterisable off-flavours, notably 'swampy', 'potatoey', 'rubbery', etc., are reasons for downgrading, blending at lower levels or total rejection.

The nature of the flavour characteristics of coffee brews has been the subject of a number of recent reviews, notably by Clifford[2] and Clarke.[40] The concentration of volatile substances, even if fully extracted by the brewing water, is necessarily some $\frac{1}{16}$ (e.g. if a coffee:water ratio of 1:16 is used) of that in the originating roasted coffee, so that levels can often be extremely small and correspondingly so in the air-space above it, according to the relative volatility of the individual compound. Some increase can be obtained by slurping the brew as it is tasted, as a result of non-equilibrium volatilisation which enhances assessment. We are now concerned with the flavour threshold of different compounds, i.e. assessment on drinking, rather than an odour threshold figure, i.e. by sniffing. There are several compounds present in coffee brews, particularly complex pyrazines with a number of substituted non-polar groups, which have extremely low threshold values (e.g. 1 μg per 100 g in water), well below the quantity present. Tressl and Silwar[41] in particular describe so-called character-impact compounds for coffee, such as furyl-2-methanethiol and kahweofuran, with coffee-like character and present in relatively high amounts compared with their threshold levels. Vernin[42] refers to the burnt-coffee-like flavour of 2-ethylfuran, thiobutyrolactone and 3-methyl-2-acetyl thiophene. Robustas are of interest in having higher levels of various phenolic compounds (e.g. guaiacols). As with roasted coffee, therefore, some of the volatile compounds present in a brew are going to be important whereas others will be of no consequence, apart from their possible synergistic effects. A useful concept in flavour assessment is that of aroma values for different compounds, defined by the actual amount present and their respective threshold values. An interesting assessment of the relative flavour importance of different volatile compounds was carried out in a comprehensive study within the USA brewing industry and reported by Meilgaard[43] in 1982. The flavour thresholds of some 250 different compounds as found in beers were very carefully established, first in very weak beer. Their relative importance was then determined in normal beers. Synergistic effects were found to be surprisingly few. A similar study is possible in coffee brews, provided individual threshold values are determined in brews from which all aromatics have been removed (e.g. by steam-stripping). Another technique, not so far reported for coffee, has been described by Parliament and Scarpellino,[44] who used the fractions obtained from gas chromatographic analysis. These were successfully compounded and then compared with total volatiles to assess their relative importance.

Summarising, therefore, for coffee brews, our current information is inadequate for the full understanding of the totality of coffee flavour, including both the aroma contribution and, more importantly, off-flavours.

4.3. Other Variants of Coffee Flavour

The foregoing has primarily concerned medium roast coffee. However, many consumers in this and other countries prefer darkly roasted coffees and others prefer the lighter. These roasting differences result in quite profound differences in the flavour of brews, though still recognisably coffee, caused by differences in bitter/acid characteristics as already described but also in the amount and proportions of volatile compounds, especially those really contributing to aromatic character, which will decline in amount and differ in proportions with increasing severity of roast. The differences resulting from the use of different coffee:water ratios and the degree of grind from coarse to fine, which directly affects the concentration and degree of extraction of both soluble solids and aromatics, are also quite marked. This is evidenced by a conventional brew at about 1·2% soluble solids in the USA and the UK, about 2% in France and Spain, and 4% in espresso coffee-making. Extraction yields of soluble solids, which can range from 15 to 28%, will also be significant. The differences arising from different types of brewing equipment, other conditions being the same, are relatively small, though these can be significant to some consumers (e.g. clear, cloudy or sedimented). Different kinds of brewing equipment require different grinds, and these factors have been examined by Pictet in Volume 2, Chapter 7, who indicates a particular panel preference for the filter brew with its more balanced aromatics.

Depending upon the actual process used, decaffeination, carried out on the green coffee, generally affects the flavour of the brews only slightly compared with that of its non-decaffeinated counterpart of the same roast colour. Instant coffees comprise a very wide range of different coffee flavour characteristics, which it would be impossible to cover in a short section. With modern sophisticated technologies (see Volume 2) many commercial instant coffees now show a very high proportion of extracted and retained aromatic volatile compounds, deducible from both instrumental and sensory analysis.

Most consumers will add sweeteners and/or milk (or milk products) to a coffee brew, depending upon the time of day or occasion of drinking. For example, in France and Spain breakfast coffee is regarded more as a nutritious drink through its hot milk content, and in the UK it is reckoned that some 70% of consumers use sugar (sucrose). The flavour implications of these various usages are again profound and little reported in respect of modifications from basic coffee flavour (i.e. tasted black). Some inferences can be drawn on taste factors as discussed, though less on aromatic factors where, for example, volatilities can be altered.

Differences can also arise from the use of different qualities of water. The water used should be at least potable (which has an EEC Directive definition), but water at different levels of hardness, non-demineralised softened water, alkalinity and chlorination, can all affect the brewing process itself (i.e. speed of filtration, extraction yield) as well as the resultant flavour and colour of the brew. In scientific flavour evaluation, distilled water is sometimes used, though lack of dissolved gases such as CO_2 through boiling may influence flavour. Many of the reported effects can be subjective; but more objective studies have been carried out by Lockhart *et al.*[45] on taste thresholds of various ions in water and coffee brew, Pangborn[46] in practical brewing experiments, and Little and Brinner.[47] The desirability of a low level of dissolved mineral salts was generally indicated, and Lockhart *et al.* set out maximum quantities for no effect (e.g. 150 ppm hardness). Pangborn showed how a softened water with a high level of Na^+ ions gave brews with much lower perceived acidity and higher pH values, and Little and Brinner showed the effect of Ca^{2+} and Mg^{2+} ions on the colour of brews.

It has also so far been assumed that fresh roasted coffee is being brewed, i.e. whether freshly roasted or roast (and ground) coffee kept in a package at low oxygen content for, say, not more than 12–18 months. If neither of these conditions is fulfilled, then roasted coffee gradually deteriorates as already noted; the perceived flavour of the brew at a slightly slower rate than that of the aroma from the dry product. The same term 'stale' is used, which may be masked to some extent by the use of milk additives and the strength of the brew. The most frequent cause of deteriorated coffee brew (e.g. in restaurants) is the holding of the prepared brew hot for several hours. This practice results in a perceived sourness and a demonstrable decrease in pH value, though in studies made by Pangborn[46] and Segall and Proctor[48] the maximum allowable time/temperature is disputed.

5. APPLICATIONS TO SENSORY ANALYSIS OF COFFEE

5.1. British Standards on Sensory Testing

A comprehensive glossary of terms used in sensory analysis has been published by the International Standards Organisation (ISO/TC34-5492/1-5) and the British Standards Institution (BS 5098:1975). In addition, there is a glossary of terms appropriate to coffee and its products under ISO/TC34-3509:1984 and a BSI equivalent under BS 5456:1985.

5.2. Environment for Sensory Assessment

The quality of roast and ground coffee, coffee brews and extracts, as well as green coffee (after roasting), depends on sensory assessments. The test methods used tend to be specific to the coffee and tea industries but are based on the general precepts which apply to all sensory judgements. The most important of these is the need for complete concentration and an absence of distraction during any test period. The environment in which sensory testing takes place is also of importance.[49,50] Even where the facilities are not optimal, it is usually possible to ensure that sensory tests are carried out in conditions that ensure that the assessor is at ease but not over-comfortable and is quiet and free from external distractions. Other conditions, such as air conditioning and variable lighting, may be advantageous but are not essential. In most cases a quiet room, in which there is an adequate bench with an impermeable top which can be kept scrupulously clean and a suitable container for rejected tastings, is all that is necessary.

5.3. Factors Affecting Sensory Judgements

The test methods to be adopted in the sensory evaluation of any given material or product will be determined by the nature of the sample and the complexity of the information sought from the test. Each calls for specific methodology aimed at reducing to a minimum any variables which may bias or adversely influence the judgement of assessors. Whatever the nature of the material under examination, it is first necessary to draw a representative sample in an approved manner as described in British Standard BS 6379/1-3, but this will to some extent be conditioned by the reasons for carrying out the test. These may include:

(a) distinguishing between one or more samples in a clearly defined way;
(b) the qualitative and quantitative characterisation of any differences that exist;
(c) the nature of any changes that have taken place (e.g. during processing or storage);
(d) establishing compliance with standards or reference material;
(e) scoring of specific attributes against a numerical index;
(f) grading according to a stated classification; and
(g) establishing a degree of acceptability based on hedonic responses.

In all of the above cases the assessors are being used as 'instruments', but, unlike mechanical or electronic instruments which have a specific data output, human assessors are capable of a high level of sensitivity and of providing a wide variety of information depending on the questions posed and the personal responses of the individual involved. One frequently made mistake is to read into results more than is justified by the test procedure and the number of judgements made. One should remember that the most consistent results are likely to derive from assessors answering simple questions directly related to the specific aims of the test. Sensory judgements may be biased and affected by hedonic reactions, but these adverse influences can be minimised by experimental design and the use of panels involving several assessors. The statistical significance of the results is related to the number of panellists used and can be determined from published tables. All sensory test methods have their faults, and the method selected should be the one that is simplest to perform, has a good degree of reproducibility of test conditions, and is most likely to provide reliable and meaningful answers which are capable of statistical analysis.

One problem facing assessors of coffee and coffee products is the need to pass judgement on a large number of samples presented simultaneously. The swallowing of large quantities of coffee, with its attendant physiological effects, may have very unfavourable consequences for the taster so that a special procedure has had to be developed. This comprises a technique called 'slurping' during which the test liquid, usually a coffee brew prepared under standardised conditions, is taken into the mouth on a spoon, the taster simultaneously inhaling through the mouth. This is rather a noisy procedure during which the liquid is thrown to the back of the mouth and on to the pharyngeal wall just as occurs in normal swallowing. A quality judgement on odour, taste and smoothness is rapidly made and the liquid is forcibly expectorated prior to repeating the process on the next-in-line sample. During slurping the aromatics in the brew reach the olfactory region via the posterior nares but more effectively than is achieved during swallowing, so that it is quite unnecessary to swallow any of the brew in order to reach a full sensory judgement. One might expect a carry-over from one sample to the next but, with training, coffee tasters learn to accommodate this effect and experts have no problem in reaching consistent judgements.

Attempts have been made to establish instrumental methods for the evaluation of odour and flavour stimuli but so far without any real success. The correlation of instrumental and sensory methods in the study of odours and flavours is the subject of a publication of the American Society for Testing and Materials (ASTM),[51] and much attention is now being paid to the use of headspace analysis and the computerised profiling of gas chromatographic data.[52,53] As yet there is nothing satisfactory to replace traditional sensory judgements.

5.4. Panels vs. Expert Judgement

There are very many occasions when the opinion of a highly trained and experienced assessor is sufficient grounds upon which to base a judgement of quality or acceptability of a given sample or set of samples. However, there are many other occasions when the combined opinion of numerous assessors, either trained or untrained depending on circumstances, is both preferable and more reliable. Specific test requirements may dictate choice, and it is often a quite emotive question to decide whether to rely on the single judgement of an expert, who may have significant status, or to insist on obtaining the views of a selected panel. There is no simple answer, and many factors must be taken into account.

In many companies, particularly those involved in coffee processing and blending which involve considerable critical sensory assessments, the role of the expert is traditional and of great importance. In the creation of new products and the selection of new or alternative raw materials for use in production, such judgements are often directly associated with ultimate success. During the often long-drawn-out period of new product development, however, there is a need for a panel of six or more assessors selected on proven ability and trained to do a specific job. When it comes to acceptance in the marketplace, there is a need for much bigger consumer panels. Such panels are carefully selected, either by representative sampling of the total population or by using specific target groups (similar to the selection method used in Gallup polls). Such panels are capable of providing a good indication of likely market response. With consumer products such as coffee, one has always to remember that the marketed item is eventually judged by randomly selected, often irrational, non-experts who constitute the large mass of consumers and who alone determine the commercial success or otherwise of the end-product.

REFERENCES

1. Teranishi, R., in *Flavor Research: Recent Advances*, Eds R. Teranishi, R. A. Flath and H. Sugisawa, Marcel Dekker, New York and Basel, 1981, 213–304.
2. Clifford, M. N., in *Coffee: Botany, Biochemistry and Production of Beans and*

Beverage, Eds M. N. Clifford and R. C. Willson, Croom Helm, London, 1985, 305–374.
3. Murray, R. G., in *Handbook of Behavioural Neurology*, Ed. R. B. Masterson, Plenum Press, New York, 1971, 91.
4. Shimamura, A., Tokunaga, J. and Toh, H., *Archiv. Hist. Japan*, 1972, **34**, 200.
5. Takeda, M. and Hoshino, T., *Archiv. Hist. Japan*, 1975, **37**, 395.
6. Scott, T. R. and Chang, F.-C. T., *Chemical Senses*, 1984, **8**(3), 91.
7. Boudreau, J. C., in *Instrumental Analysis of Foods, Vol. 1*, Academic Press, New York, 1983, 17–36.
8. Beidler, L. M., in *Flavor Chemistry*, Ed. R. F. Gould (Advances in Chemistry Series), American Chemical Society, New York, 1968, 1–28.
9. Meiselmann, H. L., *CRC Crit. Rev. Food Sci. Nutr.*, 1972, **3**(1), 89–110.
10. Meiselmann, H. L., in *Fenaroli's Handbook of Flavoring Materials, Vol. 1*, 2nd edn, Eds T. E. Furia and N. Bellanca, CRC, Cleveland, Ohio, 1972, 89–110.
11. Norgren, R., *Chemical Senses*, 1985, **10**(2), 143–161.
12. Moncrieff, R. W., *The Chemical Senses*, 2nd edn, Leonard Hill Ltd, London, 1967.
13. Yamaguchi, S., in *Food Taste Chemistry, Vol. 115*, Ed. J. C. Boudreau, American Chemical Society, New York, 1979, 33–51.
14. Shallenberger, R. S. and Birch, G. G., *Sugar Chemistry*, AVI Publishing Company, Westport, Conn., 1975.
15. Shallenberger, R. S. and Acree, T. E., *J. Agric. Food Chem.*, 1969, **17**, 701–703.
16. Kier, L. B., in *Structure–Activity Relationships in Chemoreception*, Information Retrieval Ltd, London, 1976, 101–110.
17. Lee, C. K. and Birch, G. G., *J. Food Sci.*, 1975, **40**, 784.
18. Birch, G. C., in *Structure–Activity Relationships in Chemoreception*, Ed. G. Benz, Information Retrieval Ltd, London, 1976.
19. van der Heijden, A., van der Wel, H. and Peer, H. G., *Chemical Senses*, 1985, **10**(1), 57–72, 73–88.
20. Esaki, S., Tanaka, R. and Kamiya, S., *Agric. Biol. Chem.*, 1983, **47**(10), 2319–2328.
21. Voilley, A., Sauvageot, F. and Durand, D., *Proc. 8th Coll. ASIC*, 1977, 251–260.
22. Voilley, A., Sauvageot, F., Simatos, D. and Wojcik, *J. Food Proc. Preserv.*, 1981, **5**, 135–143.
23. Cardello, A. V. and Murphy, C., *Chemical Senses Flavor*, 1977, **2**(3), 327–339.
24. Fabian, F. W. and Blum, H. B., *Food Res.*, 1943, **8**, 179–193.
25. Pilgrim, F. J., in *Physiological and Behavioral Aspects of Taste*, Eds W. S. Kane and B. P. Halpen, University of Chicago Press, Chicago, Ill., 1961, 66–98.
26. Tucker, D., *J. Gen. Physiol.*, 1963, **46**, 453.
27. Heath, H. B., *Source Book of Flavors*, AVI Publishing Co., Westport, Conn., 1981.
28. Wright, R. H., Reid, C. and Evans, M. G. C., *Chem. Ind.*, 1950 (37), 973.
29. Wright, R. H., *Ann. NY Acad. Sci.*, 1964, **116**, 552.
30. Amoore, J. E., *Proc. Sci. Sec. Toilet Goods Assoc.*, 1957, **37**, 13–23.
31. Davis, J. T. and Taylor, F. H., *Proc. 2nd Int. Congr. Surface Activity, Vol. 4*, Butterworth, London, 1957, 329.
32. Beets, M. G. J., *Amer. Perfum.*, 1961, **76**(6), 54–63.
33. McGill, J. R. and Kowalski, B. R., *Anal. Chem.*, 1977, **49**, 596.

34. Price, S., *Chemical Senses*, 1984, **8**(4), 341–354.
35. Wright, R. H., *Chemical Senses*, 1983, **8**(1), 103–106.
36. Wright, R. H., *Chemical Senses*, 1982, **6**(5), 211–213.
37. Moulton, D. G., in *Odor Quality and Chemical Structure*, Eds H. R. Moskowitz and C. B. Warren, ACS Symposium Series, No. 148, American Chemical Society, Washington, DC, 1981, 211.
38. Boeckh, J., in *Structure–Activity Relationships in Chemoreception*, Ed. G. Benz, Information Retrieval Ltd, London, 1976, 171–183.
39. McLeod, A. J., *Symp. Zool. Soc., Lond.*, 1980, No. 45, 15–34.
40. Clarke, R. J., in *Developments in Food Science 3B*, Eds I. D. Morton and A. J. Macleod, Elsevier, Amsterdam, 1987, 1–48.
41. Tressl, R. and Silwar, R., *J. Agric. Food Chem.*, 1981, **29**, 1078–1082.
42. Vernin, G., *The Chemistry of Heterocyclic Compounds in Flavours and Aromas*, Ellis Horwood, Chichester, 1982.
43. Heilgaard, M. C., *J. Agric. Food Chem.*, 1982, **30**, 1009–1017.
44. Parliament, T. H. and Scarpellino, R., *J. Agric. Food Chem.*, 1977, **25**, 97–99.
45. Lockhart, E. E., Tucker, C. L. and Merritt, M. C., *J. Food Sci.*, 1955, **20**, 598–605.
46. Pangborn, R. M., *Lebenms. Wiss. Technol.*, 1982, **15**, 161–168.
47. Little, A. C. and Brinner, L., *J. Food Sci.*, 1981, **46**, 519–522, 525.
48. Segall, S. and Proctor, B. E., *Food Technol.*, 1959, **13**, 266–271.
49. Jelineck, G., *Sensory Evaluation of Food: Theory and Practice*, Ellis Horwood, Chichester, 1985.
50. Gridgeman, N. T., in *Quality Control in the Food Industry, Vol. 1*, Ed. S. H. Herschdorfer, Academic Press, London, 1984, 299–350.
51. ASTM, Correlation of Subjective–Objective Methods in the Study of Odors and Taste. *Special Tech. Publication No. 440*, American Society of Testing and Materials, Philadelphia, PA, 1967.
52. Akesson, C., Persson, T. and von Sydow, E., *Proc. IVth Int. Congr. Food Sci. Technol.*, 1974, **II**, 183–189.
53. Wang, T. H., Shanfield, H. and Zlatkisa, A., *Chromatographia*, 1983, **17**(8), 411–417.

Chapter 7

Animal Feeding Studies with Coffee

H. P. WÜRZNER

Nestlé Research Centre, Nestec Ltd, Lausanne, Switzerland

1. INTRODUCTION

The introduction of coffee into Europe in the 17th century, as with other novel foods, immediately aroused public interest in the possible beneficial or negative effects of this new foodstuff. The first scientific report on the health effects of coffee consumption concerned its stimulatory and diuretic effects, and was presented by Antoine de Jussieu in 1713 before the French Academy of Sciences. It was really only in the last 50 years, however, concomitant with a tremendous increase in production as well as major technological developments relating to its roasting, extraction and decaffeination, that consumer groups became seriously concerned about the possible effects of long-term coffee consumption on health.

The health effects of coffee consumption are difficult to assess, mainly because of the complexity of composition of the beverage, which varies according to techniques of production and preparation, genetics, and climatic and soil conditions. Furthermore, processing of the green bean, roasting, use of blends or pure varieties and extraction methods, besides the individual's own choice of blend, cup size and strength, all result in the variation of the coffee beverage encountered from country to country and between individuals. Cultural patterns such as lifestyles and food and drinking habits are, therefore, also decisive parameters in any investigation of the physiological effects of coffee and/or its constituents.

Three main directions have been taken in the investigation of the physiology and safety of coffee:

(a) *Epidemiological studies* provide basic information and establish possible associations between long-term coffee consumption and health which in turn may result in new working hypotheses for clinical and experimental studies.
(b) *Clinical studies*, carried out normally on relatively small groups of subjects, focus on selected short-term effects and provide objective results under controlled blind or double-blind conditions. However, they do not cover long-term effects and the results are specific to a selected population.
(c) *Experimental studies*, using animal models, have become a fundamental part of many disciplines in the life sciences. In this approach cause–effect relations may be studied in a dose-dependent manner and under rigorously standardised conditions. In addition, the entire lifespan and long-term effects can be reviewed and investigated in a reasonable time.

2. CHRONIC TOXICITY AND CARCINOGENICITY

2.1. Introduction

Various published chronic studies in rodents have investigated the effects of either fresh-brewed or instant coffee, or of caffeine, the most pharmacologically active component in coffee. The objectives and consequently the design of these studies, however, have differed markedly from one another. Coffee and caffeine have either been administered via the drinking water, incorporated in a standard laboratory animal diet, or given orally by gavage. In some studies animals have even been exposed *in utero* and the resulting offspring studied for effects during early ontogenesis or on reproduction. Newer coffee processing techniques such as removal of caffeine and spray or freeze drying of extracts have also been screened in chronic rodent assays.

Certain limitations using animal models have, however, to be borne in mind. The major points include species differences in metabolism and target tissue sensitivity, potential repair mechanisms at the molecular and cellular level, immune responses, and hormonal and enzymatic profiles, as well as the route of administration, form and dose level of the test substance. Thus, effects observed in these studies should be dose-dependent, reproducible, confirmed in more than one species or strain of animals, and furthermore agree with human clinical and epidemiological evidence. Possible carcinogenic effects should be evaluated using the following criteria:

(a) higher dose-dependent tumour incidence in treated animals in comparison with controls;
(b) significance of dose-dependent earlier appearance of tumours in treated animals;
(c) unusual types of tumour observed in the treated groups; and
(d) tendency to metastasis and malignancy of the tumour types in treated animals.

2.2. Studies on Rats

There is unequivocal evidence that nutrition and feeding pattern influence longevity in experimental animals.[1-3] Although there are many studies on this topic, the actual mechanisms involved still await elucidation. In this context, when coffee drinking became established as a social habit and part of our lifestyle, several investigations aimed at assessing the effects of nutrition on lifespan also examined coffee as a dietary constituent. One of the early studies[4] reported on the effect of coffee, human diets and inheritance upon the lifespan of rats. A basal diet designed to simulate the human diet was fed with and without a modest supplement (3 ml per day) of fresh brewed coffee (1 tablespoon of coffee per cup) from 6 months of age. There was no evidence, however, that lifespan was affected by any of the diets, nor that coffee shortened the lifespan for any group.

Similar studies[5] reported on the age-related changes of feeding coffee, milk or sugar solutions to rats. Albino rats of the Yale strain, 25–64 days old and weighing 50 to 190 g, were divided into four experimental groups, each with 29–30 rats of each sex, and were maintained in air-conditioned rooms. The experimental diets were fed from weaning until natural death. Treated animals received tap water (control), 10% sucrose solution, or coffee (one tablespoon of coffee per cup) as sole beverage together with a stock diet. The coffee group was supplemented with sugar equal to the sucrose-administered group.

The growth curves differed little between the water and coffee groups with no difference for females and an insignificant trend for males. The animals in the coffee group ingested additional calories provided by the sugar supplement and yet did not attain the same body weights as those given the 10% sucrose solution. This trend towards a lower body weight with coffee at equicaloric dietary intake was again observed in later studies.

There were no striking differences in liquid consumption although female rats drank markedly more than males. As in previous studies,[4] lifespan was unaffected by the treatments and coffee consumption.

The authors[5] did not observe any effect of long-term ingestion of coffee on urinary calcium excretion, calcium per unit volume, or bone density compared with water-drinking control rats. This is of particular interest in the light of recent observations that caffeine and coffee may affect skeletal bone density and delay calcification of bones in the foetus.[6,7]

In addition, in a study in which the effects of nitrogen, phosphorus and caffeine on the calcium balance in women was investigated, it was suggested that caffeine affects the rate of loss of bone mass.[8] However, this observation was not supported by results in a longitudinal study on radial bone mass, in which caffeine, milk intake and activity were examined.[9] Very recently, two reports from one group[10,11] concluded that in healthy male and female students caffeine increased urinary calcium excretion, mainly as a result of enhanced concentration, but also by larger urinary volumes. The significance of this finding is questionable since no real calcium-balance study was performed, and it is well documented that several hormonal regulatory mechanisms may antagonise this effect by enhanced intestinal calcium absorption, renal reabsorption and bone incorporation. In any case, actual non-invasive medico-physical methods, such as mono- and dual-photon absorption, permit both in human and animals longitudinal follow-up of bone mineralisation during development and loss of bone mass in ageing processes in compact and trabecular bones.

As regards neoplasms, there was no striking difference in the total number of tumours between the treatments.[12] The most frequent neoplasm was lung lymphosarcoma, which accounted for more than 20% of the total tumour incidence. Exposure to coffee was found to reduce this particular tumour slightly in both sexes, but since coffee was administered at one dose no dose relation could be established. Chromophobe adenoma of the anterior pituitary gland accounted for 12% of the total tumours, but no relation to dietary treatment was found. The other rarely occurring tumours were of various types and in different sites and were unrelated to dietary treatment. In addition, the incidence of stomach ulcers bore no relation to dietary treatment.

Overall, these pioneering studies demonstrated certain effects of coffee such as lower body weight gain and reduced lymphosarcoma incidence, but the general pattern of disease onset and development and consequently lifespan was not significantly altered by dietary modifications.

In a two-year toxicity study,[13] diets containing soluble coffee, decaffeinated coffees, brewed coffee and caffeine were administered to rats.

The dietary levels of instant coffee ranged from 0·5 to 5·0%, and the equivalent caffeine levels from 0 to 0·177%. Weight gain was only initially impaired by the highest coffee and caffeine levels. These levels also induced liver and kidney hypertrophy. The severity of chronic nephritic lesions was linearly related to the dietary caffeine concentration. Mortality was increased at caffeine levels exceeding 0·083% of the diet.

Another investigator[14] reported no bladder tumours, hyperplasia or stones in rats given instant coffee at 5% in the diet or the human equivalent of 68–85 cups of coffee per day for up to 2 years from weaning. Obviously the study focused on bladder tumours, since no other data on tumour incidence were published.

A chronic toxicity study in rats[15] investigated the serum lipid profile after 6 to 7 months of coffee consumption. Conventional male Wistar rats were fed a standard laboratory chow and received tap water, regular or decaffeinated coffee (4 g per 100 ml water), or a caffeine solution (40 mg per 100 ml water). Each treatment group consisted of 15 rats, with an initial body weight of 120–125 g. In a second assay with the same treatments, the dose levels were doubled (8 g coffee per 100 ml water and 80 mg caffeine per 100 ml water). The coffee solutions were prepared using an extraction time of 10 min. These coffee concentrations led to the ingestion of 35–38 mg caffeine per kilogram body weight in the first study and 54–60 mg caffeine per kilogram body weight in the second, whereas those receiving decaffeinated coffee ingested only 1·6 to 2·6 mg caffeine per kilogram, respectively. Since rats ingest two to three times as much liquid as man as a proportion of body weight whereas the coffee strength corresponded to that consumed by man, these rats were exposed to much higher amounts. The studies lasted for 27 and 33 weeks, respectively, after which the animals were killed and various biochemical parameters were examined. In the first experiment the food intake of the four experimental groups was the same; however, body weights were lower than controls in both the coffee groups. The liquid consumption, organ weights (liver, heart, adrenal, testicles and spleen), haematological parameters, urine analysis, plasma cholesterol and free fatty acids showed no treatment effects. Plasma glucose levels were, however, significantly lower in both the coffee and caffeine groups as compared with controls. In the second experiment, in which dose levels of coffee and caffeine were doubled, surprisingly only the decaffeinated coffee group showed lower glucose together with a significant elevation of cholesterol levels. Total liver fat and triglycerides were comparable in all three experimental groups, and no histopathological changes were found in the liver, heart or kidneys.

Although this study again confirmed that coffee consumption resulted in lower body weights, this same effect was not observed in rats given the caffeine solution. Lipolytic effects of coffee or caffeine previously observed in acute studies were not seen in this chronic study. Lower plasma glucose values observed in both experiments were later reproduced in another chronic feeding study in rats.[16] No consistent elevation of plasma cholesterol values was noted, though later studies showed dose-dependent, elevated plasma cholesterol levels.[16,17] These differences may be related to laboratory methodology, study design, strain specificity and diet composition. Furthermore, the intestinal microflora interferes with cholesterol metabolism; thus, conventional and specific pathogen-free rats may give different responses to coffee treatment in the various studies.

Recently another group[17] studied the effect of fresh-brewed coffee in Sprague Dawley rats starting with *in utero* exposure. The offspring were provided with coffee dilutions 25, 50 or 100% as sole fluid, and were obtained from dams which had received 50% coffee for about 5 weeks prior to mating and throughout gestation and lactation. The males and control dams as well as control F_1 pups received tap water. The 100% coffee group averaged an intake of 93 ml per kilogram per day in males and about 166 ml per kilogram per day in females (respectively equivalent to 37 or 66 cups of undiluted coffee daily for a 60-kg individual). Each group consisted of 55 males and 55 females, and ten of each sex were killed after 1 year and the survivors after 2 years.

Food and liquid intake were significantly higher in the coffee group, except for the males at the highest concentration. Body weight was significantly lower in males receiving the high coffee levels, but not in females. This observation is consistent with other studies.[5,15,16] After 1 year, bone calcium content was slightly reduced in females receiving 25 or 100% coffee, whereas serum calcium level was slightly elevated in the coffee groups. This apparent bone calcium reduction returned to normal even though coffee intake continued, and is in agreement with previous observations[5] on calcium excretion and calcium content of the humerus. Elevated serum cholesterol levels were recorded in males at 2 years and at 1 and 2 years in females at the highest coffee concentration, and are again consistent with previous studies.[15,16] Mortality was not affected in males, but in females of the 50 or 100% coffee groups it was significantly increased. This appeared to be associated with tumour development, but only if it is assumed that tumours were lethal. In fact slightly greater numbers of females in the 50 and 100% coffee groups and males in the 25% group died with malignant tumours than did controls, but the actual mortality from malignant tumours was relatively small. The total incidence of primary tumours in any given organ was comparable among treated and control groups. However, chronic kidney diseases appeared to be a major cause of death in females.

Again this study shows body weight reduction, increase in liquid consumption in females, increase in plasma cholesterol levels and transitory effects of coffee on bone calcification. However, tumour incidence was not affected by the coffee concentrations administered, in contrast to some other studies in which certain tumour types were reduced at higher dose levels of coffee and caffeine.[12,18–21] This may be because mortality from other causes such as pneumonia and nephropathies complicated the picture. However, there were no unusual types of tumour or significant earlier appearance of tumours in treated animals. Since the increase of primary tumours was dose-dependent only in females, and other specific criteria for carcinogenicity were not found, no evidence for a carcinogenic effect of coffee can be derived in this study.

A two-year instant-coffee feeding study[16,18] focused on some of the main technological variables such as extraction rate, decaffeination, drying techniques and caffeine levels. The technology of instant-coffee manufacture has been under constant development and it was therefore of interest to determine whether a correlation existed between technological variables and biological effects.

Outbred male and female SPF Sprague Dawley rats received a standard laboratory chow containing soluble coffee at 6%. Although only one dose level was used, the caffeine dose–response relationship could be evaluated since a wide range of caffeine levels was provided by the various samples. The study comprised 14 experimental groups, each group consisting of 40 male and 40 female weanling rats. The basic coffee sample was a commercial blend containing arabica and robusta coffees. Decaffeination was carried out by extraction with methylene chloride. Extraction yields ranged from 23 to 50%, thus covering rates commonly used in home-brew and instant coffees. Rats were killed at 3, 12 and 24 months.

All coffee samples reduced rates of body weight gain in male rats and there was a significant negative correlation between body weight and caffeine exposure. In females this effect was considerably weaker and the correlation was not significant. Neither extraction rates nor drying processes had a significant effect on growth or food consumption. Food consumption, except at the beginning of the assay, was similar in all groups and cannot alone explain the differences in growth. It has been suggested that coffee reduces protein and energy utilisation,[22,23] and thus at least partially may be responsible for the growth inhibition. Since caffeine has been shown to increase locomotor activity and also increase basal metabolic rate and oxygen consumption,[24,25] this, in addition, may have contributed to the lower growth rates. There was a significant increase in total cholesterol in both sexes at 3 and 12 months which correlated with the caffeine content of the diet. This observation is in agreement with other chronic studies.[15,17,26] It is intriguing, however, that recent epidemiological studies in the human population have again yielded conflicting results on the effect of coffee drinking on plasma cholesterol levels as well as on low- and high-density lipoprotein profiles; see, however, chapter 3.[27–29]

Slightly higher blood urea nitrogen values observed at 12 months were unrelated to coffee dose levels but related to progressive age-related nephropathy. As in a previous study,[15] lowered plasma glucose values were recorded in both decaffeinated and regular instant coffee groups, but the biological significance of this observation with respect to human coffee consumption remains doubtful. Coronary arteriosclerosis evaluated in 10 male and 10 female rats per group at 3, 12 and 24 months was progressive from 3 months onwards. Prevalence and degree were comparable in all groups including controls, and coffee ingestion neither influenced nor accelerated these lesions.[30]

The most important data concerning the health aspects of chronic coffee ingestion were, of course, data on neoplasm incidence and appearance.[18] The most common types of neoplasm in this study were lymphosarcomas, mammary fibroadenomas, adrenal adenomas and embryonal nephromas. These accounted for 85% of the total tumours observed in the study. The frequency of spontaneous pituitary adenomas in this rat strain, as in other studies,[17,20,21] was not treatment-related and was in the range reported in the literature.[31]

In male rats coffee ingestion did not increase the incidence of benign or malignant neoplasms, but even significantly decreased them in diets containing high levels of caffeine. In female rats only one sample significantly increased tumour incidence. Without any doubt this was due to the high numbers of embryonal nephroma, which are not sex- or treatment-related, as confirmed by a high incidence in male control rats, although female controls were not affected. Thus, this very special type of tumour, characterised by very early appearance, was not correlated to coffee intake.

In general, all other neoplasms appeared simultaneously in controls and treatment groups, and there was no trend of earlier onset in any treatment. The absence of any effect with methylene chloride decaffeinated samples is noteworthy. Methylene chloride residual levels in decaffeinated coffee are very low (less than 1 ppm), and it can be argued that the potential risk was therefore not detectable. However, the objective of the study was not to test the solvent itself but rather the decaffeinated coffee so produced. Multiple regression testing suggested that the caffeine level in coffee was the only variable that influenced neoplasm incidence. It is well documented that caffeine influences repair mechanisms *in vivo*, and *in vitro* modifies radiation damage and alkylation of DNA. On the other hand, caloric restriction and the resultant decreased body weights always correlated with low neoplasm incidence in many studies.[1-3] It is therefore conceivable that the lower body weights recorded in the high caffeine groups are directly responsible for this effect, as confirmed in other long-term studies with coffee or caffeine in rats and mice.[19-21]

The effects of oral intubation (1 ml per day) of tea and coffee (25 mg per millilitre of water) were studied in rats.[32] After one year, when rats were killed, serum cholesterol (consistent with other published reports[15-17,26]) was significantly elevated compared with saline controls, but only in the coffee groups. Patchy necrosis was seen in the hearts and also in the livers of the coffee group, indicative of degenerative changes, whereas the tea group showed no changes in the heart and only slight changes in the liver.

It is doubtful, however, whether these histopathological observations were treatment-related, since the design of the study excluded a dose–response evaluation and the number of rats per treatment or sex was rather small. Furthermore, the study was too short to provide any data on tumour incidence.

2.3. Studies on Mice

A series of experiments[33] in which fresh-brewed coffee (5·2 g per 100 ml water) was administered to twenty-five C57BL6J mice from 7 months of age was reported. The coffee groups were seen to have lower weights and died earlier although they ate as much as controls. However, post-mortem examination which included gross and microscopic study of a variety of organs did not demonstrate any lesions or disease pattern that could completely explain the earlier death in the coffee groups. Indeed, most of the animals died from pneumonia both in the control and experimental group. However, no unusual occurrence of tumours was found in the test animals, and electron microscopy of the same portion of the aorta did not demonstrate any coffee-related pattern or vascular lesion in the test animals.

In yet another experiment of this series,[33] twenty-seven 5-month-old mice were given coffee at the above-mentioned concentration. The experimental animals again demonstrated lowered longevity and a gradual pattern of mortality. There was a consistently higher fluid intake in the coffee group, and they ate as much as or more food than the controls. The authors concluded that prolonged intake of brewed coffee was detrimental to the longevity of C57BL6J mice and of AJax albino mice, in which similar observations were made. However, the high incidence of pneumonia, which was lethal in both controls and treated animals, raises serious doubts about the validity of such a conclusion. Furthermore, the mortality findings were not reproduced in other studies in mice.[19] No unusual tumours or earlier appearance of tumours were, however, reported in these studies, and therefore, as reliable criteria, they suggest the overall absence of carcinogenic effects of coffee in this species.

The possible neoplastic effects of instant coffee were investigated in a standardised carcinogen bioassay.[19] Twelve-week-old SPF Swiss mice were mated and supplied with a 23–26% protein diet. After 5 days they received either a 0 or 1% coffee diet until the end of lactation. At weaning, 50 male and 50 female pups were allocated to test groups and received either 0, 1, 2·5 or 5% coffee in the diet until 720 days of age, at which time all surviving animals were killed and examined.

Body weight gain was inversely proportional to increasing coffee levels. Survival, on the other hand, showed dose-dependent increases with increasing coffee intake.

The major neoplastic entities were tumours of the liver, lung and lymphatic tissue. Hepatic tumours were mostly benign adenomas. Pulmonary tumours were either benign bronchio-alveolar adenomas or malignant carcinomas. Lymphosarcomas were observed in lymphatic tissues and other tumours in various organs singly or at low frequency. No treatment-related changes were observed either in the exocrine or endocrine pancreas and no exocrine pancreatic neoplasms were found in the whole study. Statistical analysis using a log-linear model[34] revealed that instant coffee ingestion depressed significantly ($p > 0{\cdot}001$) the total tumour incidence. Liver tumour incidence was reduced significantly ($p > 0{\cdot}001$) in males with increasing coffee doses but not in females, whereas lung carcinomas, as well as lung adenomas, were neither dose- nor sex-dependent. But lymphosarcomas of all histologic types also showed a highly significant ($p > 0{\cdot}001$) decrease in both male and female rate with increasing coffee doses. Although the proportion of benign to malignant tumours was different between the sexes, a significant dose effect was seen for tumour malignancy. Comparison of tumours of epithelial or mesenchymal origin again showed a certain dose effect, particularly at the highest dietary dose level of 5% coffee. However, coffee had no effect on general tumour latency time, and no unusual neoplasms were reported. The authors concluded that the decrease in body weight with increasing coffee in the diet was not paralleled by changes in food intake except at the highest dose level. This could possibly be explained by the thermogenic effect of caffeine.[24] In a previous trial the growth curves of rats receiving decaffeinated coffee were similar to those of controls, lending support to this theory. The dose-dependent reduction of certain neoplasms such as liver adenoma and lymphosarcoma is therefore probably not a direct effect of coffee or caffeine, but may be related to growth and body weight reduction; this is analogous to observations in food restriction studies.[1–3] Hence, increased thermogenesis and locomotor activity and overall increased energy expenditure may account for this effect.

2.4. Caffeine, Coffee and Tumour Promotion and Inhibition

Complementary to the findings of a carcinogen bioassay is the investigation of the possible promotional or inhibitory effects of dietary components on neoplasia development. Several epidemiological studies have in fact shown inverse relationships between fruit and vegetable consumption and cancer.[34–39]

The cytochrome-P450-dependent mono-oxygenase enzyme systems in the tissues and the glutathione S-transferase activity (GSH S-transferase) in the liver are of special importance in the detoxification and/or the activation of xenobiotics, and apparently respond to dietary constituents and other environmental influences.

The role of some minor dietary constituents including green coffee beans fed at 10% in the diet to rats either prior to or after DMBA administration was recently studied.[40] Mammary tumour formation was significantly inhibited by feeding of supplements of green coffee beans. GSH S-transferase activity was enhanced sixfold in the liver and sevenfold in small bowel mucosa.[41] Comparable inducing activity was found with roasted coffee beans, and also to a lesser extent with instant coffee and instant decaffeinated coffee. Two coffee diterpenes, kahweol and cafestol, together account for half of the inhibitory activity occurring in green coffee beans. While these experiments do not reflect the situation in normal coffee consumption, they reveal the presence of blocking agents, which evidently act by enhancing the activity of the detoxification system.

In the same DMBA-induced mammary gland carcinogenesis model, caffeine solutions (250 and 500 mg per litre) significantly increased mammary carcinoma incidence.[42] The increase was not significant if caffeine was administered prior to or during DMBA tumour induction. It therefore appears that under the given experimental conditions caffeine acts as a promoter and enhances tumour development only after the initiation phase. Similar tumour-promoting effects of caffeine and unsaturated fat were reported in this model.[43] The effects of caffeine and unsaturated fat in this case were synergistic.

In the azaserine-induced pancreas tumour model, coffee induced some hypertrophy of the pancreas but did not increase the mortality of rats, the incidence of exocrine pancreas tumours or tumour development.[44]

Caffeine has, however, been suggested to act as a co-carcinogen. Thus, it significantly increased the incidence of malignant skin tumours in mice treated with β-rays and 4-nitroquinoline-1-oxide,[45] but had no effect on the incidence of bladder tumours in rats treated with *N*-butyl-*N*-(4-hydroxybutyl)nitrosamine.[46]

This type of promotional study has, however, a serious drawback, namely the arbitrary selection of a specific artificial situation which cannot and should not be generalised and extrapolated to man. Carcinogenic initiation in these models is mostly obtained by a single massive dose of a carcinogen, whereas man is generally exposed to repeated, extremely low doses of carcinogens in the environment. Hence, such experiments are only helpful when investigating single components and their interaction under specific conditions, but not for establishing real cause–effect relationships.

2.5. Caffeine Studies

No discussion of possible physiological effects of coffee would be complete without reference to the effects of caffeine, its most pharmacologically active component.

Early workers[47] concluded that there was no evidence that caffeine is involved in carcinogenesis, as did others,[48] who had carried out an 18-month oral feeding study of caffeine (55 mg per kilogram per day) in C57BL6J mice.

In a subacute study,[20] groups of 10 male and 10 female 5-week-old Wistar rats were administered 0–0·8% levels of caffeine in their drinking water for 6 weeks. Mortality with 0·8% was high, but there were no deaths at levels below 0·4% caffeine. Accordingly, levels of 0·1 and 0·2% caffeine were chosen for a follow-up carcinogenicity study[20] using 50 male and 50 female Wistar rats per group. Treatment was started at 8 weeks of age and resulted in daily intakes of 28–53 mg caffeine, equivalent to 40–71 cups of coffee per day in a 60-kg man. There were no significant differences in general condition nor mortality. Although neoplastic lesions were observed in the pituitary, mammary, thyroid and uterus, their incidence was similar to that in control animals. The authors therefore concluded that caffeine was not carcinogenic in their experiment.

In another carcinogenicity study[21] caffeine was administered in drinking water to SPF Sprague Dawley rats at levels of 200, 430, 930 and 2000 mg per litre. Each group was made up of 50 males and 50 females. Two control groups received tap water. Females drank consistently more than males, resulting in higher caffeine intakes (12 to 102 mg per kilogram per day for males and 15 to 170 mg per kilogram per day for females). Survival rates were not affected, but body weight gain was considerably depressed by high caffeine intakes. Neoplasm incidence was reduced in a dose-dependent manner. This was mainly due to changes in incidence of benign tumours rather than of malignant neoplasms. This effect is consistent with observations in several studies.[12,18–20] No unusual tumours or sites were observed in the caffeine-exposed animals. Consequently the authors concluded that long-term ingestion of caffeine did not significantly increase tumour incidence even at doses exceeding the maximum tolerated dose.

2.6. Conclusions

The evaluation of the various chronic studies in rats and mice, and the application of earlier mentioned criteria, provide convincing evidence that both brewed and instant coffee, as well as caffeine alone, have no carcinogenic potential in the species examined and at the dose levels used. Growth depression was frequently observed and appears to be related to inhibitory effects on certain tumour types. However, interaction and promotion studies using established carcinogens as initiators and coffee or caffeine as promoters have yielded conflicting results up to now, and need deeper and more extensive investigation to be useful and relevant.

Since elevated cholesterol levels were found in several experimental studies and also in human investigations, this effect merits further in-detail examination using specific approaches, particularly prospective clinical studies. However, the experimental and human studies do not support the hypothesis that coffee is a major causal factor in the development of ischaemic heart disease and arteriosclerosis. In none of the studies were striking treatment- and dose-related lesions in other organs reported (Table 1).

Table 1
Overview feeding studies with coffee and caffeine chronic toxicity and carcinogenicity

Authors	*Species*	*Number*	*Dose and administration*	*Time of treatment*	*Observations*
Sperling *et al.*[4]	Yale albino rat	20 per group per sex	3 ml coffee per day per rat orally as supplement	From 6 months to end of life	No effect on longevity
McCay *et al.*[5]	Rat	30 per group per sex	Coffee brew *ad lib.* (one tablespoon per cup)	From weaning to end of life	No effect on longevity
Saxton *et al.*[12]	Rat	30 per group per sex	Coffee brew *ad lib.* (one tablespoon per cup)	From weaning to end of life	Slight reduction of lymphosarcoma in coffee-drinking rats
Daubert[13]	Rat	Not reported	0·5–5·0% in diet instant coffee, brew, decaff. coffees as sole fluid	2 years	Growth impairment
Zeitlin[14]	Sprague Dawley rat	Control 40 per sex	5% instant coffee in diet	2 years from weaning	No bladder tumours observed
Strubelt *et al.*[15]	Wistar rat	15 males per group	4 and 8 g per 100 ml H_2O brew of regular and decaff. coffee sole fluid	27 and 33 weeks	Lowered body weight and effect on plasma chemistry
Palm *et al.*[17]	Sprague Dawley rat	55 per group per sex; two control groups	Brewed coffee dilutions 25, 50, 100% sole fluid *in utero* exposure	2 years	No evidence of carcinogenic effect
Würzner *et al.*[16,18]	SPF Sprague Dawley rat	40 per group per sex	Regular and decaff. instant coffee 6% in diet	2 years	No evidence of carcinogenic effect
Patel and Venkatakrishna-Bhatt[32]	Rat	25 per sex per group	Intubation 1 ml per day of coffee 25 mg per ml	1 year	Elevated plasma cholesterol
Bauer *et al.*[33]	Mice C57BL6j	25 per sex per group	Brewed coffee *ad lib.*	103 + 105 and 80 weeks	No evidence of carcinogenic effect
Stalder *et al.*[19]	SPF Swiss mice	50 per sex per group	Regular instant coffee at 1, 2·5 and 5% in diet *in utero* exposure	2 years	No evidence of carcinogenic effect
Takayama and Kuwabara[20]	SPF Wistar rats	50 per sex per group	Caffeine in drinking water, 0, 0·1, 0·2%	2 years	No evidence of carcinogenic effect
Mohr *et al.*[21]	SPF Wistar rats	50 per sex per group	Caffeine in drinking water, 200, 430, 930, 2000 mg per litre	2 years	No evidence of carcinogenic effect

3. TERATOLOGY AND REPRODUCTION

3.1. Introduction

Concern has repeatedly been expressed by the public as to the possible health hazards of coffee and caffeine ingestion during pregnancy. Several epidemiological studies examined the question as to whether heavy coffee drinking influenced birth weight and pregnancy outcome.[49–54] Since coffee drinking is often associated with smoking and alcohol consumption, these two recognised embryotoxic factors may confuse the results of the studies.[51,55] In addition, it is well documented that caffeine metabolism slows down in the third trimester of pregnancy and hence may lead to accumulation of circulating caffeine under certain circumstances.[56] A careful evaluation of currently available human studies, however, provides no support for the conclusion that normal caffeine consumption as a food and beverage constituent causes birth defects in humans.

Although many experimental studies have investigated the teratogenic potential of coffee and caffeine, their evaluation, assessment of risk and extrapolation to humans is complicated for a variety of reasons. The length of gestation and time course of development differ from species to species. Placental morphology and consequently placental and fetal kinetics, and metabolic profiles as well as general metabolic capacity, show considerable variation. Therefore the design, route of administration, dose levels and possible maternal toxicity must be considered before meaningful evaluations and conclusions can be made. In this context, studies in which coffee or caffeine was administered by injection or gavage (total daily dose at one time) are not considered, since pharmacokinetics, circulating caffeine levels and consequently resulting teratogenic effects differ considerably from those encountered in feeding studies. These latter studies mimic much more closely human consumption patterns, where coffee is normally consumed at intervals of several hours.

3.2. Studies on Rats

In an early longevity study,[4] the influence of moderate amounts of coffee upon growth, reproduction and lactation was also investigated. Three groups of adult female rats (about 4 months old) were fed stock diet and provided with, as sole liquid, either strong coffee (one tablespoon of coffee per cup), weak coffee (half strength) or water. Their average daily consumption was 29, 32 and 26 ml, respectively. After 5 weeks the females were mated, and mating was continued at intervals of 3 months. There was no evidence that coffee affected the lifespan of these animals, and no untoward effect of coffee consumption on reproduction, lactation and growth was detected in three generations of rats.

In an extensive generation study,[7] weanling male and female rats were given either brewed or instant coffee solutions (100, 50, 25 or 0%) only. This amounted to caffeine contents of 0·057% and of 0·045% at the 100% concentration for the brewed and instant coffee, respectively. Average daily caffeine intakes of males consuming brewed coffee plateaued at about 50, 30 and 17 mg per kilogram and those drinking instant coffee at 40, 28 and 13 mg per kilogram. Corresponding values for females were 80, 45 and 30 mg per kilogram per day from brewed coffee and 70, 40 and 25 mg per kilogram per day from instant coffee. The rats consuming the full-strength coffee received the equivalent of 50–60 cups of coffee per day for a 60-kg man.

Weights of kidneys and livers in both sexes were increased at the high coffee levels in both instant and brewed coffee drinking groups. Testicles in males and ovaries in females had lower weights. None of the coffee solutions affected the percentage of conceptions nor adversely affected fetal or neonatal viability. Mean pup weights at weaning in coffee groups were lower than controls, but only significantly so for 100% solutions. This is in agreement with other reports[57] in which caffeine below 60 mg per kilogram per day had no effect on birth weight.

There were no differences in conception rates at the second matings nor in numbers of corpora lutea or implantations in dams killed at day 13 or 21. A difference in resorptions was seen only with 25% instant coffee at day 13, but as this was not dose-dependent and was not seen at day 21 it can be concluded that it was not coffee-related.

No differences were seen in the number of fetuses with soft-tissue defects nor any frank abnormalities in skeletons, although the incidence of pups with developmental variations in the sternebrae (missing or hypoplastic 2nd and 5th) was significantly increased in the two groups fed full-strength and 50% brewed coffee. Thus there seemed to be a retardation in the ossification and calcification of bone in the fetuses exposed to high coffee levels, although no cleft palates were observed. It can therefore be stated that relatively high coffee (full strength) consumption in this study showed no harmful effects on implantation, growth or viability of embryo or fetus, and no soft-tissue or specific skeletal abnormalities. In a similar study,[58] either brewed or instant decaffeinated coffees (25, 50 or 100%) were administered to weanling rats for about 6 months. Rats were bred twice after 90 days to assess reproduction and possible teratogenicity. The levels administered are equivalent to human consumption of 12, 25 and 50 cups of coffee per day. Again in this study none of the coffee treatments had a significant effect on reproduction parameters, and no significant fetal abnormalities were observed in either soft-tissue or skeletal examinations carried out at days 13 and 21 of pregnancy.

In a generation study,[6] the teratogenic potential of both fresh-brewed coffee and caffeine in the rat was evaluated. Dilutions of fresh-brewed coffee at 12·5, 25 and 50% (providing approximately 9, 19 and 38 mg caffeine per kilogram per day) were offered to female rats (25 per group) as beverage for 5 weeks prior to mating, throughout gestation and until 27 days post-partum. Aspirin (125 mg per kilogram) served as positive control and was administered by intubation as well as caffeine at 30 mg per kilogram body weight. On day 19 of pregnancy the dams were killed and the fetuses were thoroughly examined. Calcium levels in bones were measured in adult rats. At least five dams from each treatment group and ten from the water control were left to rear their pups. After weaning, the F_1 pups were provided with water only, and at 100 days of age they were randomly mated and their reproductive capacity was followed.

Coffee-treated females showed no difference in behaviour, other gross signs, weight gain or food consumption. After killing on day 19, the mean values for days to pregnancy, number of resorptions, number of fetuses per litter, number of dead fetuses per litter, male:female ratio and fetal weight were comparable between groups. The fetuses collected were comparable in size and organ weight although at the 50% coffee level mean weights of fixed brains, lungs and liver appeared slightly but significantly lighter as a percentage of body weight. This was also the case for the aspirin-treated group. In the 25 and 50% coffee, aspirin and caffeine in water groups, there was a significantly increased incidence of kidney–pelvis underdevelopment in fetuses. Cleft palate was more common in coffee and aspirin groups than in water controls, but in the coffee groups the incidence was inversely related to coffee and caffeine levels. Consequently this observation cannot be attributed to the coffee consumption. In addition, this effect was not observed in other studies[7,58] in which full-strength 100% coffee was administered. A significant effect of shortened fibulae was seen in 2·8% of the coffee group but this was not dose-related. However, a statistically significant and dose-related reduction in bone density was noted with coffee. Apart from a slightly lower body weight in the caffeine-intubated group the numbers of surviving pups and their weight were similar in all groups. No difference was noted in any treatment group in the mean number of pups per litter at birth, the mean body weight of the pups, the number of pups dead at birth, the percentage surviving or mean body weight at day 28.

The authors[6] accordingly concluded that the administration of up to 50% coffee (27 cups per day for a 60-kg person) during gestation and lactation appeared to have no effect on the reproductive performance of the F_1 offspring. Furthermore, they concluded that although there was some apparent delay in sternebral bone ossification, no dose-related frank teratological effects could be linked with coffee consumption. Therefore the data on resorption, sex ratio, fetal weight and incidence of cleft palate, which were all unrelated to coffee consumption, are fully in agreement with similar studies,[7,58] despite the difference in coffee strength.

In a multigeneration study,[59] possible reproductive effects of instant coffee and the specific effects of some technological variables such as drying technique, caffeine content and decaffeination were investigated. All instant coffee samples were fed at 3% in the diet. Caffeine, the positive control, was fed in amounts equivalent to the highest level of caffeine in the instant coffee samples. This exposure corresponds approximately to an intake ranging from 40–60 cups per 60-kg adult. Circulating plasma caffeine levels varied enormously according to sampling time but reached extreme values of 40 μg per millilitre. In humans, levels of 5–10 μg per millilitre have been reported in pregnancy.[56] Whereas fertility and gestation index were not affected by the treatment, for the viability index a negative correlation with the caffeine content in the instant coffee samples appeared by applying multiple regression analysis. Likewise the lactation index was affected in the first two generations, but disappeared in the third. There was no treatment effect on the number of pups born alive whereas survival till weaning was negatively correlated with caffeine content of the administered coffee samples. Only plasma triglyceride levels were positively correlated with the caffeine content in both litters and sexes of each of the three generations. Therefore one can conclude that the technological parameters studied did not significantly influence the biological data and outcome of this study. Caffeine content was the only factor which, at higher dose levels, affected birth weight, plasma triglycerides, organ weights of liver, kidney and testicles, and viability after birth. Birth-weight reductions at high coffee dose levels were also observed in other feeding studies.[6,7,58]

3.3. Studies on Mice

In a recent trial,[60] the effect of a range of coffee doses (4·8–24 mg coffee per day) on pregnant mice, equivalent to 4–20 cups per day per human, was examined. There was no significant difference for litter size or gestation length. Average weights and lengths of neonate groups decreased in a dose-related manner. No gross malformations were observed nor alimentary abnormalities. The authors postulated that, since caffeine causes systemic vasoconstriction and increases blood pressure as well as catecholamine production, uteroplacental circulation becomes impaired.

In another study[61] mice received daily in their drinking water caffeine up to 39 mg per kilogram. No dose-related teratogenic effects were observed, although at 12–18 mg per kilogram some offspring of treated animals were under weight at weaning and one litter was completely ectrodactylous.

3.4. Caffeine Studies

In fact, very many studies have been undertaken to investigate the teratological potential of caffeine in mice, rat, hamster, rabbit and chick embryo, and were recently reviewed.[62] The results obtained are highly variable and again show that species, route of administration, dose and time of the treatment are important factors which decisively influence the outcome of the study.

Two extensive studies were recently carried out[63,64] in which the possible teratogenic potential of caffeine was investigated by administering it to Osborne Mendel rats either by oral intubation or in the drinking water. They show clearly the difference in effects related to the route of administration.

Since single doses administered by intubation result in high plasma peak levels and therefore affect placental circulation and fetal caffeine exposure, only the data from the drinking study are discussed here, which are also comparable to those from the coffee-drinking studies in rats.

Caffeine was dissolved in the drinking water at levels of 0, 0·007, 0·018, 0·036, 0·07, 0·10, 0·15 or 0·20% from days 0–20 of gestation. The corresponding intakes were 0–204·5 mg per kilogram body weight. Extreme doses of 160·9 and 204·5 mg per kilogram were associated with decreased implantation efficiency and increased resorptions at doses of 115·8–204·5 mg per kilogram per day. Fetal body weight and length were decreased and numbers of oedematous fetuses increased at a dose level of 86·6 mg per kilogram per day and above. No dose-related gross anomalies were observed. Sternebral ossification deficiencies were observed at all dose levels except 10·1 mg per kilogram per day, and skeletal ossification deficiencies were increased in a dose-dependent fashion for 86·6–204·5 mg per kilogram body weight. Similar observations on delayed ossification in mice were made at 150 mg per kilogram body weight.[65] This finding, also reported in previous studies, appears to be transient and is not considered to be a teratogenic response.[66,67]

Table 2

Overview feeding studies into coffee and caffeine. Effects on reproduction and teratogenesis

Authors	*Species*	*Number*	*Dose and administration*	*Type of study, killing date*	*Observations*
Sperling *et al.*[4]	Yale albino rat	20 per group per sex	50, 100% coffee brew as sole fluid intake	Reproduction study, three generations	No effect on reproduction and lactation
Nolen[7]	Sprague Dawley rat	20 M, 30 F per group	25, 50, 100% coffee brew; 25, 50, 100% instant coffee	Reproduction and teratogenesis; killing on days 13 and 21 of pregnancy	No significant soft-tissue or skeletal abnormalities; unossified sternebrae at 50, 100% coffee and 100% instant coffee
Nolen[58]	Sprague Dawley rat	20 M, 30 F per group	25, 100% coffee brew; 25, 50, 100% decaff. coffee brew; 25, 50, 100% decaff. instant coffee	Reproduction and teratogenesis; killing on days 13 and 21 of pregnancy	No significant soft-tissue or skeletal abnormalities; unossified sternebrae at 100% coffee brew
Palm *et al.*[6]	Sprague Dawley rat	25 F per group	12·5, 25, 50% coffee brew; 30 mg/kg b.w. caffeine; 125 mg/kg b.w. aspirin	Reproduction and teratogenesis; killing on day 19 of pregnancy	No dose-related teratogenic effects in coffee groups; delayed ossification in all treated groups; cleft palate incidence inversely related to coffee dose
Würzner *et al.*[30]	Sprague Dawley rat	10 M, 10 F per group per generation	Instant coffee 3% in diet; decaff. instant coffee 3% in diet	Reproduction study, three generations; killing at 3 months	Negative correlation to caffeine dose for viability and lactation index; high triglyceride plasma values related to caffeine dose
Murphy and Benjamin[60]	Mice	6 F per group	4·8, 9·6, 14·4, 19·2, 24·0 mg instant coffee in diet	Reproduction and teratogenesis; examination at birth, day 20–21	No gross abnormalities; resorptions at high dose levels
Thayer and Kensler[61]	Mice	20 M, 20 F per group	4–5, 12–18 and 25–39 mg caffeine/kg/day	Reproduction and teratogenesis; examined at birth and day weaning	No dose-related teratogenic effect; one litter ectrodactylous
Collins[63]	Osborne Mendel rat	61 F per group	10·1, 27·4, 50·7, 86·6, 115·8, 160·9, 204·5 mg/caffeine kg b.w. in drinking water	Teratology study; killing day 20 of pregnancy	No dose-related gross anomalies, dose-related skeletal ossification deficiencies at the four higher dose levels; resorptions and reduced viability at 160 and 204 mg caffeine/kg b.w.

In conclusion, the effects of caffeine observed in the *ad libitum* sipping trial[63] were less severe than by intubation,[64] but are in agreement with other studies in rats,[68] in which doses of 40–50 mg per kilogram in drinking water produced no teratogenic effects, if similar dose levels are compared.

3.5. Conclusions

There is no doubt that coffee and caffeine administered to experimental animals in single massive doses already toxic to the dams can cause specific defects in the fetus and offspring (generally more than 75 mg per kilogram body weight). This way of exposure is, however, quite different from normal human coffee consumption. Several carefully conducted experimental studies indicate that there is indeed a threshold for dose-related irreversible teratogenic effects. In addition, this threshold appears to be significantly higher in feeding studies in comparison with single dose administration. None of the coffee and caffeine feeding studies reported provides evidence for any coffee-related irreversible skeletal or visceral gross anomalies. Reversible dose-related effects such as reduced birth and organ weights and delayed skeletal ossification were apparent only at very high levels of coffee and caffeine (generally above 50 mg caffeine per kilogram body weight). The exact mechanism by which caffeine and coffee may lead in experimental models to these effects during gestation is presently unclear and needs further careful investigation. However, the reported studies strongly suggest that caffeine or its metabolites alone account for them. The determination of circulating caffeine plasma levels in the experimental models in comparison with reported human values may offer more precise possibilities for risk assessment than extrapolating mere dose levels.

The combined existing results from epidemiological and mutagenicity studies, as well as experimental feeding studies, provide, however, strong evidence for the absence of coffee-related birth defects in humans at current consumption levels (Table 2).

4. FINAL REMARKS AND OUTLOOK

The coffee/caffeine feeding studies in experimental animals necessarily have to be viewed in conjunction with the preceding chapters on metabolic, clinical and physiological studies, as well as epidemiological studies, for a comprehensive overview of the effects of coffee consumption and its possible risk to human health. This will lead also to the identification of still existing research gaps and controversial issues and indicate the possible outlook of future research in this area.

Concerning experimental studies it would appear that investigations on tumour promotion and inhibition in relation to coffee consumption and interactions with other food constituents represent a topic where valuable insight into the molecular mechanisms could be gained. The mechanism by which coffee may affect adenosine receptors represents a research topic of great current interest. Moreover, studies on the mechanism by which coffee consumption influences bone mineralisation at various ages needs to be pursued using modern non-invasive methods and computer-assisted image analysis. There is also little work reported on specific functional aspects of certain organ systems such as the central nervous and cardiovascular systems, investigated in adult offspring exposed *in utero* to coffee and caffeine.

These research endeavours require a concerted effort by pertinent scientific disciplines in order to discover new avenues, to resolve certain uncertainties and to confirm definitely the absence of any risk of coffee drinking to the health of humans.

ACKNOWLEDGEMENT

The author would like to thank Ms R. Acheson for editorial help.

REFERENCES

1. Ross, M. H. and Bras, G., *J. Natl Cancer Inst.*, 1971, **47**, 1095.
2. Roe, F. J. C. and Tucker, M. J., *Proc. Eur. Soc. Study of Drug Toxicity*, 1974, **15**, 171.
3. Tannenbaum, A., *Ann. NY Acad. Sci.*, 1947, **49**, 5–17.
4. Sperling, A., Loosli, J. K., Barnes, L. L. and McCay, C. M., *J. Gerontol.*, 1946, **1**, 426–432.
5. McCay, C. M., Lovelace, F., Sperling, G., Barnes, L. L., Lin, C. H., Smith, C. A. H. and Saxton, J. A., *J. Gerontol.*, 1952, **7**, 161–172.
6. Palm, P. E., Arnold, E. P., Rachwall, P. C., Leyczek, J. C., Teague, K. W. and Kensler, C. J., *Toxicol. Appl. Pharmacol.*, 1978, **44**, 1–16.
7. Nolen, G. A., *Toxicol. Pharmacol.*, 1981, **58**, 171–183.
8. Heaney, R. P. and Recker, R. R., *J. Lab. Clin. Med.*, 1982, **99**(1), 46–55.
9. Eliel, L. P., Smith, L. C., Ivey, J. L. and Baylink, D. J., *Calcified Tissue Int.*, 1983, **35**(4–5), 669.
10. Massey, L. K. and Wise, K. J., *Nutr. Res.*, 1984, **4**, 43–50.

11. Massey, L. K. and Berg, T. A., *Nutr. Res.*, 1985, **5**, 1281–1284.
12. Saxton, J. A., Sperling, G. A., Barnes, L. L. and McCay, C. M., *J. Gerontol.*, 1953, **8**, 255–263.
13. Daubert, B. F., *Proc. 3rd Int. Coll. ASIC*, 1967, 387.
14. Zeitlin, B. R., *Lancet*, 1972, **1**, 1066.
15. Strubelt, O., Siegers, C. P., Breining, H. and Steffen, J., *Z. Ernährungs. Wiss.*, 1973, **12**, 252–260.
16. Würzner, H. P., Lindström, E., Vuataz, L. and Luginbhühl, H., *Food Cosmet. Toxicol.*, 1976, **15**, 289–296.
17. Palm, P. E., Arnold, E. P., Nick, M. S., Valentine, J. R. and Doerfler, T. E., *Toxicol. Appl. Pharmacol.*, 1984, **74**, 364–382.
18. Würzner, H. P., Lindström, E., Vuataz, L. and Luginbhühl, H., *Food Cosmet. Toxicol.*, 1977, **15**, 7–16.
19. Stalder, R., Luginbhühl, H., Bexter, A. and Würzner, H. P., *Proc. 10th Coll. ASIC*, 1982, 339–345.
20. Takayama, S. and Kuwabara, N., *Gann*, 1982, **73**, 365–371.
21. Mohr, U., Althoff, J., Ketkar, M. B., Conradt, P. and Morgareidge, K., *Food Chem. Toxicol.*, 1984, **22**(5), 377–382.
22. Eggum, B. O., Pedersen, B. and Jacobsen, I., *Br. J. Nutr.*, 1983, **50**, 197–205.
23. Silbald, I. R., *Nutr. Rep. Int.*, 1982, **25**, 133–137.
24. Acheson, K. J., Zahorska-Markiewicy, B., Pittet, Ph., Anantharaman, K. and Jéquier, E., *Am. J. Clin. Nutr.*, 1980, **33**, 989.
25. Miller, D. S., Stock, M. J. and Stuart, J. A., *Proc. Nutr. Soc.*, 1974, **33**, 28A.
26. Naismith, D. J., Akinyanju, D. A. and Yudkin, J., *J. Nutr.*, 1969, **97**, 375–381.
27. Thelle, D. S., Arnesen, E. and Forde, O. H., *N. Engl. J. Med.*, 1983, **308**, 1454–1457.
28. Heyden, S., Bartel, A., Lassel, J. C., Hames, C. G., Tyroler, H. A. and Meier, R., *Z. Ernährungs. Wiss.*, 1972, Suppl. **14**, 1.
29. Friedman, G. D., in *Coffee and Health*, Eds B. MacMahon and T. Sugimura, Banbury Report 17, Cold Spring Harbor, 1984, 207–217.
30. Würzner, H. P., Lindström, E., Vuataz, L. and Luginbhühl, H., *Proc. 8th Coll. ASIC*, 1977, 319–323.
31. Melby, A. and Altman, R. (Eds), *Handbook of Laboratory Science, Vol. III*, CRC Press, Cleveland, Ohio, 1976, 234.
32. Patel, V. K. and Venkatakrishna-Bhatt, H., *Asian Med. J.*, 1982, **25**(10), 737–742.
33. Bauer, A. R., Rank, R. K. and Kerr, R., *Life Sci.*, 1977, **21**(1), 63–70.
34. Everitt, B. S., *The Analysis of Contingency Tables*, Chapman & Hall, London, 1977.
35. Graham, S. H., Dayal, H., Swanson, M., Mittelman, A. and Wilkinson, G., *J. Natl Cancer Inst.*, 1978, **61**, 709–714.
36. Haenszel, W., Correa, P., Cuello, C., Guzman, N., Burbano, L., Lores, H. and Munoz, J., *J. Natl Cancer Inst.*, 1976, **57**, 1021–1026.
37. Haenszel, W., Kurihara, M., Segi, M. and Lee, R. K. C., *J. Natl Cancer Inst.*, 1972, **49**, 969–988.
38. Maclennan, R., DaCosta, J., Day, N. E., Law, C. H., Ng, Y. K. and Shanmugaratnam, K., *Int. J. Cancer*, 1977, **20**, 854–860.
39. Phillips, R. L., *Cancer Res.*, 1975, **35**, 3513–3522.
40. Wattenberg, L. W., *Cancer Res.*, 1983 (Suppl.), **43**, 2448s–2453s.

41. Sparnins, V. L., Lam, L. K. T. and Wattenberg, L. W., *Proc. Am. Assoc. Cancer Res.*, 1981, **22**, 114.
42. Welsch, C. W., Scieszka, K. M., Senn, E. R. and DeHoog, J. V., *Int. J. Cancer*, 1983, **32**, 479–484.
43. Minton, J. P., Abou-Issa, H., Foecking, M. K. and Sriram, M. G., *Cancer*, 1983, **51**, 1249–1253.
44. Stalder, R. and Muller, A. M., *Proc. 11th Coll. ASIC*, 1985, 113–120.
45. Hoshino, H. and Tanooka, H., *Toxicol. Lett.*, 1979, **4**, 83–85.
46. Nakanishi, K., Hirose, M., Ogiso, T., Hasegawa, R., Arai, M. and Ito, N., *Gann*, 1980, **71**, 490–500.
47. Kuhlmann, W., Fromme, H., Heege, E. and Ostertag, W., *Cancer Res.*, 1968, **28**, 2375–2389.
48. Macklin, A. W. and Szot, R. J., *Drug. Chem. Toxicol.*, 1980, **3**, 135–163.
49. Borlee, I., Lechat, M. F., Bouckaert, A. and Misson, C., *Louvain Med.*, 1978, **97**, 279–284.
50. Weathersbee, P. S., Olsen, L. K. and Lodge, J. R., *Postgrad. Med.*, 1977, **62**(3), 64–69.
51. Mau, G. and Netter, P., *Geburts. Frauenheilkde*, 1974, **34**, 1018–1022.
52. Linn, S., Schoenbaum, S. C., Monson, R. R., Rosner, B., Stubblefield, P. G. and Ryan, K. J., *N. Engl. J. Med.*, 1982, **306**(3), 141–145.
53. Rosenberg, L., Mitchell, A. A., Shapiro, S. and Slone, D., *J. Am. Med. Assoc.*, 1982, **247**(10), 1429–1432.
54. Kurppa, K., Holmberg, P. and Saxen, L., *N. Engl. J. Med.*, 1982, **306**(24), 1548.
55. Van den Berg, B. J., in *The Epidemiology of Prematurity*, Eds D. M. Reed and F. J. Stanley, Urban and Schwarzenberg, Baltimore, MA, 1977, 157–176.
56. Knutti, R., Rothweiler, H. and Schlatter, C. H., *Env. J. Clin. Pharmacol.*, 1981, **21**, 121–126.
57. Aeschbacher, H. U., Milon, H., Poot, A. and Würzner, H. P., *Toxicol. Lett.*, 1980, **7**, 71–77.
58. Nolen, G. A., *J. Toxicol. Environ. Health*, 1982, **10**, 769–783.
59. Würzner, H. P., Stalder, R., Bexter, A. and Vuataz, L., *Proc. 10th Coll. ASIC*, 1982, 355–362.
60. Murphy, S. J. and Benjamin, C. P., *Microbios. Lett.*, 1981, **17**, 91–100.
61. Thayer, P. S. and Kensler, C. J., *Toxicol. Appl. Pharmacol.*, 1973, **25**, 169–179.
62. Collins, T. F. X., *FDA By-Lines*, 1979, **9**, 352–373.
63. Collins, T. F. X., *Food Chem. Toxicol.*, 1983, **21**(6), 763–777.
64. Collins, T. F. X., Welsh, J. J., Black, T. N. and Collins, E. V., *Regul. Toxicol. Pharmacol.*, 1981, **1**, 355–378.
65. Elmazar, M. M. A., McElhatton, P. R. and Sullivan, F. M., *Toxicology*, 1982, **23**, 57–71.
66. Kimmel, C. A. and Wilson, J. G., *Teratology*, 1973, **8**, 309–316.
67. Fritz, H. and Hess, R., *Teratology*, 1968, **3**, 331–338.
68. Bachmann, G., Haldi, J., Wynn, W. and Ensor, C., *J. Nutr.*, 1946, **32**, 239–247.

Chapter 8

Mutagenicity of Coffee

H. U. AESCHBACHER
Nestlé Research Centre, Nestec Ltd, Lausanne, Switzerland

1. INTRODUCTION

Numerous environmental substances, including those occurring in our food, are known to cause DNA damage and are therefore associated with a health risk for man.[1,2] In fact DNA alterations have been shown to be causal factors in birth defects and cancer.[3–5] These observations led to the development of several mutagenicity/carcinogenicity screening tests[6,7] which can detect substances interacting with the DNA. Actual validation and standardisation of these tests has only recently been achieved with the accumulation of a sufficiently large data bank which revealed that only a few of the *in vitro* screening tests are reliable for risk prediction.[8,9] It has also become clear that knowledge of the metabolic fates is imperative in such risk assessment especially for carcinogenicity.[10,11] Furthermore, the action of single substances or compounds must be considered in context especially in the case of complex mixtures such as coffee, which contains more than a thousand compounds (though many of these are present at very low levels).

However, despite these complications, experimental evidence is an important factor in estimations of the health risk of coffee, especially since some epidemiological studies on human coffee drinkers have yielded results that are still unclear.[12]

2. MUTAGENICITY EVALUATION OF WHOLE COFFEE

Mutagenicity evaluation of coffee has been limited to the bacterial Ames mutagenicity tests[13–20] until recently when a few studies using mammalian cells *in vitro*[21,22] or whole mammals[23] have been reported. Evidence was provided that coffee roasting led to the formation of reactive compounds which cause an effect in *in vitro* mutagenicity test systems.[16,19] These compounds were, however, shown to be readily deactivated by liver metabolising enzymes in all *in vitro* tests,[15,18,21] and this was further confirmed by the non-mutagenicity observed in mammalian tests.[23]

2.1. Bacterial Tests

Several publications have shown a direct-acting effect of instant or home-brew coffee in bacterial mutagenicity test systems.[13–19,24] Generally in these studies home-brew coffee was prepared by dissolving 5 to 10% w/v ground coffee in hot water, as is usually done for home preparation. These home-brew coffee solutions were lyophilysed for concentration and then rediluted in water or other solvents, e.g. dimethylsulphoxide, prior to testing. In the case of instant coffee several different brands have been used including decaffeinated coffees, and the powders were directly dissolved prior to testing.

In one case[24] extracts of overheated coffee syrups prepared from roast and ground beans were tested. For the majority of the mutagenicity studies with either instant or home-brew coffee, maximal dose levels of 15–50 mg per plate were used. These dose levels were limited by bactericidity and are considered as very high since potent mutagens are already active at microgram levels in bacterial test systems.[25] Despite these high dose levels, home-brew or instant coffees caused, with some exceptions, only relatively weak mutagenic effects *per se* in Ames *Salmonella* or *E. coli* bacterial tester strains.

In fact a wide range of effects has been reported in the Ames *Salmonella* mutagenicity test. Freshly brewed coffee showed no effect in two studies,[17,24] and a weak mutagenic induction (mutation factors (MF) between 2 and 4),[13–15] or somewhat higher effects (MF:7)[16] or MF:11[18] in other studies. In the latter study, which reported relatively high mutation factors, the spontaneous levels were at the upper limit which might increase sensitivity but also the tendency to the occurrence of artefacts. Coffee at doses of 30–60 mg per plate also caused a clear-cut effect in *E. coli* K12, inducing mutation factors which were estimated as between 5 and 10.[16] The above variations cannot be attributed to differences between methods since the direct Ames plating test[14] and a preincubation test[13] gave identical results whereas the results within the same preincubation test varied considerably.[13,16,18] Inter-laboratory variations in bacterial tests are known to occur and are attributed to a variety of factors.[26]

Other factors directly related to the test substance might also lead to inter- or intra-laboratory variation. It was, for instance, calculated that, at the highest dose level of coffee tested, 70 μg of total histidine was added per plate.[15] Such a level of free histidine can interact in the Ames *Salmonella* test to yield non-mutant colonies.[27] However, the majority of histidine in coffee solutions is present in the bound form[15] and therefore cannot interact with the Ames test.[27] This point is confirmed by the observation that liver enzymes negate the effect of coffee.[13,15,18] The mode of sterilisation may also be a cause of inter-laboratory variations, since autoclaving tends to produce mutagenic heat reaction products[28] or else destroy biologically active compounds whereas filter sterilisation could result in mutagenic substances being retained. It seems, however, that the various sterilisation procedures had no effect on the mutagenic outcome of coffee.[15] Coffee test solutions at a pH of about 5·0 to 5·5 were also suspected of interacting with bacteria mainly owing to cytotoxicity. However, a pH much lower than 5 did not affect bactericidity[15] nor mutagenicity.[18] Accordingly artefacts in the bacterial test systems owing to the preparation or chemistry of the coffee solution can be excluded.

Despite the inter-laboratory variation it can be concluded that all coffees caused similar relatively weak effects in bacterial test systems. When both home-brew and instant coffee were evaluated in the same study, no differences in mutagenic potential were observed.[13,15] There was also no difference in the mutagenic outcome between different brands of instant coffees,[13,16] between caffeine-containing or decaffeinated coffee,[13,15,16] or between different blends.[16,19] The latter observation was confirmed with Arabica Santos, Arabica Columbia, Robusta Indonesia and Robusta Cameroun varieties, which gave comparable effects in the Ames mutagenicity test upon roasting.[19]

In all the above bacterial mutagenicity studies, with one exception where *E. coli* was also used,[16] the Ames *Salmonella* tester strains were employed. The results are summarised in Fig. 1. In this figure effects induced by the various Ames tester strains are given as mutation factors: number of induced mutants (revertants) at the highest non-bactericidal coffee concentration per number of mutants (revertants) with untreated bacteria (spontaneous rate). The mutation factors were calculated from values given by various authors (numbers of the corresponding reference are given in parentheses). For Ames tester strain TA 104 not enough information was available to calculate the mutation factor. However, a slightly higher

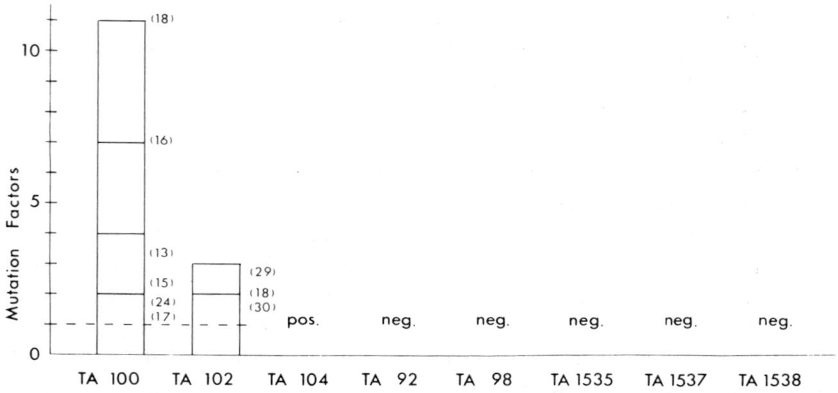

Fig. 1. Mutagenicity of coffee in Ames tester strains. ---, Negative control (spontaneous level).

response was reported with TA 104 than with TA 102.[30] Instant coffee was reported to give comparable results to home-brew coffee within the same working group, and therefore mutation factors given in Fig. 1 refer to both coffees (where compared). Coffee was generally only significantly mutagenic *per se* to base pair Ames tester strains (TA 100, TA 102 and TA 104) but not to all the other Ames tester strains which respond to a variety of other specific mutations. All the studies revealed that coffee only exerted a clear-cut effect in the sensitive 'base pair' Ames tester strains TA 100,[13,15,18,19] TA 102[18,29,30] and TA 104[30] (see mutation factors in Fig. 1). Other Ames tester strains such as TA 1535, TA 1537 and TA 1538,[15,30] the sensitive frameshift strains TA 98[14–16,24] or TA 92[16] did not respond to coffee. The above observations indicate that coffee causes a very specific effect on the genome of the bacteria. Coffee was not mutagenic in *E. coli* WP2 or base pair *Salmonella* tester strain TA 1535 which both contained no plasmid, but exerted an effect in the same tester strains which contained the plasmid. This observation suggests that the presence of the plasmid is essential to obtain an effect with coffee.[16] This is not, however, the only factor which is required to obtain a mutagenic effect with coffee. *Salmonella* tester strain TA 92, which also contains the plasmid but has a normal lipopolysaccharide barrier of the cell wall and excision repair capacity, was not affected by coffee either.[16] Hence bacteria require a deficient cell wall or excision repair in order to respond to coffee. It has also been speculated that the effects of coffee might be efficiently repaired and that major mutagens in coffee might not have DNA crosslinking activity.[30] Such repair of coffee-induced lesions had also been suggested[16] since coffee did not induce prophage activity in a repair-competent *E. coli* strain. Hence coffee was revealed to have a very specific and possibly reparable effect on the DNA, at least in bacteria.

Coffee has also been seen to be able to act directly since the bacterial strains were only affected in the absence of the liver enzyme fractions. In studies where liver microsomal enzymes, which partly mimic metabolism, were added to bacterial cultures, the effect of coffee was efficiently reduced or even totally removed[13–15,18] (see Fig. 3). This observation is very important for risk evaluation since substances which are deactivated in *in vitro* mutagenicity tests are usually not mammalian carcinogens.[10] It was therefore essential to study metabolic fates in mutagenic systems more thoroughly to determine whether coffee is metabolically deactivated or bound to protein. It was shown that the deactivation was due to metabolic activity since the presence of bovine protein or heat-deactivated liver supernatant S-9 did not have a 'deactivation' capacity.[15,18] No additional NADPH generating system was required for the 'metabolic activation system' to deactivate coffee,[18] hence suggesting that even suboptimal metabolising capacity can deactivate coffee *in vitro*. This efficiency of coffee deactivation by liver enzymes was again confirmed in a study where as little as 0·6% of S-9 liver supernatant readily decreased mutagenicity,[18] although optimal deactivating capacity was observed with an S-9 mix containing 10% liver supernatant whereas no further improvement was observed for levels of 10–40%. It has been suggested that the inactivating enzymes are situated in the heat-labile cytosol fraction of the S-9 liver supernatant, and that an oxidative process might be responsible for the deactivating activity.[18] Enzymes such as glyoxalase I and II, together with reduced glutathione as well as catalase activity which occur in mammalian cells, suppressed, to some extent, the 'mutagenicity' of coffee,[18,20] hence suggesting a deactivating capacity in mammals also.

2.2. Mammalian Tests *in vitro*

Coffee has caused mutagenicity in some *in vitro* mammalian cell test systems. It exerted mutagenic activity in cultured Chinese hamster lung cells[22] when tested without the metabolising enzyme system. In this study two different instant coffees were used, both at concentrations up to 6 or 8 mg per millilitre of culture (see Fig. 2). The highest dose of one instant coffee (A) increased the mutation factor to about eight times the

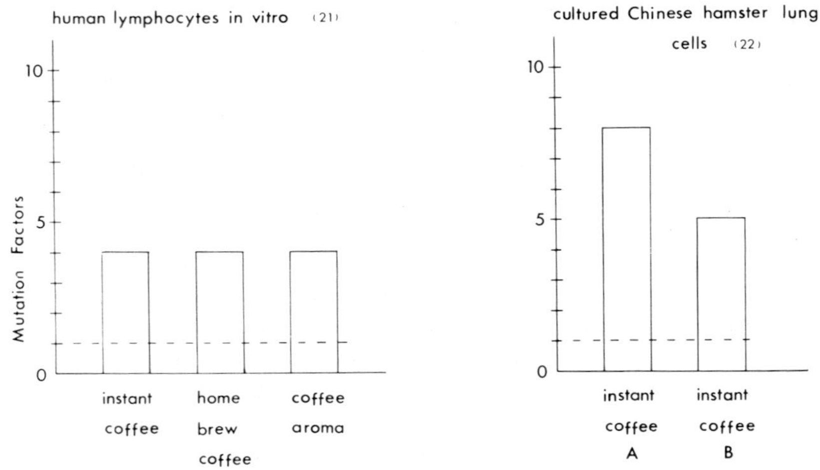

Fig. 2. Mutagenicity of coffee in mammalian cells *in vitro*. ---, Negative control (spontaneous level).

spontaneous level whereas the other coffee (B) had only about half of this activity.[22] No systematic difference between various instant or home-brew coffees was, however, observed with the other *in vitro* tests (Ames test and *in vitro* mammalian tests) which are reported here. The values given in Fig. 2 are mutation factors (induced with highest coffee dose/spontaneous) of total chromosome aberrations for the lymphocyte test or diphtheria toxin-resistant mutants for cultured Chinese hamster lung cells. These results were obtained without the 'metabolic activation system'.

Coffee also caused dose-dependent chromosome aberrations in cultured human lymphocytes when tested at doses up to 4·3 mg per millilitre of culture in the absence of the 'metabolic activation system'.[21] Both home-brew coffee, prepared from 50 g ground beans per litre by the Melitta filter method and then freeze dried, and regular instant coffee caused similar effects (see Fig. 2). In fact both home-brew coffees prepared from either decaffeinated or caffeine-containing beans as well as caffeine-containing or decaffeinated instant coffee caused about a fourfold increase of total chromosome aberrations for the highest concentrations over the spontaneous level when tested *per se*. In this same experiment[21] it was shown that the principal chromosome breaking agents were also present in coffee aroma as concentrated coffee aroma (prepared by a standard method)[31] at concentrations up to 50 μl per millilitre of culture induced a fourfold increase in total chromosome aberrations when compared with the spontaneous level of untreated lymphocytes. To obtain these effects specific preincubation procedures were required to render the test system sensitive enough to detect such slight chromosome damage. About half of the chromosomal aberrations observed with coffee and coffee aroma were of the 'gap' type. However, these gaps induced by coffee do not necessarily indicate DNA damage,[32] and hence the real mutagenic effect might be attributed to only half the total chromosome aberrations. It can therefore be concluded that in mammalian cell *in vitro* tests, as for the bacterial test, high coffee concentrations are required and even then only a relatively weak effect is observed. Furthermore, it has been shown in cultured lymphocytes that the chromosome damage observed with coffee or coffee aroma *per se* was removed in the presence of the liver metabolising enzyme system (see Fig. 3). In Fig. 3 the mutation factors indicated for the Ames test (tester strain TA 100) are average values obtained with different coffees in the same report (references are given in brackets). The results were obtained either without 'metabolic activation' (− S-9) or with 'metabolic activation' (+ S-9). For both the Ames and human-lymphocyte test the results given with or without S-9 were obtained under the same experimental conditions. These results show a complete deactivation of coffee mutagenicity in the presence of S-9 because the dotted line indicates the spontaneous level (untreated negative control), and a mutation factor below 2 is generally considered as the criterion for a non-significant effect. This deactivation property was also observed for the other base pair Ames tester strains, TA 102 and TA 104, which responded to coffee. The same deactivation property of coffee with liver 'metabolic activation' enzyme systems was also observed

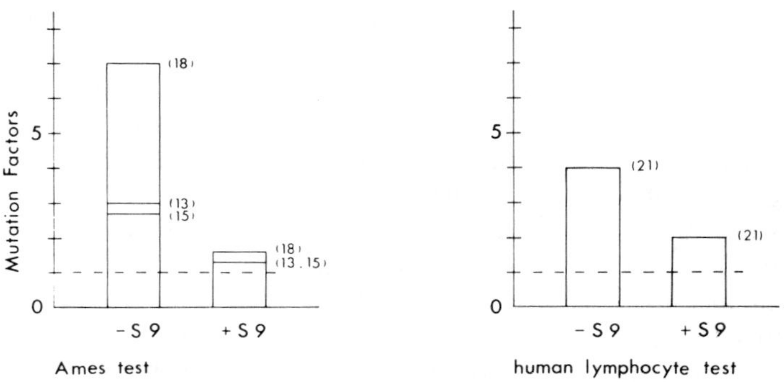

Fig. 3. Metabolic deactivation of coffee mutagenicity *in vitro*.

with human lymphocytes in culture. The number of total chromosome aberrations was significantly lower for all coffees in the presence of S-9 (as compared with effects without S-9) and generally no or only weakly significant increases of chromosome aberrations were observed with S-9. The dotted line again represents the zero level (spontaneous level). Since the composition of this microsomal S-9 enzyme system is comparable to that used in bacterial tests it can be hypothesised that similar deactivation mechanisms occur in cultured mammalian cells as described for the bacterial test systems.

2.3. Tests under *in vivo* Conditions

Mutagenicity test systems which are near to mammalian conditions, e.g. *Drosophila*, which contain a metabolising system comparable with that of mammals, or tests where tissue or body fluids from rodents (host mediated) or human coffee drinkers (urine) are evaluated in *in vitro* mutagenicity tests (Ames, CHO), are usually referred to as 'tests under *in vivo* conditions' (Table 1).

Instant and home-brew coffees were evaluated in different mutagenicity assays in germ and somatic cells in *Drosophila melanogaster*. For germ cells sex-linked recessive lethals which detect mainly point mutations and small chromosome aberrations, dominant lethals scoring for cytotoxic and genotoxic effects, and sex-chromosome losses detecting chromosome breaks and non-disjunction, were considered. Possible induction of mutation and recombination in somatic cells of the wing imaginal disc was studied separately.

High concentrations, up to 4% (w/v) for instant and 3% for home-brew coffee, which were near the lethal dose, did not cause consistent genotoxic effects in the different *Drosophila* cell systems.[33] This observation was also confirmed by the same workers in excision repair deficient cells which increased the sensitivity of the test.

In another test where doses up to 3 or 6 g of instant coffee were orally administered per kilogram body weight in mice in the 'intrasanguine mutagenicity test', no mutagenic effect was observed.[14] In this assay mutagenic agents present in the blood would readily affect the tester organisms (Ames strain TA 1537 or *E. coli* K-12) previously intravenously injected into the host. This test system, although in theory suitable for detecting possible reactive compounds, is considered to be relatively insensitive.

Attempts were also made to investigate the possible mutagenic activity

Table 1
Effect of coffee in *in vivo* mutagenicity test

Dose levels of coffee	*Test system*	*Results*	*Reference*
In vivo conditions			
3–4% in diet	*Drosophila*	No effect	33
1–6 g/kg body weight	Host-mediated test in mice	No effect	14
12 g/day (5 cups)	Human coffee drinkers' urine extracted with XAD-2 → Ames test	No effect	34
5 cups	Human coffee drinkers' urine extracted with reverse-phase HPLC → CHO cells	Equivocal effect	35
In vivo			
1·0–2·5 g/kg body weight	Sister chromatid exchange test (SCE)	No effect	23
0·5–3·0 g/kg body weight/day	Micronucleus test	No effect	23

Under '*in vivo*' the results were obtained with two widely used *in vivo* mutagenicity tests in mice (micronucleus test) or hamsters (sister chromatid exchange test). Dose ranges of instant coffee administered (which was generally used in these trials) are given in the table.

The results indicated as 'no effect' also include those results which were considered as not giving consistent evidence for significant mutagenic induction by the authors.

of human coffee drinkers' urine and to compare them with that of the urine of smokers as, based on bacterial test results, the risk of coffee drinking was thought to be similar to that of cigarette smoking.[13] Human coffee drinkers who received 12 g of coffee per day during 4 days or 12 g within 2 hours and whose urine was purified by XAD-2 column chromatography showed no signs of mutagenicity in Ames tester strains TA 98 and TA 100.[34] However, the urine of cigarette smokers smoking 20–30 cigarettes per day showed clear-cut mutagenic activity in Ames tester strain TA 98 upon metabolic activation.[34] In another study with human coffee drinkers the urine was extracted and separated into three fractions by preparative reversed-phase high-pressure liquid chromatography and tested in Chinese hamster cells (CHO). In this experiment, in which the subjects drank at least five cups of coffee, a significant increase of chromosomal aberrations was reported.[35] These authors observed similar effects with subjects who smoked at least 20 cigarettes but also reported that cigarette smoking did not potentiate the effect of coffee drinkers' urine. They suggested the involvement of active oxygen species in the clastogenic response, but some discrepancies in their results cast doubt on their hypothesis. In fact clastogenicity of fractions one and two was abolished by catalase or peroxide dismutase but not that of fraction three. In contrast, there was apparently a correlation between the ability to generate hydrogen peroxide and induction of clastogenic action with fractions two and three but not one.[35] Due to these equivocal results this study is inconclusive. Since the CHO cells system is very sensitive, it is also subject to possible artefact. Furthermore, some recent evidence suggests that highly reactive components of coffee aroma provoke chromosome breaking (clastogenic effects) in mammalian cells *in vitro*,[21] and therefore it is not very likely that those reactive species would readily occur in urine as active oxygen species.

2.4. Mammalian Tests *in vivo*

So far only two studies have been reported on coffee mutagenicity in target cells of intact mammals (Table 1). In a widely used micronucleus test, mice received instant coffee by gavage for 5 days consecutively at dose levels between 0·5 or 3 g per kilogram body weight by gavage. This 5-day treatment has been found to be the most efficient in the detection of mutagens in complex mixtures of unknown pharmacokinetic properties. In addition to the evaluation of whole coffee, a study was also carried out on coffee aroma at doses up to 50 ml per kilogram body weight. With coffee aroma a single treatment was required owing to the reactivity of the aroma constituents. The results from both instant coffee and coffee aroma in the micronucleus test revealed that *in vivo* they had no chromosome breaking capacity.[23] In addition to the micronucleus test which detects the induction of chromosome aberrations and is considered appropriate for cancer prediction,[36] the so-called sister chromatid exchange test (SCE) was also used in the mutagenic evaluation of coffee. This latter test detects DNA damage which is not detectable in the micronucleus test, and it is therefore generally considered to be more sensitive.[37] In the sister chromatid exchange test, when hamsters received single doses of 1 to 2·5 g of instant coffee per kilogram body weight, no consistent DNA damaging effect was observed.[23] Based on these results it can be concluded that coffee is not mutagenic in the widely used, relatively sensitive mutagenicity tests in rodents.

3. MUTAGENICITY EVALUATION OF COFFEE COMPONENTS

The identification of compounds present in complex mixtures which might be biologically reactive, e.g. mutagens, is essential for toxicological evaluation. Pure compounds are necessary to investigate factors such as metabolic fates and specificity of action, e.g. repair, attainment of target organs, etc., which need to be known for risk evaluation. The objective of identifying mutagenic compounds in coffee, which contains numerous very reactive compounds, is, however, a formidable task even if between-compound interaction is excluded. In fact as many as 660 volatile substances such as sulphur compounds, pyrazines, pyridines, pyrroles, oxales, furanes, aldehydes, ketones, phenols, etc., have been identified in coffee.[38] Actually some of these volatile compounds, which are prone to interact with the DNA because of their reactivity, are suspected of being responsible for the positive effect observed in *in vitro* mutagenicity tests. This hypothesis is supported by the fact that coffee only becomes mutagenic *per se* upon roasting.[16,19] Furthermore, some advanced Maillard reaction products which might occur in coffee in traces were found to be mutagenic in the Ames bacterial test.[28] The presence of chromosome-damaging substances in coffee aroma[21] also suggests the involvement of reactive advanced Maillard reaction products. Some of the compounds that occur during the roasting of coffee, such as methylglyoxal,[39] glyoxal, maltol, diacetyl[40] and hydrogen peroxide,[41] were shown to be active in the Ames mutagenicity test. Other substances present in coffee, such as caffeine,[2] chlorogenic acid[42] and flavonols, were suspected of contributing to the direct-acting 'mutagenic' effect of coffee. There is, however, some evidence that these latter compounds are not causal factors, as will be discussed below.

Attempts have repeatedly been made to concentrate and isolate mutagenic substances from coffee via extraction.[13,17,24] Different extraction procedures such as XAD-2,[13] Soxhlet[24] or concentration upon evaporation and direct solvent extraction[17] have been used. XAD-2 extraction of coffee solution where a first water extraction was discarded has been claimed to yield about a sixfold purification in the methanol eluate.[13] In another study where coffee was concentrated through evaporation, no effect was observed except for the chloroform extraction upon partition into *n*-butanol, methanol or hexane, where a borderline effect was reported in the Ames test.[17] In contrast, with the Soxhlet extraction no mutagenic activity was reported in the dichloromethane nor in the ethanol extracts of coffee solid residues or from freshly brewed coffee.[24] The difficulties in concentrating mutagenic compounds in coffee as illustrated above have to be attributed to the high reactivity and volatile character of the suspect aroma components which are mainly also water-soluble. Therefore extraction and concentration methods which are generally efficiently used to extract mutagens from heated food products do not seem suitable in the case of coffee. Another approach using coffee components which have already been chemically characterised was shown to be more appropriate for mutagenicity testing.[20,39,40]

3.1. Aliphatic Carbonyls

The few coffee compounds occurring at very low but still significant concentrations that have so far been found to have mutagenic activity all belong to the class of aliphatic carbonyls, e.g. methylglyoxal, glyoxal, diacetyl, maltol. Their mutagenic property is attributed to their alkylating activity. Methylglyoxal, which is the most active compound in the Ames tester strain TA 100,[18,20,39] was initially considered as being the major mutagen of coffee in which it is present at levels of 100–150 μg per gram. It was estimated that it accounts for over 50% of coffee mutagenicity.[39] However, further detailed studies on methylglyoxal in the Ames test provided evidence that methylglyoxal was only a minor factor.[20] Cysteine suppressed the mutagenicity of methylglyoxal but did not alter the mutagenicity of coffee, whereas catalase suppressed most of the coffee mutagenicity but not that of methylglyoxal. The conclusion that methylglyoxal only accounts for a minor part of the mutagenicity of instant coffee was also drawn from a mutagenicity study using Chinese hamster lung cells.[22]

The other aliphatic carbonyl compounds caused lower mutagenic activity than methylglyoxal and can be ranked in declining order of activity as glyoxal, diacetyl and maltol.[40,43] Their mutagenic activity was again shown to be specific to Ames tester strain TA 100. Some attempts have been made to establish a structure–mutagenic activity relationship. For instance, it has been observed that only dicarbonyl compounds in the acyclic series with an oxidation capacity equivalent to that of aldehydes and ketones such as glyoxal and diacetyl were mutagenic, whereas dicarbonyl compounds with the carboxyl oxidation level (dimethyl oxalate and pyruvic acid) were inactive. Of the other compounds such as maltol, non-quininoid, 1,2-diketo compounds six-membered rings were required for mutagenic activity.[40]

3.2. Methylxanthines

Caffeine is the only methylxanthine present in coffee in substantial amounts. Its content varies considerably between and within coffee species but it is estimated to range from about 1 to 2·5% in roasted coffee and from 3 to 5% in instant coffee.[44] Caffeine was one of the first substances found to cause chromosome damage *in vitro*, and therefore a very large body of literature exists on its effect and interaction with DNA which has been thoroughly reviewed.[45–47]

Caffeine was shown to be negative in the Ames test,[48,49] which is the most widely used bacterial mutagenicity test. Otherwise mutagenic effects in bacteria were only observed under very specific conditions.[47] The negative effect in the Ames test was further confirmed by the fact that caffeine-containing and decaffeinated coffees showed identical mutagenic properties in the Ames test.[13,15] This observation also indicates that caffeine does not potentiate the mutagenic effect of other coffee constituents. In *Drosophila* caffeine caused equivocal results,[47] but on the other hand it caused chromosome damage in various mammalian cells although generally only at high dose levels.[46,47] This observation was extended to cultured human HeLa and lymphocyte cells. However, only concentrations at least 10 to 100 times higher than the peak levels experienced in man caused DNA damage.[46,47] Furthermore, the amounts of caffeine present in the maximal dose levels of coffee used in a lymphocyte test did not contribute to the induction of chromosomal aberrations.[21] It has never been conclusively proved that caffeine causes DNA damage in experimental animals, or that it represents a mutagen-related risk for man.[45–47]

Other methylxanthines such as theophylline and theobromine are also present in coffee, but only in traces.[44] Therefore these other methylxanthines do not seem to be important in coffee mutagenicity, especially since their mutagenicity is comparable to or only slightly higher than that of caffeine.[47]

3.3. Phenolic Compounds

Total chlorogenic acid levels are about 2% in roasted arabica, 4% in roasted robusta and 6% in instant coffee powder.[50] Chlorogenic acid, caffeic acid and to a lesser degree quinic acid were shown to have some mutagenic activity in *Saccharomyces cerevisiae* and cultured Chinese hamster ovary cells, but none of the substances caused an effect in Ames tester strains TA 100 and TA 98.[42] Considering these results it is difficult to evaluate the role of chlorogenic acid in coffee mutagenicity since coffee was positive in both the Ames test and in cultured mammalian cells. Different blends of coffee, the chlorogenic acid content of which varied between 2 and 6%,[50] also caused similar mutagenic effects in the Ames test.[19] Furthermore, it has been suggested that other phenolics including flavonoids, which are widely found in plants and also in coffee beans at low concentrations, might be involved in its mutagenicity.[42] It is known that a large number of plant flavonols exert a mutagenic effect in the Ames test and also in mammalian cells.[51] Their implication in coffee mutagenicity is not, however, very likely since flavonoids are more efficient at inducing frameshift than 'base pair' mutation and are generally metabolically activated, whereas coffee causes only specifically 'base pair' mutations and is metabolically deactivated. There is thus quite a lot of evidence that phenolic compounds are not likely to be involved in the mutagenicity of coffee.

3.4. Hydrogen Peroxide

Formation of hydrogen peroxide has been reported upon roasting coffee. It was suggested that a cup of coffee of about 2·2 g of instant coffee contains 500–750 μg of hydrogen peroxide and that this concentration increases with time after preparation.[41] Such an increase in hydrogen peroxide over time is, however, astonishing since such reactive compounds are generally very labile and mutagenicity of coffee solutions has been reported to decrease with time.[18] Hydrogen peroxide itself was found to be significantly but very weakly mutagenic in the Ames test. Hence the mutagen risk of hydrogen peroxide present in coffee is not yet clear, especially since it is constantly being formed by enzymatic reactions in the human body which possesses an efficient defence system, e.g. catalase or glutathione peroxidase, against free radical toxicity.

4. INTERACTION

The great number of different and mainly highly reactive compounds present in coffee makes it highly probable that a variety of interactions may occur. Such interactions were shown to occur between coffee constituents but also with other compounds present in the food. It is therefore essential to consider these interactions in the assessment of the mutagenic risk of single coffee constituents. Such an approach is, however, extremely complex and additionally complicated by the fact that different mutagenic outcomes are obtained in the various mutagenicity test systems.

4.1. Interactions between Coffee Constituents

The main interaction in terms of mutagenicity was reported to occur between methylglyoxal and hydrogen peroxide. It has been proposed that the mutagenicity generated through interaction between these two compounds accounted for the majority of coffee's mutagenicity in the Ames test.[41] The mode and biological significance of such an interaction is, however, far from clear as it is known that methylglyoxal is converted by hydrogen peroxide to pyruvic acid which is not mutagenic by itself nor in combination with hydrogen peroxide. Furthermore, plant phenolics such as chlorogenic or caffeic acids which are present in coffee at substantial concentrations were shown to have an antimutagenic activity. This activity was attributed to the interaction of phenols with oxygen radicals.[52] This observation leads to the proposition that the mutagenicity of hydrogen peroxide or other advanced Maillard reaction products is diminished by other coffee constituents such as phenolics.

4.2. Interactions of Coffee Constituents with Other Food Components

Plant phenols of coffee might be of importance not only in inhibiting other coffee constituents but also in their interactions with other mutagens found in food. It was found, for instance, that phenolics occurring in coffee and tea had inhibitory properties on mutagenicity resulting from nitrosation.[53] There are also possible complex interactions between caffeine and other compounds in terms of mutagenicity. In fact many hypotheses have been advanced to explain the mechanisms involved in the interaction of pure caffeine and other biologically active pure compounds.[46] Caffeine was shown to act either as inhibitor or enhancer of mutagens/carcinogens in a variety of *in vitro* test systems.[46,47] Claims have also been made that sodium sulphite suppressed the mutagenic activity of coffee in the Ames test and inactivated phage-inducing activity whereas ascorbic acid enhanced the mutagenicity of coffee.[22,43] In this context it must, however, be remembered that there is also some experimental evidence that ascorbic acid is a cancer inhibitor in mammals.[54]

Although there is quite a lot of evidence that interaction between coffee components or other food components alters the mutagenic outcome, the picture is not yet clear because of the complexity of the problem.

5. RISK ASSESSMENT

To obtain a significant mutagenic effect with coffee or its constituents in *in vitro* mammalian and bacterial mutagenicity test systems, specific and sensitive conditions as well as high doses are required. This fact suggests that coffee is only relatively weakly active under *in vitro* conditions. Several of the *in vitro* experiments provide evidence that chemically reactive compounds are responsible for the direct-acting activity of coffee. This explains the fact that a deactivation or complete abolition of the mutagenic activity occurred in the presence of liver microsomal enzymes, as demonstrated in all the *in vitro* mutagenicity tests on coffee to date. The mechanism of metabolic deactivation of coffee or coffee compounds was shown to be due to enzymatic activity, e.g. glyoxylase, glutathione transferase, catalase, etc.,[18,20] which suggests the ready deactivation of direct-acting mutagens such as aliphatic carbonyls or hydrogen peroxide. In this context it was also suggested that the conjugate of the aromatic carbonyl maltol was unlikely to be mutagenic as a result of the loss of its α-dicarbonyl functionality.[40] The use of results of interaction studies for risk evaluation is not yet feasible, because of the complexity of interaction between coffee components and the fact that interaction between coffee constituents can lead to either potentialisation (methylglyoxal/hydrogen peroxide) or inhibition (phenolics) of the mutagenic activity. There have also been some contradictions in the mutagenic response within *in vitro* tests which additionally complicate the issue. In this context it is important to mention that the reliability and predictability, especially for carcinogenic risks, is much better with some *in vitro* tests than others.[8,55]

Therefore in the present situation it is most reliable to use results of whole coffee obtained in well validated mutagenicity tests for risk assessment. Taking into account this fact and considering the mechanisms of metabolic deactivation *in vitro* and the suggestion that substances which are deactivated in *in vitro* mutagenicity tests are usually not carcinogens in mammals,[10] it can be concluded that coffee does not represent a mutagenic/carcinogenic risk. This conclusion is confirmed by the results obtained with coffee in mutagenicity tests carried out *in vivo*. Therefore coffee does not appear to be associated with mutagenicity-dependent health risks.

Recently, various published work has also shown an antimutagenic[56,57] and anticarcinogenic[58] activity of coffee or coffee constituents. The most likely candidates for this beneficial effect seem to be plant phenolics, e.g. chlorogenic acid and its derivatives, or Maillard reaction products with antioxidant properties.

6. SUMMARY

1. Coffee causes a relatively weak direct-acting activity in some bacterial and mammalian *in vitro* mutagenicity tests.
2. Mutagenicity is mainly attributed to chemically reactive coffee components such as aliphatic carbonyls.
3. Interaction between single coffee components is very complex and results in either potentiation or inhibition of mutagenic activity.
4. Coffee mutagenicity is readily deactivated by detoxifying mammalian enzyme systems.
5. No conclusive evidence for mutagenic effects of coffee was observed when tested under *in vivo* conditions.
6. Coffee was not mutagenic in widely used *in vivo* mutagenicity tests.
7. It can be concluded from the experimental results currently available that coffee does not represent a mutagen-dependent health risk for man.

REFERENCES

1. Sugimura, T. and Sato, S., *Cancer Res.*, 1983, **43**, 2415s–2421s.
2. Ames, B. N., *Science*, 1983, **221**, 1256–1264.
3. Childs, B., Miller, S. M. and Bearn, A. G., in *Mutagenic Effects of Environmental Contaminants*, Eds H. Sutton and M. I. Harris, Academic Press, New York, 1972, 3–14.
4. Hartman, P. E., *Environ. Mutagen.*, 1983, **5**, 139–152.
5. Ames, B. N., *Science*, 1979, **204**, 587–593.
6. Hollstein, M. and McCann, J., *Mutat. Res.*, 1979, **65**, 133–226.
7. Anderson, D. and Longstaff, E., *Analyst*, 1981, **106**, 1–22.
8. Sobels, F. H., *Mutat. Res.*, 1985, **147**, 1–4.
9. Drake, J. W. and De Serres, F. J., in *Screening Assays for Carcinogens*, Eds R. Montesano, H. Bartsch and L. Tomatis, IARC Monographs, Lyon, 1980, 295–308.
10. De Flora, S., *Toxicol. Pathol.*, 1984, **12**, 337–343.
11. Wood, A. W., Bartsch, J., Fishbein, L., Garattini, S., Gielen, J., Malaveille, C., Roberfroid, M., Teichman, B. and Vesell, E., in *Screening Assays for Carcinogens*, Eds R. Montesano, H. Bartsch and L. Tomatis, IARC Monographs, Lyon, 1980, 277–294.
12. MacMahon, B., in *Coffee and Health*, Eds B. MacMahon and T. Sugimura, Banbury Report 17, Cold Spring Harbor, 1984, 243–244.
13. Nagao, M., Takahashi, Y., Yamanaka, H. and Sugimura, T., *Mutat. Res.*, 1979, **68**, 101–106.
14. Aeschbacher, H. U. and Würzner, H. P., *Toxicol. Lett.*, 1980, **5**, 139–145.
15. Aeschbacher, H. U., Chappuis, C. and Würzner, H. P., *Food Cosmet. Toxicol.*, 1980, **18**, 605–613.

16. Kosugi, A., Nagao, M., Suwa, Y., Wakabayashi, K. and Sugimura, T., *Mutat. Res.*, 1983, **116**, 179–184.
17. Stoltz, D. R., Stavric, B., Krewski, D., Klassen, R., Bendall, R. and Junkins, B., *Environ. Mutagen.*, 1982, **4**, 477–492.
18. Friederich, U., Hann, D., Albertini, S., Schlatter, Ch. and Würgler, F. E., *Mutat. Res.*, 1985, **156**, 39–52.
19. Albertini, S., Friederich, U., Schlatter, Ch. and Würgler, F. E., *Food Chem. Toxicol.*, 1985, **23**, 593–597.
20. Fujita, Y., Wakabayashi, K., Nagao, M. and Sugimura, T., *Mutat. Res.*, 1985, **142**, 145–148.
21. Aeschbacher, H. U., Ruch, E., Meier, H., Würzner, H. P. and Munoz-Box, R., *Food Chem. Toxicol.*, 1985, **23**, 747–752.
22. Nakasato, F., Nakayasu, M., Fujita, Y., Nagao, M., Terada, M. and Sugimura, T., *Mutat. Res.*, 1984, **141**, 109–112.
23. Aeschbacher, H. U., Meier, H., Ruch, E. and Würzner, H. P., *Food Chem. Toxicol.*, 1984, **22**, 803–807.
24. Blair, C. A. and Shibamoto, T., *Food Chem. Toxicol.*, 1984, **22**, 971–975.
25. Ames, B. N., McCann, Y. and Yamasaki, E., *Mutat. Res.*, 1975, **31**, 347–364.
26. Seiler, J. P., Herbold, B. A., Aeschbacher, H. U. and Göggelmann, W., *Teratogen. Mutagen. Carcinogen.*, 1983, **3**, 183–217.
27. Aeschbacher, H. U., Finot, P. A. and Wolleb, U., *Mutat. Res.*, 1983, **113**, 103–116.
28. Shibamoto, T., *Instrum. Anal. Foods*, 1983, **1**, 229–277.
29. Aeschbacher, H. U., Meier, H., Ruch, E., Wolleb, U. and Würzner, H. P., in *Coffee and Health*, Eds B. MacMahon and T. Sugimura, Banbury Report 17, Cold Spring Harbor, 1984, 89–97.
30. Nagao, M., Suwa, Y., Yoshizumi, H. and Sugimura, T., in *Coffee and Health*, Eds B. MacMahon and T. Sugimura, Banbury Report 17, Cold Spring Harbor, 1984, 69–77.
31. Pintauro, N. D., in *Coffee Solubilization*, Ed. N. D. Pintauro, Noyes Data Corp., London, 1975, 156–175.
32. Brogger, A., *Hereditas*, 1975, **88**, 131–136.
33. Graf, U. and Würgler, F. E., *Food Chem. Toxicol.*, 1986, **24**, 835–842.
34. Aeschbacher, H. U. and Chappuis, C., *Mutat. Res.*, 1981, **89**, 161–177.
35. Dunn, B. P. and Curtis, J. R., *Mutat. Res.*, 1985, **147**, 179–188.
36. Jenssen, D. and Ramel, C., *Mutat. Res.*, 1980, **75**, 191–202.
37. Evans, H. J., in *Long-term and Short-term Screening Assays for Carcinogens*, Eds R. Montesano, H. Bartsch and L. Tomatis, IARC Monographs, Lyon, 1980, 227–244.
38. Clifford, M. N., in *Coffee. Vol. 1: Chemistry*, Eds R. J. Clarke and R. Macrae, Elsevier Applied Science Publishers, London, 1985, 153–197.
39. Kasai, H., Kumeno, K., Yamaizumi, Z., Nishimura, S., Nagao, M., Fujita, Y., Sugimura, T., Nukaya, H. and Kosuge, T., *Gann*, 1982, **73**, 681–683.
40. Bjeldanes, L. F. and Chew, H., *Mutat. Res.*, 1979, **67**, 367–371.
41. Fujita, Y., Wakabayashi, K., Nagao, M. and Sugimura, T., *Mutat. Res.*, 1985, **144**, 227–230.
42. Stich, H. F., Rosin, M. P., Wu, C. H. and Powrie, W. D., *Mutat. Res.*, 1981, **90**, 201–212.

43. Suwa, Y., Nagao, M., Kosugi, A. and Sugimura, T., *Mutat. Res.*, 1982, **102**, 383–391.
44. Macrae, R., in *Coffee. Vol. 1: Chemistry*, Eds A. J. Clarke and R. Macrae, Elsevier Applied Science Publishers, London, 1985, 115–149.
45. Thayer, P. S. and Palm, P. E., in *Critical Reviews in Toxicology*, Ed. L. Goldberg, CRC, Lyon, 1975, 345–369.
46. Kihlman, B. A., in *Caffeine and Chromosomes*, Ed. B. A. Kihlman, Elsevier, Amsterdam, 1977, 249–415.
47. Timson, J., *Mutat. Res.*, 1977, **47**, 1–52.
48. Ishidate, M., Sofuni, T., Yoshikawa, K., Hayashi, M., Nohmi, T., Sawada, M. and Matsuoka, A., *Food Chem. Toxicol.*, 1984, **22**, 623–636.
49. McCann, J., Choi, E., Yamasaki, E. and Ames, B. N., *Proc. Natl Acad. Sci. USA*, 1975, **72**, 5135–5139.
50. Smith, A. W., in *Coffee. Vol. 1: Chemistry*, Eds R. J. Clarke and R. Macrae, Elsevier Applied Science Publishers, London, 1985, 1–41.
51. MacGregor, J. T., *Adv. Exp. Med. Biol.*, 1984, **177**, 497–526.
52. Huang, M. T., Chang, R. L., Wood, A. W., Newmark, H. L., Sayer, J. M., Yagi, H., Jerina, D. M. and Conney, A. H., *Carcinogenesis*, 1985, **6**, 237–242.
53. Stich, H. F., Rosin, M. P. and Bryson, L., *Mutat. Res.*, 1982, **95**, 119–128.
54. Shamberger, R. J., *Mutat. Res.*, 1984, **133**, 135–159.
55. Zeiger, E. and Tennant, R. W., Abstract, *4th Int. Conf. Environ. Mutagen.*, Stockholm, 1985, 42.
56. Obana, H., Nakamura, S. and Tanaka, R., *Mutat. Res.*, 1986, **175**, 47–50.
57. Yamaguchi, T. and Iki, M., *Agric. Biol. Chem.*, 1986, **50**, 2983–2988.
58. Nishikawa, A., Tanaka, T. and Mori, H. J., *Nutr. Growth Cancer*, 1986, **3**, 161–166.

Chapter 9

Allergens and Mould Toxin Contaminants

R. G. K. STROBEL

The Procter & Gamble Company, Cincinnati, Ohio, USA

1. INTRODUCTION AND BACKGROUND

1.1. General

This chapter considers the allergies caused by green coffee bean dust, and discusses the moulds and mycotoxins found on green coffee beans. The next chapter deals specifically with polycyclic aromatic hydrocarbons present in roasted coffee beans as a result of pyrolysis. An effort has been made to focus on historical developments in all these three areas. A perspective is also provided to put these areas into context with allergies caused by specific and general environments, and the presence and physiological actions of moulds and resulting mycotoxins in general, as well as of those physiologically active compounds generated by natural and industrial pyrolytic processes.

It is important to note that, as regards each of these issues, the coffee industry does not contribute an increased hazard beyond what is naturally present in the human environment. This statement results from and finds support in the abundant data provided in these chapters.

Man, since his existence, has been living in an environment that abounds with compounds of biochemical activity. Such compounds are naturally produced by plants; by microorganisms such as bacteria and moulds; or by chemical, pyrolytic or catalytic reactions.

It is the duty of the food-producing industry to provide and distribute foods and beverages that are safe for human consumption. In order to achieve such a goal, the consumer-oriented industries have to be knowledgeable of the toxins produced by plants and microbes, and by the chemical reactions that occur during industrial processing of foods and beverages.

In this chapter attention will be directed to the potential toxins and allergens associated with agricultural commodities in general, and special attention will be given to coffee beans and the coffee beverage. Qualitative and quantitative analytical procedures will also be described. The results of these data in the various areas will provide the proper perspective necessary to comprehend that, in spite of the ubiquitous presence of pathogenic microbes and of microbial and chemical toxins, commercially produced foods and beverages are practically free of microbial and chemical toxins and without detriment to human physiology. The continued co-operation and collaboration of the scientists employed in industrial research with those in governmental, private and academic institutions is, however, necessary to generate, maintain and disseminate the knowledge necessary to provide the consumer with foods and beverages safe for consumption. But, regardless of however safe the commercially produced food may be, there is still the hazard of food spoilage in the homes of the consumers. An effort should therefore be made to increase the level of consumer knowledge and awareness not only to practise proper food storage and home preparation in order to avoid food spoilage and deterioration but especially to prevent ingestion of such food, thereby avoiding any further consequences.

The Earth on which we live and the air we breathe are not so clean and pure as we once believed. The soil abounds with microorganisms, predominantly moulds (fungi), many of which are beneficial to plants, animals and humans. However, some directly cause toxicoses and some produce toxins, which are toxic upon ingestion. Like the pollen of plants, the spores of fungi, called conidia, can cause allergies. Mycotic diseases can ensue when humans are exposed to overwhelming numbers of conidia or when inoculated traumatically, or when the fungi find the human body in a state of malnutrition, stress, pregnancy, etc. When only the outer areas of the body are involved, e.g. skin and mucous membranes, then these diseases are relatively benign. However, when dissipation of the fungi to subdermal or vascular regions or organs occurs, they can be rather severe. Active growths of fungi in man with ensuing disease are called *mycoses*. Ingestion of fungal toxins and the resulting diseases are called *mycotoxicoses*, and the disease resulting from ingestion of poisonous mushrooms (macrofungi) is called *mycetism*.

1.2. Importance of Fungi

Fungi, when exposed to the proper conditions of temperature, humidity, moisture and nutrient availability, will actively grow on nearly any organic carbon source. Some of the fungi will produce, mostly at the point of exhaustion of the carbon source, so-called secondary metabolites. Some of these secondary metabolites can be very beneficial to man, animals and plants (like those from specific *Penicillium* spp.); some, however, like aflatoxins, when inhaled or ingested as a contaminant with the foodstuff, can be very noxious. There are many fungal toxins, but none more than aflatoxin had a more dramatic effect in popularising the subject of the 'forgotten disease' and in bringing mycotoxicoses into the forefront of industrial, scientific, medical and veterinary attention. Which of the various fungal effects are important to the coffee industry?

(a) the allergies that can be contracted by mainly atopic, sensitive workers exposed to green coffee beans or, more specifically, to green coffee bean dust; and
(b) the potential contamination of coffee beans with mycotoxins as a result of incorrect conditions of transport or storage of coffee beans.

Mycoses, fungal diseases caused by moulds, may be contracted by the traveller who ventures into coffee-producing areas where such diseases are known to be endemic. Fungal diseases will more often be harboured by humans in an occult stage and will not be noticed by outright disease symptoms. However, immunological testing will reveal that an infection of the unsuspecting native or traveller in an endemic area has occurred. This is further described in Section 1.4.

Most stored food is susceptible to spoilage by moulds. Surveys[1,2] have shown that the world loss in food commodities amounts to about 20% of the production; more than half of this loss is as a result of spoilage by moulds, which, in the case of coffee, are responsible for a loss of an estimated $3 billion in a total net value of some $15 billion.

When conditions of climate (e.g. temperature, rain and relative humidity) and nutrient availability are ideal for specific microorganisms to multiply, a striking increase in microbial growth will occur. If such epidemic growth of microorganisms is also accompanied by the production of powerful toxins, the results can be dramatic for man, animals or plants. The elimination of entire plant species in certain regions of the world has resulted from such microbial epidemics. The near total elimination of grapes by downy and powdery mildew (*Peronospora viticola* and *Oidium*) at the turn of this century in the United States, especially in the Ohio Valley, is one such example.[3] The total elimination of coffee trees by the coffee rust *Hemileia vastatrix* in 1868 in Ceylon (now Sri Lanka) and India in 1869 is another example (see Volume 4 in this series). Potato blight in Europe from 1840 to 1850 caused by the phytopathogenic fungi *Phytophthora infestans* and *Fusarium solani* destroyed an important staple food and led to famines. According to Bischoff,[4] about 800 000 people died in Ireland from 1846 to 1849 because of famine and typhoid fever. About three million people emigrated between 1846 and 1857 from Europe to the United States mainly because of the famine caused by the potato blight. Men and animals can also be directly and severely affected by the epidemic growth of microorganisms and the formation of mycotoxins. The diseases (mycotoxicoses) caused by the ingestion of mycotoxins are exemplified by the large-scale epidemics of ergot, known as St Anthony's fire, that occurred in Europe during the Middle Ages. Ergot is caused by the infection by the parasitic fungus *Claviceps purpurea* of rye or wheat grown for bread-making, and the subsequent production of toxins.[5] The resulting symptoms are initial sensations of cold hands and feet, followed by intense burning sensations. In advanced cases, the extremities become gangrenous and necrotic; patients also suffer from convulsions. All these symptoms are caused by the complex mixture of ergot alkaloids which are for the most part derivatives of lysergic acid.[6]

Joffe[7] reported that sporadic outbreaks of alimentary toxic aleukia (ATA), also called septic angina, have been recorded in Russia since 1913 and have been known to have occurred before that time in eastern Siberia as a food intoxication. During the spring of 1932, ATA reappeared in several districts of western Siberia. The symptoms included fever, haemorrhagic rash, bleeding from the nose, throat and gums, necrotic angina, and exhaustion of the bone marrow. Mortality was about 60% as entire families and villages in agricultural areas were affected. The same disease reappeared more broadly and severely during the period from 1942 to 1947. However, it was not until 1960 that the disease was understood to be caused by fungal toxins, mainly produced by *Fusarium poae* and *F. sporotrichioides* growing on millet, wheat and other grains kept through the winter in the fields and subsequently used for the production of flour for baking bread.

In Japan, research on mycotoxins was stimulated by the assumption made by Sakaki in 1891 that disease beri-beri was caused by mouldy rice. Miyaki *et al.*[8] and Tsunodai and Haruki[9] started in 1940 to work with mouldy rice, and later, through animal feeding tests, it became evident that artificially mould-infected rice was indeed toxic.

According to Graham,[10] more than 5000 horses died of nervous disorders (necrosis of the spinal cord) in Illinois, USA, in the winter of 1933–34 after being fed mouldy grain.

Balkan endemic nephropathy, first described over 30 years ago, which was confined to the areas of Yugoslavia, Romania and Bulgaria close to the River Danube and its tributaries, claimed over 20 000 victims. In 1960 Dimitrov suggested that fungal toxins were involved. An epidemiological study of this disease by Barnes *et al.*[11] led to the suggestion that the mould *Penicillium verrucosum* var. *cyclopium* was responsible for these outbreaks.

There are other examples of crises of mycotoxicoses, such as the aflatoxin poisoning in Thailand[12] and in Uganda,[13] but none had a more stimulating effect on the awareness of the continuous existence and sporadic outbreaks of microbe-caused toxicoses than turkey X-disease in England in 1960.[14] This disease caused the deaths of more than 100 000 turkeys that had been fed peanut meal imported from Brazil and various regions of Africa that was strongly contaminated with aflatoxins produced mainly by the fungus *Aspergillus flavus*.

According to Lillehoj *et al.*,[15] 'If a disease can be said to have a value, aflatoxicosis may be credited with drawing attention to the need for increased scrutiny of fungal contamination of the world's food and feed supply.' Vigorous activity in mycotoxin research developed as a direct result of concurrent outbreaks of disease in poultry and in fish.[16]

The 1985 world crop of coffee, according to USDA estimates, amounted to 93·6 million bags. An estimated 70 million bags are exported from the grower countries to all parts of the world. Systematic examination and screening of coffee beans for fungal toxins as a precautionary measure were started in 1964 by Levi and Borker.[17] Earlier reports on the presence of moulds in coffee were given in 1946 by Heim,[18] in 1961 by Natarajan *et al.*[19] and Majumder *et al.*,[20] and in 1953 by Nicot.[21] Levi and others continued this work, which is more fully discussed in Section 3 and is also described in Chapter 1. Levi[22] concluded her studies of mycotoxins in coffee beans in 1980. She stated that the probability of mycotoxins occurring in commercial coffee beans is very low because infected coffee beans produce strong off-flavours which will affect product quality adversely. Moreover, mould proliferation in green coffee beans could be readily prevented by controlling temperature, humidity and access of air.

The survey of nearly 1500 green coffee bean samples (see Table 12 on p. 300) showed that the frequency of aflatoxin, ochratoxin and sterigmatocystin contamination was extremely low, and when contamination did occur the level was low and was reduced further by roasting. Therefore the probability of finding these mycotoxins in the coffee beverage was also very low. Certain countries, such as the USA through the Food and Drug Administration (FDA), have rigorous procedures, based on sampling and minimal counts, to prevent importation of green coffee consignments containing mouldy coffee beans, though not necessarily carrying toxins. It is relatively easy to detect the presence of mycotoxins and to analyse for down to nanogram (10^{-9}) levels per gram.

1.3. Growth of Fungi

1.3.1. Nature of Fungi

The fungi are a group of small, living, eukaryotic organisms. In contrast to the usually even smaller and more primitive prokaryotic organisms such as bacteria, they have membrane-bound true nuclei, mitochondria and chromosomes that contain complex structures of deoxynucleic acids (DNA) and smaller amounts of ribonucleic acids (RNA). Prokaryotic cells usually have only a single molecule of DNA. Fungi are different from plants because they lack the anatomical organisations of roots, stems and leaves. They are also different from other eukaryotes such as algae in being heterotrophic, i.e. they lack the chlorophyll that enables eukaryotes, like plants, to utilise the energy of light to convert CO_2 and water, in an autotrophic fashion, to carbohydrate. Fungi, according to the classifications of Whittaker,[23] Rippon[24] and Brock,[25] belong to the third kingdom of living organisms after Protista (II) and Monera (I), and before Plantae (IV) and Animalia (V).

In order to obtain food and to maintain and disseminate their species, fungi, similar to the bacteria, have highly specialised enzyme systems. With the help of these enzyme systems, they decompose cellulose, lignin, chitin and protein of their hosts, and often even synthetic organic materials to use as their food. The transport of the food into the organism is carried out by diffusional processes (active transport) or by capture of sufficiently small food particles (such as emulsion droplets) by the cell walls of fungal hyphae and the indentation or invagination of the cell wall (phagocytosis) at the attachment site. This is followed by total enclosure of the food particle by the cell wall, and subsequent separation from it, to form an independent vesicle. By detaching from the cell wall and moving to the interior of the cell, the independent vesicle becomes a digestive system. After enzymatic transformation of the food particle into cellular carbon components, the digestive membrane finally disbands as indicated by the striking decrease in phospholipid and sterol ester content and also by electron microscopic observations. This ability to ingest food through the cell wall membrane differentiates fungi from animals, which are equipped with a permanent digestive system. When fungi live on dead matter, they are described as saprophytic or, preferably, saprobic. Fungi can be single-celled and can live under aquatic conditions. They may be multi-celled and may be parasites of plants (for example as rusts, wilts and blights) and animals including man. They then become the subject for study by the plant pathologist or by veterinary or medical mycologists.

Fleshy, edible macrofungi (mushrooms) are used by man as food. Other fungi (microfungi) are used to improve the texture, flavour, palatability and digestibility of foods such as cheese and in the production of many oriental fermented foods[26] such as shoju, tempeh, otjom, hamanatto and miso, to name but a few. Alcohol fermentation of grains and fruits by selected fungi (yeasts) preserves these materials from other fungal or bacterial attacks, thus transforming food materials into storage-stable goods.

Antibiotics are microbial antagonists produced by certain fungi on an industrial scale that can then be used to fight bacterial and pathogenic fungal infections in man and other animals. There are about 175 pathogenic fungi among the approximately 100 000 species known. There is a far greater number of fungi that are physiologically adapted to function, either in localised or systemic fashion, as plant parasites than there are fungi pathogenic to man or other animals. Among fungi and actinomycetes, there are about 20 species that cause systemic and potentially fatal disease, and 35 that cause systemic disease of less severity, but yet may cause severe, localised, cutaneous, subcutaneous or lymphatic infections. About 45 cause superficial infections of the skin, the mucosa and the keratinised skin layer.

The predilection of the saprobic fungi for specific types of soil enriched by excreta and for decaying organic matter (e.g. coffee beans when they either become wet or are exposed to air with a relative humidity giving rise to a moisture content above 14% w/w for prolonged periods of time), and the relationships which some of the fungi have with specific animals, are essential to an understanding of the epidemiology of the mycoses for man. It is also essential to place the potential involvement of mainly green coffee bean derived potential mycoses (Section 1.4) or mycotoxicoses into proper perspective. Most fungi are saprobes, and therefore are important as the decomposers of the ecosystem. Satchell[27] estimated the biomass of different soil microbes and litter found in deciduous woodland. For bacteria and actinomycetes he gave 2 kg as the dry weight per hectare; for fungi 454 kg (!); and finally, for soil and litter fauna, 36 kg. Lacey[28] reviewed about 200 reports and tabulated the presence of air spora in different parts of the world, which he then grouped together into climatic regions. The total number of spores in the air varies from fewer than 200 per cubic metre of air to more than two million spores per cubic metre of air. There are more spores in the air in temperate and tropical regions than at high latitudes and over dry deserts. In humid, subtropical climates (corresponding to coffee-growing regions), the conidial fungi were composed of *Cladosporium* 39%, *Alternaria* 15%, *Fusarium* 5%, *Epicoccum* 4%, *Helminthosporium* 3% and *Apsergillus/Penicillium* 1%. Spores of Ascomycetes and Basidiomycetes comprised 6% and 10%, respectively, and those of rusts and smuts 2% and 4%, respectively. The very great number of spores in the air indicates that it is actually impossible to prevent mould spores from settling on to a substrate, no matter where in the world the substance may be located.

1.3.2. Governing Factors in Growth

The germination and growth of saprobic fungi is governed by seven factors:[29]

(1) water activity,
(2) temperature,
(3) oxygen tension,
(4) hydrogen ion concentration,
(5) nutritional status,
(6) specific solute effects, and
(7) preservatives (fungistats, fungicides).

The effect of all these factors is well described in the literature, and those of the first four are summarised in Table 1. Examination of the water sorption isotherms for different food commodities will indicate the maximum percentage water content allowable for safe storage against mould growth; for example, green coffee beans show a moisture content of 13% w/w at an equilibrium relative humidity of 67% at 28°C (see Volume 1, Chapter 2, and see Table 1).

In respect of nutritional status for the growth of moulds, Pitt[29] in his 1981 review on food spoilage and biodeterioration of materials utilised or manufactured by man states:

> The more successful (in survival and expansion) a conidial fungus is, the less nutritionally demanding it must be. Materials not containing effective preservatives and subjected to high humidity will almost all eventually support mould growth. It may be of little consequence what mould [species] it is, but for the mycologist attempting to understand the

Table 1
Governing factors in the growth of moulds

Factor in growth	*Range of conditions*	*Examples*
Water activity (a_w)	Xerophilic	*Aspergillus flavus*
	—germination, <0·87–0·62	*Chrysosporium festidium*
	—optimal growth, 0·95	*Aspergillus chevalieri*
	Safe upper limit of 70% ERH for storage for food products	
	Mesophilic	*Aspergillus fumigatus*
	—germination, 0·8–0·9	
	—optimal growth, 0·95–1·0	
	Hydrophilic	*Mucor circinelloides*
	—germination, >0·9	*Trichothecium roseum*
	—optimal growth, ~1·0	
Temperature tolerance	Thermophilic	*Humicola lanuginosa*
	—maximum growth, ~50°C	
	—but also growth <20°C	
	Psychrotolerant thermophiles	*Aspergillus fumigatus, A. candidus, Absidia ramosa*
	Mesophilic	Most fungi
	—growth range, 10–40°C	
	—optimal, 25°C	
	(NB: below human blood temperature)	
	Psychrophilic	*Botrytis cinerea*
	—optimal growth, 5–10°C	
	Cryophilic	*Fusarium sporotrichioides*
	—growth possible, <5°C	*Cladosporium herbarum*
Oxygen requirement (% O_2 in headspace)	Conidial fungi	*Mucor* and
	—from 2·1% upwards high requirement	*Penicillium roquefortii*
	—no change in growth to 4·2%	*Aspergillus flavus*
	—stimulation of growth, 15–20% CO_2	
Hydrogen ion concentration (pH)	Conidial fungi and yeasts	Most fungi
	—growth range, 3–8	*Aspergillus niger*
	—optimal at 5	
	—pH 2 possible	

parameters governing mould growth, the carbon source can be of great significance.

Perlman[30] has listed more than 150 different carbon sources known to be utilised by fungi, many as sole carbon sources. Missing from Perlman's list are cellulose, lignin, rubbers and hydrocarbons, compounds which, according to Pitt, are of great importance for biodeterioration. For fungal attack on coffee beans, hydrocarbons in the form of cuticle wax as the outermost protective layer (see Volume 1, Chapter 6) and cellulose-like materials as the next layer constitute the most accessible carbon sources for fungal infestation by fungal spores attached to the outside of the coffee bean. According to Pitt,[29] conidial fungi will utilise almost all sources of nitrogen whether it comes from nitrate, ammonium ions or organic nitrogen compounds. In support of this view, Strobel *et al.*[31] showed that certain *Fusarium* spp. did not show any growth increase or retardation when supplying the moulds with nitrate or nitrite. As an exception to the general rule, Ormerod[32] showed that a few xerophilic mould species are somewhat more finicky, preferring yeast autolysate, an amino acid mixture, or tryptone. Proline was identified as the specific nitrogen compound responsible for growth stimulation.[32] In 1975 Pitt[33] also defined additional xerophilic species (*Chrysosporium*) as nutritionally fastidious. However, the critical nitrogen compounds were not identified.

Fries,[34] in his 1965 review of organic growth factors and vitamins, stated that very few biodeteriorative and food spoilage fungi have any specific growth factor requirement. In 1981 Pitt wrote that this fundamental fact had remained essentially unchanged.

In respect of specific solute effects, whereas slow or no growth of moulds can mostly be attributed to the water activity conditions (a_w; see Table 2), some solutes present in foods can have additional effects on growth rate, according to Pitt.[29] When Scott, in 1957,[35] used sucrose or glucose for water activity adjustment rather than glycerol or inorganics such as sodium chloride or magnesium chloride, *Aspergillus amstelodami* grew 50% faster at its optimal a_w of 0·96. Similar growth inhibitions were observed on *A. chevalieri* and *Chrysosporium fastidium* by Pitt and Hocking.[36] On the other hand, the same publication, and also Hocking and Pitt,[37] showed that neither germination nor growth of a number of *Aspergillus* and *Penicillium* species indicated that they had a negative sensitivity to glycerol or the two inorganic compounds.

When reviewing all the above parameters and their specific influences on mould germination and growth, it is not too surprising that, in a given environment, there are only relatively few moulds that are governed by a specific environment. Such specific growth of only a few mould species or, in some instances, of virtually only one species, occurs in alcoholic fermentation, where candida yeasts hold the ruling position so long as an anaerobic environment is maintained. It also occurs on rotting leaves where almost exclusively *Aspergillus niger* is found. Of course, the formation of alcohol during the fermentation process is a compelling factor for the prevention of germination and growth of many other microorganisms.

1.3.3. *Formation of Secondary Metabolites*

This section deals with the phenomenon of why some fungi, under certain circumstances, produce antibiotics which are beneficial to man and other animals, as well as toxins (called mycotoxins) which are noxious to man,

Table 2
Minimum water activities (a_w) permitting germination of conidial fungi[29,38–40]

Fungus	*Minimum reported* (a_w)	*Time (days)*	*Temperature (°C)*
Fungi (in general)	0·95–0·75	ns[a]	ns
Alternaria citri	0·84	ns	ns
Aspergillus candidus	0·75	14	25
Aspergillus echinulatus	0·62	730	25
Aspergillus flavus	0·78	ns	33
Aspergillus fumigatus	0·82	ns	40
Aspergillus nidulans	0·78	ns	37
Aspergillus niger	0·77	ns	35
Aspergillus ochraceus	0·77	57	25
Aspergillus oryzae	0·85	14	25
Aspergillus terreus	0·78	ns	37
Aspergillus versicolor	0·78	24	25
Botrytis cinerea	0·93	2	20
Cladosporium herbarum	0·88	7	20
Fusarium solani	0·90	56	20
Fusarium sporotrichioides	0·88	56	20
Fusarium oxysporum	0·89	56	20
Geotrichum candidum	0·90	15	20
Penicillium aurantiogriseum (= *P. cyclopium*)	0·81	120	25
Penicillium brevicompactum	0·78	24	25
Penicillium citrinum	0·80	14	25
Penicillium cyclopium	0·81	ns	ns
Penicillium expansum	0·82	22	25
Penicillium herquei	0·88	6	25
Penicillium implicatum	0·78	10	25
Penicillium islandicum	0·83	ns	ns
Penicillium oxalicum	0·86	5	23
Penicillium simplicissimum	0·86	11	25
Penicillium viridicatum (= *P. verrucosum*)	0·80	21	25
Rhizopus nigricans	0·93	ns	ns
Rhizoctonia solani	0·96	ns	ns
Sporendonema	0·77	ns	ns
Trichothecium roseum	0·90	4	20
Wallemia sebi	0·75	ns	22
Xeromyces bisporus	0·61	ns	ns
Yeasts (in general)	0·95–0·86	ns	ns
Osmotolerant species	As low as 0·62	ns	ns
Bacteria (in general)	>0·90	ns	ns
Halophilic species	As low as 0·75	ns	ns

[a] ns = Not stated.

other animals and plants. Special attention, however, is directed to the growth of fungi and fungal toxins in general, and to the growth of moulds and the formation of secondary metabolites (mycotoxins) in coffee beans. Attention is also paid to the quantitative determination of fungal toxins. An important fact has to be emphasised: mycotoxins and antibiotics produced by fungi are only produced when certain circumstances are met.

As noted by Moss,[41] primary metabolites are absolutely necessary for the growth and functionality of organisms, ranging from prokaryotes, to eukaryotes, to plants and animals. Encompassing this diverse range and size of organisms, there are but a few basic cellular building blocks involved in their primary metabolic functionality. These include about 20 different amino acids involved in protein biosynthesis, about 15 pyrimidines (e.g. cytosine, uracil, thymine, 5-methylcytosine and 5-hydroxymethylcytosine) and purines (inclusive of guanine, adenine, xanthine, hypoxanthine and uric acid, all of which are involved in the structure of DNA and RNA), a finite number of carbohydrates and derivatives thereof, and so on.

The primary metabolites and their pathways of origin and destination are reasonably well known; the production of all these diverse components is well regulated and essentially identical for all organisms. Primary metabolites have therefore been termed by Martin and Demain[42] 'general metabolites'. In contrast, secondary metabolites, according to Bu'Lock,[43] are genotypically and phenotypically specific. The same author[44,45] also made the differentiation that the primary metabolites are produced during the tropophase, the growth phase which sometimes occurs in an exponential fashion. The production of secondary metabolites, however, occurs during the 'peculiar phase' or 'idiophase', which follows the tropophase. There is some possible overlap of the two phases during which growth of secondary metabolites occurs, as defined by Aharonowitz and Demain.[46] The general pathways of secondary metabolites from primary metabolite precursors are shown in Fig. 1.

Two main questions come to mind:

(a) Why do some organisms produce secondary metabolites and others do not?
(b) Under what circumstances and when are secondary metabolites produced by those organisms capable of secondary metabolite production?

Smith and Moss[51] state categorically that 'it has not generally been possible to rationalise the biological function of secondary metabolites, although some are very active against other microorganisms (antibiotics),

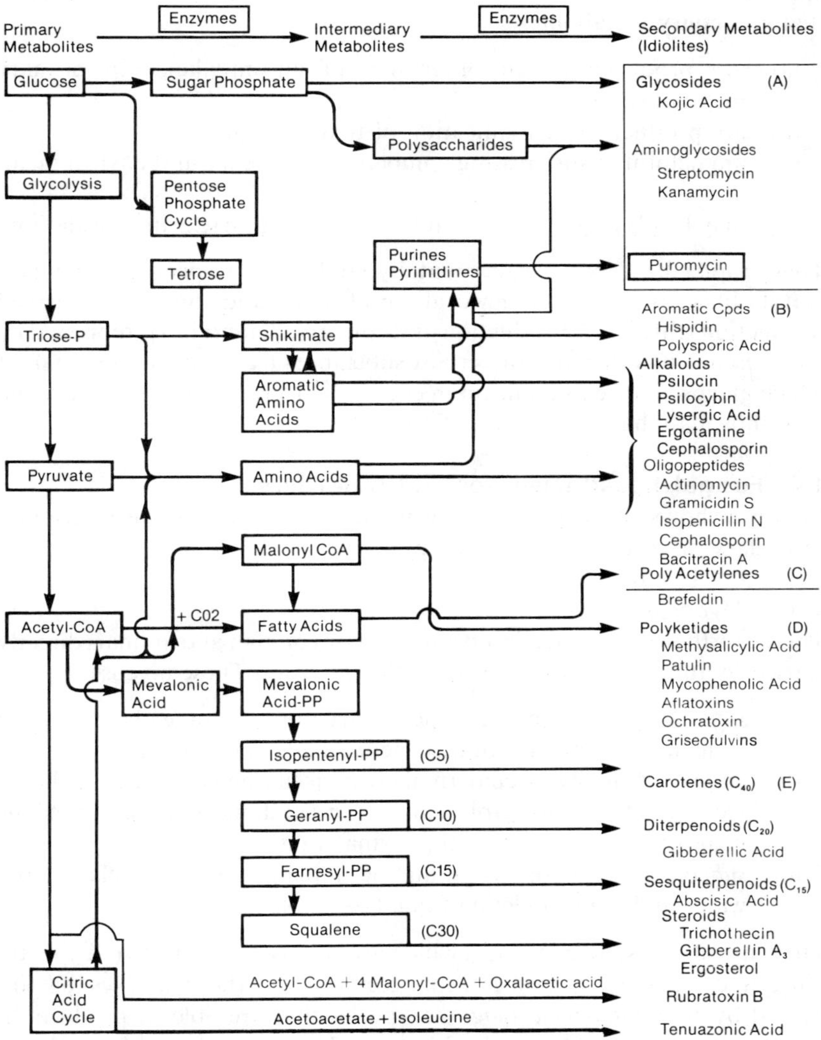

Fig. 1. General pathways of secondary metabolites from primary metabolites.[47–50]

plants (phytotoxins) or animals and humans (mycotoxins)'. Zähner *et al.*[47] approach the same theme from a somewhat different angle. They speculate that secondary metabolites:

(a) are produced as an outgrowth of an imbalance in primary metabolism;
(b) are products of detoxification of toxic intermediates;
(c) are helpful in suppressing competitors in the natural environment; and
(d) may be playing a role in cellular differentiation (e.g. in sporulation).

These topics have been experimentally studied, in respect to secondary metabolites, both under natural conditions and on an industrial production scale.[52–56] Production of secondary metabolites is repressed by the presence or otherwise of certain substances (catabolic repressors), of which glucose is often significant in effect.[57,58] Certain sources of nitrogen and phosphate have a negative effect.[59–61]

1.4. Fungal-based Diseases and Disorders

Figure 2 shows the sequence of events leading to mycosis, primary and secondary mycotoxicosis, and allergy.

1.4.1. Mycoses

There are about two dozen important diseases of man and animals directly caused by fungi; these diseases are called *mycoses*. These mycoses can be

(a) *superficial*, involving only 'dead' structures or outer layers of the skin such as hair, stratum corneum, nails and feathers;
(b) *cutaneous*, involving epidermal layers as in ringworm of scalp, body, feet (athlete's foot), groin, etc., or in candidiasis of skin, mucous membranes (tongue, mouth, vagina, anus); or
(c) *subcutaneous*, involving traumatic implantation of unrelated soil, water and plant-infecting fungi (see Table 3).

There are also systemic mycoses which are subdivisible into two distinctly different categories. The first category contains only those diseases that are caused by *true pathogenic fungi*, i.e. species that are able to produce an infection process in a normal, healthy host. These are tabulated in Table 4. Coccidioidomycosis (Table 4) is known to occur in severe forms in Filipinos and pregnant women. Severe disseminated histoplasmosis (Table 4) has been recognised in patients suffering from hypogammaglobulinaemia.

In the second category, caused by *opportunistic or adventitious fungi*, are

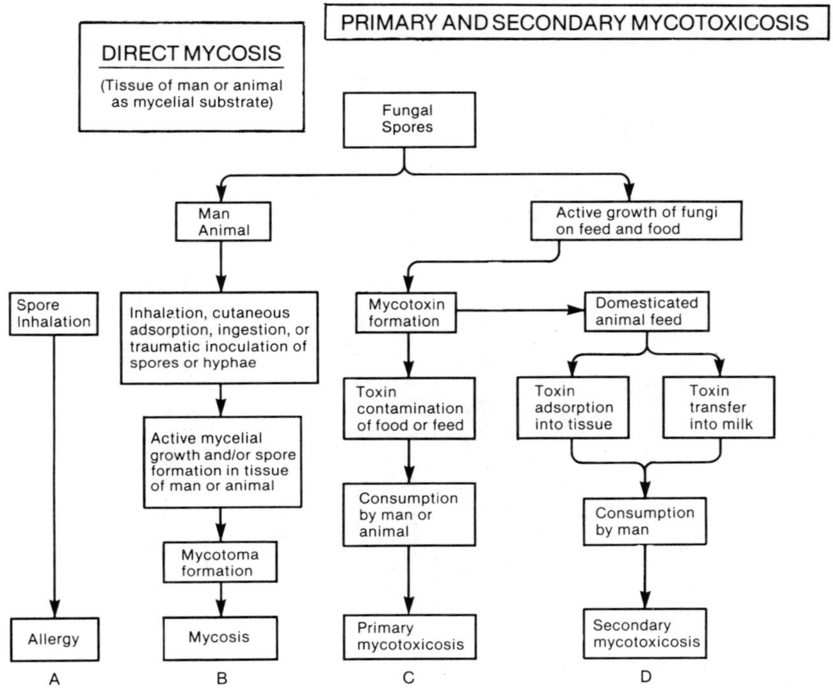

Fig. 2. Events leading to allergy, mycosis and primary and secondary mycotoxicosis. A, Spore inhalation without ensuing spore germination may cause allergy. B, If inhalation, etc., or traumatic implantation of spores leads to spore germination and mycotoma formation, then a mycosis exists. C, A primary mycotoxicosis results if mycotoxin is directly ingested. D, A secondary mycotoxicosis results when, for example, a domesticated animal ingests mycotoxin-contaminated feed and the mycotoxin is then excreted into milk or deposited into meat tissue and consumed by man.

four common types of infections, as tabulated in Table 5. The disease-causing fungi of these mycoses do not possess sufficient virulence to infect the normal host, but they are capable of attacking the debilitated, disease-weakened host or the host under severe stress. Most of the systemic, lymphatic or subcutaneous mycoses are caused by fungi that are normally saprobes living in decaying vegetation, humus, bird or animal excreta, or humus or soil enriched with bird or animal excreta. Such saprobic fungi, which can suddenly become pathogenic, are called by Ahearn (Scientific Publication, PAHO, No. 205, 1970) 'adventitious' fungi; more conventionally, they are called opportunistic fungi (Table 5).

Table 3
Characteristics of some subcutaneous mycoses[24]

Medical term	*Symptoms*	*Causative agent*	*Region of occurrence*
Chromoblastomycosis (verrucous dermatitis)	Wart-like growths or hypoplasia on the skin	From implantation of *Phialophora verrucosa, Cladosporium carrionii*	Tropics, among barefoot-walking population
Mycotic mycetoma	Tumour-like swelling and sinus tracts which rupture and extrude	Bacteria with fungi-like growths. *Nocardia brosithensis, Actinomedint streptomyces somcliensis*	As above
Entomophthoromycosis	Subcutaneous nodules	Fungal: *Madurella mycetomatis, Leptosphaeria senegalesis*, etc., *Basidiobolus haptosporus* via insect bites	Uganda, Sudan, Kenya, Nigeria, Iraq, India, Indonesia
Rhinosporidiosis	Lesions in the nose, forming polyp-like growths; also in or on other organs, e.g. anus, vulva, penis, larynx, pharynx	*Rhinosporidium seeberi*, aquatic and also present in sandstorms	India, Ceylon, Argentina, Central America, Brazil, mainly India and Ceylon, >2000 cases reported
Lobomycosis	Painless nodules, mostly on feet first, then spreading to other sites, even lymph nodes	*Loboa loboi* from insect sting and other skin trauma (mostly feet)	Brazil (Mato Grosso), South and Central America (agricultural workers)
Sporotrichosis	Wide spectrum of disease types	*Sporothrix schenkii* by various traumatic skin infections, also inhalation of conidia. Present in dry *and* humid areas	Mexico, Guatemala, Central America, Columbia; in 1900s in France, South Africa

Table 4

Characteristics of mycoses from true pathogenic fungi[24]

Medical term	*Symptoms/treatment*	*Causative agent*	*Region of occurrence*
Histoplasmosis	95% of infections are subclinical. Pulmonary and some wider effects, leaving immunity and possibility of hypersensitivity. Symptomatic disease treated with amphotericin B	*Histoplasma capsulatum* growing on starling droppings, and on bat and chicken manure. Agent distributed by winds and inhaled	Widely distributed, including Mississippi and Ohio Valley areas
Blastomycosis (Gilchrist's disease, Chicago disease)	Pulmonary primary stage and osseous stages (ribs, etc.). Often also verrucous lesions in skin exposed to sunlight. Amphotericin B is drug of choice	*Blastomyces dermatitis* growing on tree bark and pigeon manure. Disease caused by inhaling conidia	North America (especially Kentucky), but many other areas in South America and Africa
Coccidioidomycosis (Pesode's disease)	Usually pulmonary and may cause lung damage, but when dissemination occurs almost every tissue of the body may be involved; in most cases recovery occurs. Amphotericin B is drug of choice	*Coccidiodes immitis* in soil of carbonised organic material, high salt concentration, which is usually prohibitive for other microorganisms which prefer rich, decomposing fauna in humid areas. Inhalation but also direct percutaneous inoculation	California, but other areas with desert-like (hot/dry) environments. Caught by archaeology students digging in endemic areas
Paracoccidioidomycosis	Pulmonary lesions and spreading to visceral organs ranging from subclinical to chronic and sometimes fatal	*Paracoccidioides brasiliensis*. Has large hyphal diameter (30–60 μm) and easily detected by microscope. Inhalation	Endemic areas in Brazil (over 3000 cases) but also coffee-producing regions ranging from central Mexico to southern Brazil. High, sylvatic regions with high rainfall (600–2000 mm)

Table 5
Characteristics of mycoses from adventitious fungi[24]

Medical term	*Symptoms*	*Causative agent*	*Region of occurrence*
Mucormycosis	Infection may begin in paranasal sinuses but can spread as cellulitis to brain and meninges. Mainly associated with malnutritions, diabetes, etc.	Mainly *Hyphomyces destruens*, *Basidiobolus ranarum*, present in dung of herbivorous animals, decaying plant material (such as sugar-containing fruit)	Tropical and subtropical zones
Candidiasis	Wide disease spectrum from thrush to, more rarely, serious meningitis, etc., but generally only in debilitated hosts for various reasons	*Candida albicans* (a yeast). Member of normal flora of the surface of the body and some cavities/tracts. Infection may be cutaneous, involve mucous membranes, subcutaneous or systemic	General
Cryptococcosis	Pulmonary, but generally subclinical. Only occurs when other diseases are present. Has, however, a predilection for central nervous system	*Cryptococcus neoformans* associated with filth of old buildings and where pigeons roost	Europe and USA, 200–300 cases per year of cerebral meningitis
Aspergillosis (1) Allergic aspergillosis —asthmatic alveolitis (2) Colonising aspergilloma (3) Invasive	Coughs, wheezing (similar to effect of other conidia, pollen, etc.) in allergic individuals, immediate. In normal individuals, delayed, from massive insult. See Section 2. Formation of fungus balls in the lung, following other diseases. Uncommon, except for patients with leukaemia, etc.	*Aspergillus* spp. (some 600 spp.) with *Aspergillus fumigatus* the most prevalent	Ubiquitous

It has become quite obvious that diseases due to fungi occur more frequently under rather special conditions. People afflicted with Hodgkin's disease will relatively frequently contract cryptococcosis (Table 5). Patients with acute leukaemia quite often have severe systemic fungal infections, mostly aspergillosis and candidiasis (Table 5). Orbital–cerebral mucormycosis (Table 5) has been diagnosed in patients with diabetic acidosis. Organ transplant patients requiring prolonged treatments with immunosuppressive drugs and with corticosteroids will be prone to aspergillosis, and occasionally to cryptococcosis and histoplasmosis. X-Ray therapy has been known to predispose towards severe microbial infections, including fungal infections. Stress is considered a causative factor for mycoses. Aspergillosis is an important agent of opportunistic infections. Systemic disseminated diseases caused by aspergilli are termed by Rippon[24] 'diseases of medical progress' since they have essentially been brought into being in the antibiotic era.

The two major physiological barriers to fungal growth within tissue, according to Rippon,[24] are temperature and redox potential. Most fungi are mesophilic (see Section 1.3) and have an optimal growth range considerably below body temperature. Also most fungi are saprobic and their enzyme pathways function more efficiently at the oxidation–reduction potential of non-living substrates. Furthermore, the body has a highly efficient set of cellular defences to guard against fungal proliferation. Rippon[24] states that thousands of cases of true fungal infection by *Histoplasma* and *Coccidiodes* occur each year, but only a few ($\sim$1%) result in a definite disease process. For opportunistic or adventitious fungi of lesser virulence, such as *Cryptococcus*, *Aspergillus*, *Mucor*, etc., the defect in the host's defence and/or the exposure must be even greater.

1.4.2. Mycotoxicoses

Mycotoxicoses are diseases caused by a wide range of mycotoxins after inhalation, ingestion or skin contact, and afflict a broad range of body organs. Pier *et al.*[62] in 1980 categorised the effects of mycotoxins into three different areas, depending on dose-related exposure of the disease:

(a) primary *acute* mycotoxicoses,
(b) primary *chronic* mycotoxicoses, and
(c) secondary diseases caused by exposure to relatively small quantities of mycotoxins mostly over longer periods of time.

Primary acute mycotoxicoses are produced when relatively large quantities of mycotoxins are consumed over a very short period of time.

The ensuing disease is very marked and relatively easily identifiable, and rapid death often occurs.

Primary chronic mycotoxicoses are produced by sub-lethal ingestion of mycotoxins. The diagnosis of such mycotoxicoses is rather difficult since the symptoms are not nearly as clear and obvious as in the acute type.

In farm animals (cattle, pigs, poultry) the response to chronic mycotoxicoses is expressed in a range of symptoms such as reduction of weight gain, of milk production, of feed intake and of egg production, and also coagulopathy, immunosuppression and non-functionality of vaccines.[63]

Secondary diseases caused by mycotoxins: Smith and Moss,[64] in their review of mycotoxicoses, state that 'consumption of low levels of certain mycotoxins can lead to the impairment of the native and acquired resistance to infectious diseases'. Thus the diminished resistance to infectious diseases to which animals (and humans) are naturally exposed leads to greatly increased disease susceptibility. (It would be of great interest to study whether or not the epidemic-like outbreaks of viral or bacterial diseases are preceded and/or initiated by increased exposure to low levels of mycotoxins.)

Immunodeficiency caused by low levels of, for example, aflatoxin is apparently associated with a suppression (failure) of the cell-mediated immune system. At slightly higher levels of mycotoxin, the production of antibodies is also assumed to be impaired. Aflatoxin consumption has been associated with increased opportunistic susceptibility to candidiasis, Marek's disease, coccidiosis and salmonellosis in chickens; pasteurellosis and salmonellosis in turkeys; erysipelas and salmonellosis in pigs; fascioliasis and clostridial infections in calves; and intramammary infections in cows.

A deadly horse disease, initially thought to be caused by the bacterium *Pyosepticum viscosum*, was found to be a secondary disease occurring after the horses were fed with barley straw heavily overgrown with *Stachybotrys alternans*.[65]

In addition to the aflatoxins, which belong to the decaketides (aflatoxin G_1 included), other polyketides are also carcinogenic, namely:

(a) the tetraketides (patulin and penicillic acid),
(b) the pentaketides (ochratoxin A (if associated with sterulic acid) and citrinin),
(c) the octaketides (luteoskyrin and rugolosin), and
(d) the decaketide (sterigmatocystin).

Table 6
Comparison of bactericidal, mutagenic and carcinogenic activity of aflatoxins, aflatoxin metabolites and biosynthetic precursors of aflatoxins[66–68]

Compound	*Carcinogenicity*	*Relative mutagenicity (%)*	*Bactericidal properties*
Aflatoxin B_1	Most potent liver carcinogen in trout and rats	100	++++
Aflatoxin B_2	No carcinogenicity (in the rat, weak for trout)	0·2	—
Aflatoxin G_1	In trout and rats, not as active as B_1	3·3	+++
Aflatoxin G_2	No tumours in trout	0·1	+
Aflatoxin M_1	About 33% as active as B_1 in trout	3·2	
Aflatoxin Q_1	No tumours in trout	1·1	
Aflatoxin R_0 (Aflatoxicol)	About 50% as active in B_1 in trout		
Aflatoxin B_{2a}	No activity when administered per os (but see Figs 3 and 4)	0·0	—
Aflatoxin G_{2a}	—	0·0	—
Biosynthetic precursors			
Sterigmatocystin	Strong liver carcinogen	10·7	
Versicolorin A		5·8	
Versiconal hemiacetal acetate		0·2	
Averufin		0·2	
Norsolorinic acid		0·2	

The various aflatoxins are of vastly differing carcinogenic activity. This was pointed out by Wong and Hsieh,[66] Wong *et al.*[67] and Reiss,[68] as shown in Table 6, where the carcinogenicity, mutagenicity and bactericidal properties of aflatoxin, aflatoxin metabolites and aflatoxin biosynthetic precursors are tabulated.

Aflatoxin B_1, after biochemical transformation to the 8,9-epoxide derivative, shows a reactivity with nucleic acids in decreasing order:[69] DNA > RNA > poly-guanilic acid > poly-adenilic acid > poly-uridilic acid > poly-cytidilic acid. Mononucleotides and proteins showed only weak interaction.[70]

In 1977 Essigman *et al.*[71] determined by *in vitro* experiments that, for each 60-DNA-nucleotide residue, one aflatoxin B_1-epoxide molecule is attached. Based on *in vivo* experiments, it is proposed that by binding covalently to the guanine residue the double helix of the DNA, held together by hydrogen bonds, can dissociate and thus mutations can arise.[72]

According to Miller[73] in 1970, many carcinogens are actually pre-carcinogens. After biochemical transformation in the host cell, the carcinogenic properties are created. Namely, the carcinogenic biochemical transformation always produces compounds which are of pronounced negatively polarised/charged nucleophilic character and therefore are ideally capable of interacting in covalent fashion with positively polarised/charged electrophilic compounds present in the host cells. Such electrophiles are proteins (enzymes), DNA and RNA. See Chapter 10 for further information on this subject.

1.4.2.1. Primary, acute human aflatoxicosis. In 1971 Crawford calculated that in order to produce acute aflatoxicosis a person would have to eat about 16 kg of peanuts heavily infested with aflatoxin. In those countries with progressive agricultural methods where agricultural commodities are properly harvested, dried, stored prior to distribution and transported under dry conditions to the consumer, there have been relatively few fatal cases (four) of acute aflatoxicosis.[68,74]

1.4.2.2. Chronic aflatoxicosis. There are certain areas in the world where aflatoxin-producing moulds find nearly ideal conditions for growth. Such regions are usually the warmer and more humid ones. From epidemiological studies it appears quite certain that the ingestion of small quantities of aflatoxins over a longer period of time will cause chronic diseases. Some of these diseases are discussed in subsequent paragraphs.

Primary liver cancer is directly caused by the action of the mixed function oxidases in the liver microsomes, which transform non-carcinogenic aflatoxins into carcinogenic oxidation products (e.g. aflatoxin B_1-8,9-epoxide and aflatoxin G_1-8,9-epoxide), as illustrated in Figs 3 and 4. The reaction of these carcinogens is direct (not via metastases), and therefore the organ (the liver) that is capable of transforming xenobiotic aflatoxins into carcinogens is directly attacked and becomes the cancerous site.

Primary liver tumours are rarely found in the colder or intermediate climates, but are frequently found in Africa (south of the Sahara Desert, south of Egypt). As shown in Table 7, in the less humid, less warm climates, only about 1–5 persons per 100 000 suffer from primary liver cancer, whereas the warmer, more humid climates show a drastically increased rate of liver cancer.

It was found that the increase in primary liver cancer was directly correlated with the quantity of aflatoxin ingested via aflatoxin-contaminated food. In Kenya, with about 0·5 ng aflatoxin daily uptake per kilogram body weight, only one person in 100 000 contracted primary liver cancer. At the high end of the scale, in Mozambique, where the daily uptake is about 2·5 ng kg^{-1} body weight, the incidence of primary liver cancers is very much higher.[79,80] There is fairly good consensus that the quantity of

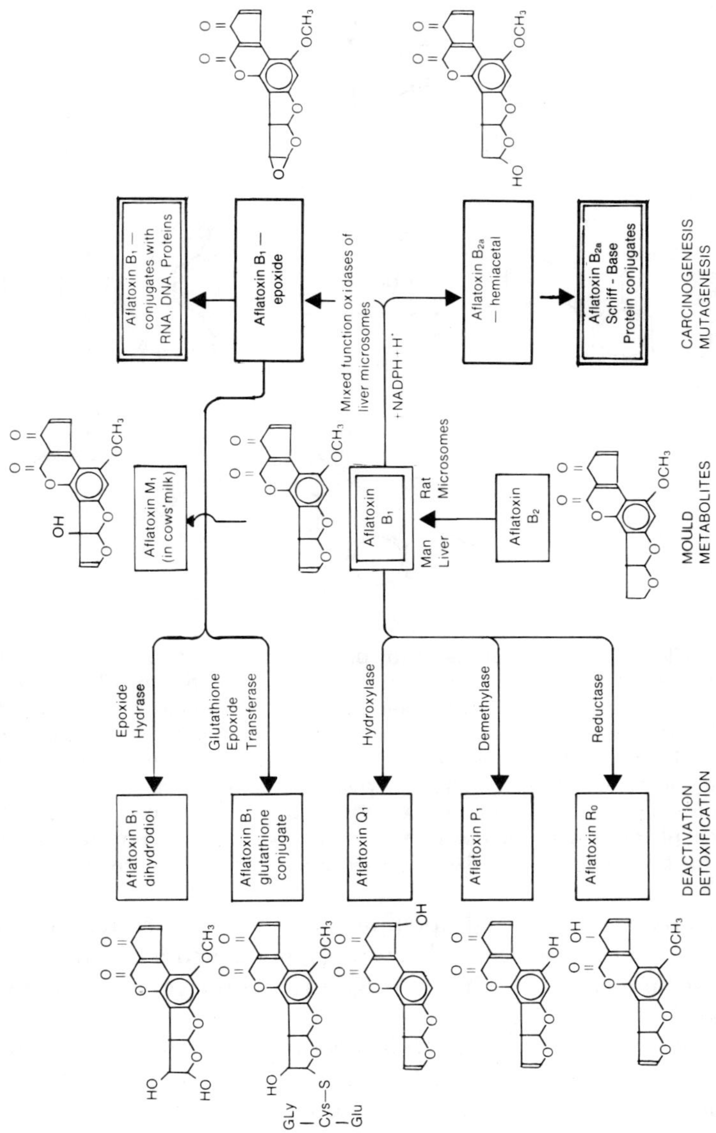

Fig. 3. Transformation of aflatoxins by liver enzymes (modified from refs 68, 75 and 76).

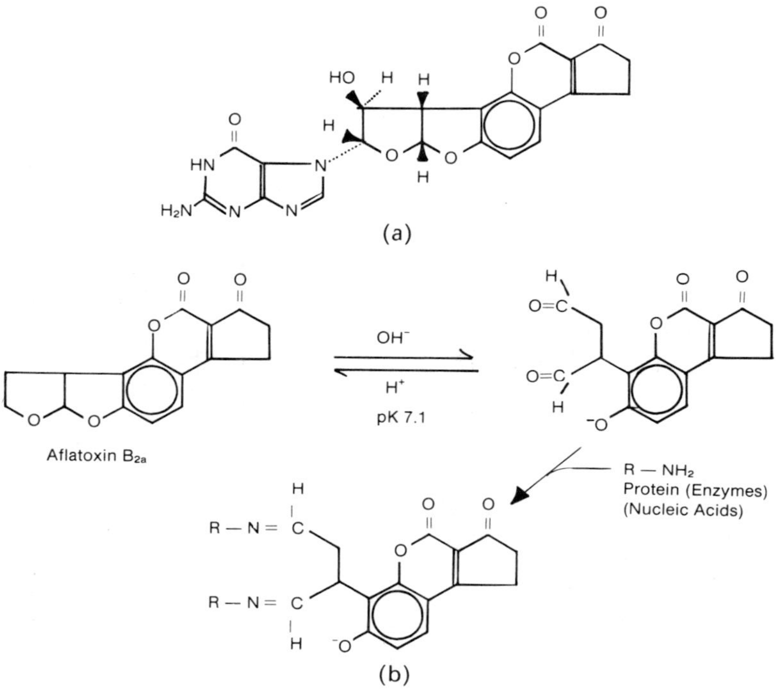

Fig. 4. Aflatoxin B_1—guanine and protein conjugates. (a) Guanido-conjugate with aflatoxin B_1 epoxide.[77] It is the transformation of aflatoxin B_1 to the corresponding epoxide by the mixed oxidases in the liver microsomes that causes the pronounced formation of the hepatic carcinoma. (b) Campbell and Hayes[76] in 1976 and Patterson[78] in 1977 called the aflatoxin the *acute toxin* since they postulated that the strong, covalent interaction of the dialdehyde with the free amino groups of amino acids, peptides, proteins and especially enzymes at physiological reaction condition are perhaps even more responsible for severe acute liver damage than the aflatoxin B_1 epoxide.

daily aflatoxin-contaminated food intake is directly correlated with primary liver cancer incidence;[80] however, a multifactorial aetiology that includes the presence of the hepatitis B virus as a prerequisite is strongly indicated.[81]

Indian childhood cirrhosis (ICC) and hepatitis: According to Shank,[74] ICC is a disease restricted to specific regions of India, south-east Asia and the Middle East, and is most likely to occur in direct connection with aflatoxin-contaminated food. The disease peaks in children three years of age. It is characterised by fatty infiltration, degeneration and fibrosis. In

Table 7
Distribution of primary liver cancer[50]

Primary cancer cases per 100 000 people	*Location*
~1–5	Canada, Norway, Holland, Denmark, USA
9	Africa (Nigeria)
14	Thailand
19·2	Africa (Bantus in Johannesburg)
31	Far East (Chinese males in Singapore)
98·2	Mozambique (males)

advanced stages jaundice and ascites set in; the disease usually ends in hepatic coma.

Indian children suffering from kwashiorkor had 30–60 g of peanut meal per day per child added to their diets. Twenty children, 1·5 to 5 years old, consumed the supplement for 5 to 30 days before it was discovered that the peanut meal was contaminated with 300 μg aflatoxins per kilogram supplement. On this discovery, 16 of the children were found to have developed soft hepatomegaly. The diet was stopped at that stage. Two months thereafter, the soft hepatomegaly had progressed to firm hepatomegaly. Four months after withdrawal, perilobular fibrosis was detected. Ten months after withdrawal, complete loss of lobular morphology, faecal necrosis and bile-duct proliferation were evident from liver biopsies. These morphological liver changes described above are identical to those in the Indian childhood disease, but are not observed in victims of kwashiorkor.

Reye's syndrome is a liver disease. It reaches epidemic proportions in north-eastern Thailand, but has also been reported from Taiwan, New Zealand, Czechoslovakia, Uganda, Puerto Rico, Canada and the southern and central United States. It mainly affects children. Since, in this disease, relatively low levels of aflatoxin (e.g. 1·5 mg) cause acute illness, Bourgeois *et al.*[82] suggest that Reye's syndrome is initiated first by a toxic, nutritional or infectious insult. Such an insult causes subclinical liver injury. When a second insult by relatively small quantities of aflatoxin occurs, *acute* aflatoxicosis is experienced. The morphological changes involve cerebral oedema and fatty degeneration of the liver, heart and kidneys. The symptoms are malaise, followed by vomiting, hypoglycaemia, convulsions, hyperammoniaemia and coma. In severe cases death may set in from 6 to 48 h after onset.

1.4.2.3. Ergotism. This is the oldest known mycotoxicosis and is thought to have been recognised by the Chinese 5000 years ago. In 1480 BC the armies of Athens and Sparta were afflicted by this disease, as were the troops of Peter the Great of Russia in 1710–12 when he lost the second war with Turkey. During the ninth and tenth centuries, the disease was particularly rife in France where limbs of the afflicted became gangrenous and fell off. It was not until the seventeenth century that it was realised that the ergots, the black, hardened masses of mycelia growing on female sex organs of grasses, contain numerous ergot alkaloids, most of which either directly constrict the smooth muscles of the vascular system and thus prevent sufficient blood supply into the extremities and/or affect the nervous system, thus leading to convulsions and spasms. The oxytoxic effect of ergot alkaloids was utilised much before that time.[83]

The ergot alkaloids have a wide spectrum of pharmacological activity, which includes antagonism towards adrenaline and 5-hydroxytryptamine, impairment of lactation, and smooth muscle stimulation of uterus, the walls of arterials and sphincter pupillae. The toxic effects, e.g. anoxia, death of peripheral tissue and dry gangrene, result from these muscle-contracting activities. The uterine contractions caused by ergometrine are used for involution of the uterus and in controlling post-partum haemorrhage.

1.4.2.4. Fusarium *mycotoxicoses.* (a) *Scab.* According to Mirocha,[84] it has long been recognised that food grains, especially barley, rye and wheat, when exposed to wet or humid weather conditions during maturation, become unwholesome. One of the first written descriptions of ear rot of maize (caused by *Fusarium moniliforme*), according to Booth,[85] was derived from the Aztecs as reported by a Franciscan monk in the sixteenth century. Woronin in 1891 reported on a disease in humans present in Siberia and Sweden that caused dizziness (*Taumelgetreide*), headache, shivering, vomiting, disturbance of vision, general malaise and staggering as a result of eating bread made from flour of grain infected with *Fusarium* (scab).

Fusarium scab (blight) broke out in 1980 as a serious epidemic of wheat in Ontario and Quebec, Canada. In America two years later, the hard red winter wheat of Kansas, Missouri, was strongly infected. There are similar outbreaks (red mould) described for mainland China and for Japan. Ueno *et al.*[86] reported on a red mould disease, 'Akakabi-byo', which occurs in wheat and barley in times of excessive rainfall and cold weather in spring, and which is apparently similar to the wheat scab found in the Americas. In man, the symptoms are vomiting and diarrhoea, and in horses and other farm animals, refusal of feed, congestion and haemorrhaging in the lung, adrenals, intestine, uterus, vagina and brain are observed. In 1926

Dounin[87] reported that, in eastern Siberia, when bread was made from grain infected with *Fusarium graminearum* and *F. roseum*, it caused vomiting and the sensation of drunkenness. Deoxynivalenol (vomitoxin) was the main cause of the emetic effects.

(b) *Gastrointestinal haematotoxicosis* (alimentary toxic aleukia [ATA], formerly called septic sore throat). As pointed out in the introduction, Joffe[7] reported on sporadic outbreaks of ATA in Russia before the turn of the century, in 1913 and 1932, and from 1942 to 1947. These outbreaks of ATA were associated with grain (wheat, rye, millet, oats, barley and buckwheat) that was left standing in the fields during wintertime and was not harvested until spring. The combination of low temperatures and humidity encouraged the growth of *Fusarium* moulds, mainly *F. sporotrichioides* and *F. poae*, on the various grains covered with snow. According to Mirocha,[88] *Fusarium* spp. also grow prolifically in the USA and Canada. Corn (kernels and stalk) becomes infested while standing in the field during growth and ripening. It is not just the colder regions of the US that are afflicted with *Fusarium* spp., but also the relatively warmer regions of the Mid-West (Ohio, Missouri, Illinois, Kansas, Indiana), and a wide spectrum of climates, for example those of France, Germany, the United Kingdom, Tunisia, India, South Africa and Japan, will permit the proliferation of *Fusarium* spp. Although about 40 *Fusarium*-derived 12,13-epoxy-trichothecene-derived tetracyclic mycotoxins can be isolated in laboratory cultures, only five are consistently found occurring naturally in foods: T-2 toxin, nivalenol, deoxynivalenol (DON, as in *Fusasium* scab), monoacetoxyscirpenol and diacetoxyscirpenol (DAS). The trichothecenes belong to the most bioactive small molecules among natural products. Application to pigs of 0·05 mg kg^{-1} of deoxynivalenol (intraperitoneal) or 0·1 mg kg^{-1} (oral) induced vomiting. Individuals exposed to *Fusarium* mould-infested food experience first a burning sensation in the mouth, oesophagus and stomach. This is followed by severe emesis, diarrhoea and abdominal pain. In the second stage there is a marked decrease of leucocytes and the development of agranulopenia and lymphocytosis. In a third stage, petechial haemorrhages are noticed on the chest, arms, thighs and face. At this stage necrotic areas develop in the throat (septic angina), and there is sepsis, exhaustion of the bone marrow, bleeding from the nose and gums, and fever, all symptoms typical of ATA. The patient at this point may either expire or recover. According to Mirocha *et al.*,[89] 'The signs (symptoms) of ATA have been recognised in man under other circumstances: T-2 toxin has been found to be the major toxic component used in Laos and Kampuchea in chemical warfare directed against the Hmong people.'

(c) *Kashin–Beck disease.* In the eastern regions of Russia, northern China and North Korea a disease is found that afflicts the developing skeletal system. Children, especially, suffer from chronic arthritis and degenerative skeletal malformation. These symptoms are brought about by narrowing of the blood vessels leading to the joints. The visceral organs are apparently unaffected. According to Nesterov,[90] the cause of this disease is derived from cereal grain infested with *Fusarium sporotrichioides.*

It should be mentioned at this point that there exist many different strains classified under the same fungus name. When 110 different *F. sporotrichioides* strains were tested for the degree of toxicity, they were separated into five groups ranging from non-toxic to severely toxic. Three strains proved non-toxic, 15 strains showed a toxicity of 1st degree, 24 strains one of 2nd degree, 55 one of 3rd degree, and 13 one of the highest (4th) degree.[91] The main mycotoxins of the virulent strains were zearalenone and T-2.

1.4.2.5. Acute cardiac beri-beri. Cardiac beri-beri has claimed numerous victims in Japan and other countries of the Far East. The compound responsible for this disease is citreoviridin, which is abundantly produced on rice at temperatures ranging from 12 to 18°C by *Penicillium citreo-viride* (Biourge). The LD_{50} is 20 mg kg^{-1} (in mice subcutaneously injected). At chronic *low doses,* citreoviridin causes: (1) congestion of veins, (2) haemorrhage of liver and spleen, (3) excessive swelling of kidney. At *higher doses* it causes: (1) paralysis (ascending type) affecting the motor nerves of the spinal cord, (2) difficulties in breathing, (3) death. The compound is a neurotoxin that exhibits an intense yellow fluorescence under UV light. Post-mortem examination under UV light has revealed that the compound is specifically enriched in the central nervous system, the adrenal cortex and the kidneys.

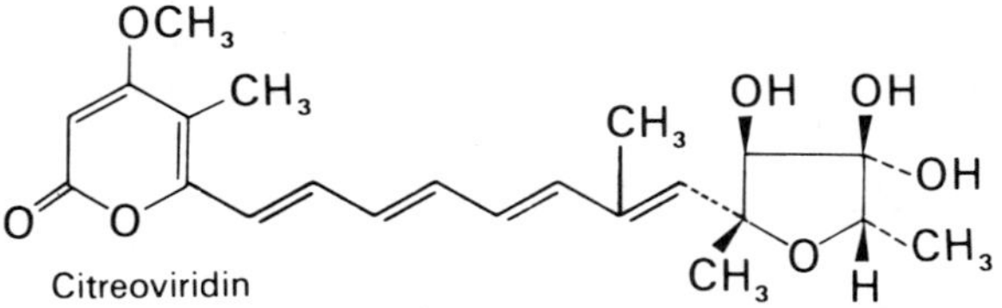

Citreoviridin

There are additional fairly severe mycotoxicoses in man. It is not, however, the intention of this chapter to recount all of them. This discussion was undertaken with the intention of reiterating that the growth of moulds on food products is almost unavoidable. Since many moulds grow in the home on food products, the education of the population in general on the significance of mycotoxins should be attempted. It is this writer's opinion that mycotoxicoses in man are perhaps unduly neglected and warrant more serious consideration as the primary cause of many maladies.

1.4.2.6. Endemic Balkan nephropathy. This occurs in 16 areas of the Danube Basin in south-eastern Europe. The areas are distributed over Yugoslavia, Rumania and Bulgaria. There is a highly significant correlation between high levels of precipitation in the months from August to November (the time of the grain harvest and the critical storage period) and outbreaks of nephropathy, i.e. the number of people dying in the course of two successive years after high rain precipitation.

Barnes *et al.*[11] in 1977 were able to identify one of the moulds responsible for the toxin formation, namely *Penicillium verrucosum* var. *cyclopium*. This mould produces, besides other toxins, mainly ochratoxin A. Scott *et al.*[92] in 1970 and van Walbeek *et al.*[93] in 1969 identified *Penicillium viridicatum* as a producer of ochratoxin A. Krogh and Elling[94] pointed out the similarity of nephropathy in man with porcine nephropathy in Danish pigs. There is little doubt that porcine nephropathy is caused by maize contaminated with, mainly, ochratoxin A. (See Fig. 5 for chemical formula and biosynthesis pathway.)

Ochratoxin A is produced not only by *Penicillium* spp., but also by *Aspergillus* spp., especially *A. ochraceus*. However, the latter require a higher temperature (up to 37°C) than *Penicillium* spp., which will still produce the toxin down to 5°C.

1.5. Immunology

1.5.1. Introduction

Since the allergic reactions provoked by green coffee bean dust belong in the field of immunology, some very brief and fundamental description of the main factors involved in immunological reactions is presented for the benefit of those readers not familiar with the field of immunology. For a more detailed insight into the very complex and wide area of immunology, the following books are recommended: *Cellular Functions in Immunity and Inflammation* by J. J. Oppenheim, D. L. Rosenstreich and M. Potter, published by Elsevier/North-Holland, 1981; *Clinical Aspects of Immunology*, Volumes 1 and 2, 4th edition, P. J. Lachmann and D. K. Peters (Editors), Blackwell, 1982; and *Essential Immunology*, 5th edition, by I. Roitt, Blackwell, 1984). There are new publications appearing every day in the area of immunology, demonstrating the enormous scientific attention that is given to this very important aspect of human physiology.

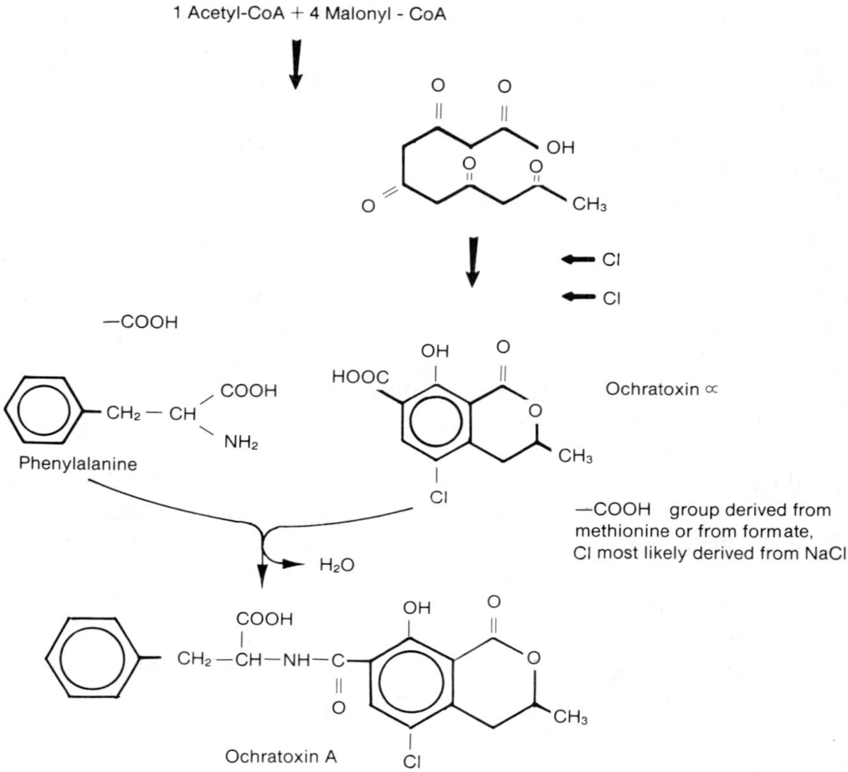

Fig. 5. Formation of ochratoxin A.[95,96]

Immunology is the study of the defence mechanism activated by the body against infection. Infections may range from bacterial, viral or mould infections (e.g. leprosy, tuberculosis; smallpox, rabies, polio, measles, chickenpox, mumps; coccidioidomycosis, blastomycosis, histoplasmosis; etc.) to macromolecules, foreign to the body. Such macromolecules may be composed of proteinaceous, polysaccharide or lipid complexes.

1.5.2. Abbreviated History of Immunology

The development of immunology, according to Roitt, started in India and China in ancient times where protection against smallpox was achieved by inoculating with live organisms from disease pustules.

Noting that milkmaids retained a pox-free facial skin, the English physician Edward Jenner in 1798 started to vaccinate successfully against smallpox with non-virulent cowpox.

Robert Koch in 1877 and Pasteur in 1879 succeeded in isolating the pathogens of ovine anthrax and fowl cholera. In 1879, while studying the causative agent of fowl cholera, some aged cultures of the gram-negative coccobacillus unexpectedly failed to kill the fowl (chickens). On re-inoculation with fresh, active bacillus, the chicken previously inoculated with aged non-virulent bacillus survived the inoculation while fresh chickens died in due course. Pasteur realised that he had rediscovered a different means of inducing immunity in close relationship to Jenner's discovery 81 years earlier.

Pasteur in 1881 produced and used successfully in a dramatic animal experiment (sheep, horse, cows) an attenuated vaccine against anthrax disease that was ravaging among farm animals at this time. Metchnikoff in 1883 pointed out the role of phagocytes in immunity. In one of his experiments, he placed rose thorns under the skin of transparent starfish larvae. He observed that phagocytes migrated to the thorn, attached to it and attempted to destroy it by means of phagocytosis. In 1890 von Behring and Kitasato recognised antibodies to the diphtheria toxin in the blood serum. Koch in 1882 isolated the tubercle bacillus. By 1891 he found that re-inoculation of previously inoculated guinea pigs with tubercle bacillus no longer caused non-healing ulceration at the site of inoculation. Also inoculation with killed bacilli or a protein isolated from tubercle bacilli, which he called tuberculin, produced the same subdued phenomenon. The Koch phenomenon is an example of delayed hypersensitivity. In 1895 Denys and Leclef discovered that phagocytosis is greatly enhanced by immunisation. Paul Ehrlich in 1897 proposed the receptor theory in immunoglobulins (which he called 'Antikörper', i.e. antibody) for explaining the astounding antibody specificity. Bordet discovered in 1899 that the lysis of invading foreign cells by antibodies requires the co-operation of a factor contained in the serum. This serum factor is now collectively termed *complement*. Landsteiner in 1900 discovered the human specific blood types A, B, O and the natural isohaemagglutinins.

Richet and Portier described, in 1902, anaphylaxis in dogs sensitised with eel serum.

Maurice Arthus in 1903 sensitised rabbits by giving multiple injections of horse serum several days apart. After subcutaneous injection of horse serum, an inflammatory lesion developed at the injection site. The lesion was accompanied by considerable swelling and sometimes was necrotic. This phenomenon is now known as the Arthus reaction. The Arthus reaction can be passively provoked by injecting the serum of a sensitised animal into a normal animal. On antigen injection, such an animal will now produce a localised 'Arthus reaction' at the antigen injection site. In this reaction an antibody–antigen interaction occurs. This interaction is the basis for the immune complex disease.

Von Pirquet and Schick described in 1905 the symptoms of serum sickness following the injection of foreign (e.g. horse) serum, causing fever, swollen lymph nodes and skin rash.

Pirquet in 1906 recognised the relationship of immunity and *hypersensitivity*; he introduced the term 'allergy'.

Fleming in 1922 discovered that by mixing nasal mucus with bacteria the bacterial cell walls were lysed. This, he found, was due to the presence of lysozyme. Tears are also rich in lysozyme. Lysozyme is capable of hydrolysing specifically the polysaccharide component of the bacterial cell wall.

Zinsser in 1925 described contrast between immediate- and delayed-type hypersensitivity. *Immediate-type hypersensitivity* (anaphylaxis) is an allergic response occurring within minutes after exposure to antigen. It is antibody mediated.

In *delayed-type hypersensitivity*, special subsets of T-cells (see Fig. 6) of the T-delayed-type hypersensitivity (T-DTH) and TK-cells (cytotoxic T-cells) are involved in reactions after the body is invaded by certain microorganisms or after skin contact with certain sensitising chemicals, e.g. poison ivy.

Heidelberger and Kendall, from 1930 to 1935, carried out their classical quantitative precipitin studies on antigen–antibody interactions. These precipitin reactions occurred when mixing solutions of antigen with antiserum. The precipitin reaction is caused by antigen interacting via various non-covalent bonding modes (electrostatic, hydrophobic, Van der Waals and hydrogen bonding) with the antibody to form a *precipitate*. Upon adding more and more antigen, less and less precipitate is formed, and antigen and serum antibody form soluble complexes. Depending on the 'valency' (di- or polyvalent) of the serum antibody, a corresponding number of antigens can be bound to one antibody molecule.

Tiselius and Kabat discovered that the antibody activity of the blood serum resides in the gamma globulin fraction of the blood serum. Electrophoretic separation of blood serum into an albumin and eventually six different globulin fractions (α_1, α_2, β_1, β_2, γ_1 and γ_2) showed a decreased γ-globulin fraction after mixing blood serum and antibody. Eventually the γ-globulin fraction was separated via combined electrophoretic separation and ultracentrifugation into five major structural classes called the *immunoglobulins* (see Fig. 6) with different molecular weights and present at different concentrations. It also was found that about 10% of people who suffer from allergy (called atopic individuals) have a highly increased quantity of immunoglobulin E (IgE).

Astrid Fragaeus actually showed for the first time in 1948 that antibodies are produced by plasma cells. In 1955 Niels Jerne proposed the natural selection theory for antibody formation. This theory was immediately expanded by Sir Macfarlane Burnet with the theory that the precursors of antibody-forming cells are selected by antigen, and cloning of such specific cells occurs. Nossal and Lederberg supported Burnet's theory by demonstrating that a small number of genetically determined precursor cells are activated by antigen to mature to immunoglobulin-secreting plasma cells.

James Gowan's lymphocyte recirculation experiments in 1964 provided the evidence that lymphocytes are the precursors of antibody-secreting cells. He showed that some of the circulating lymphocytes entered tissue sites to transform into mature plasma cells. Identification of immunoglobulin-producing lymphocytes became possible by specific fluorescent antibody techniques.

Karl Landsteiner and Merril Chase demonstrated in 1943 that the delayed hypersensitivity caused by a specific antigen was transplantable to untreated animals by cell population rich in lymphocytes but not by cell-free serum. This experiment indicated that aside from cell-free, antibody-produced immunity there was a second mode of immunity, namely the delayed type produced by lymphocytes. When Miller thymectomised neonatal mice and these mice, as a result of the removal of the thymus gland, showed deficiencies in cell-mediated immunological functions, the missing thymus-derived lymphocytes were implicated for the resulting immune deficiency. Later Miller and Mitchell, and Claman, demonstrated with irradiated mice (to inactivate all lymphocytes in these animals) that in these mice, upon re-supply with thymus-derived lymphocytes (obtained from healthy animals), the *cell-mediated immunity* was re-established. When re-supplying the animals with non-thymus-derived lymphocytes, the *humoral immunity* was re-established. These important experiments were the basis for differentiating lymphocytes into thymus-derived T-lymphocytes and bursa-derived (in birds) B-lymphocytes.

Radio-isotope labelling of the lymphocytes permitted the tracing of the lymphocyte path in the body. The lymphocytes were found to pass from the blood into (1) lymph nodes, (2) spleen and (3) other tissues. The lymphocytes

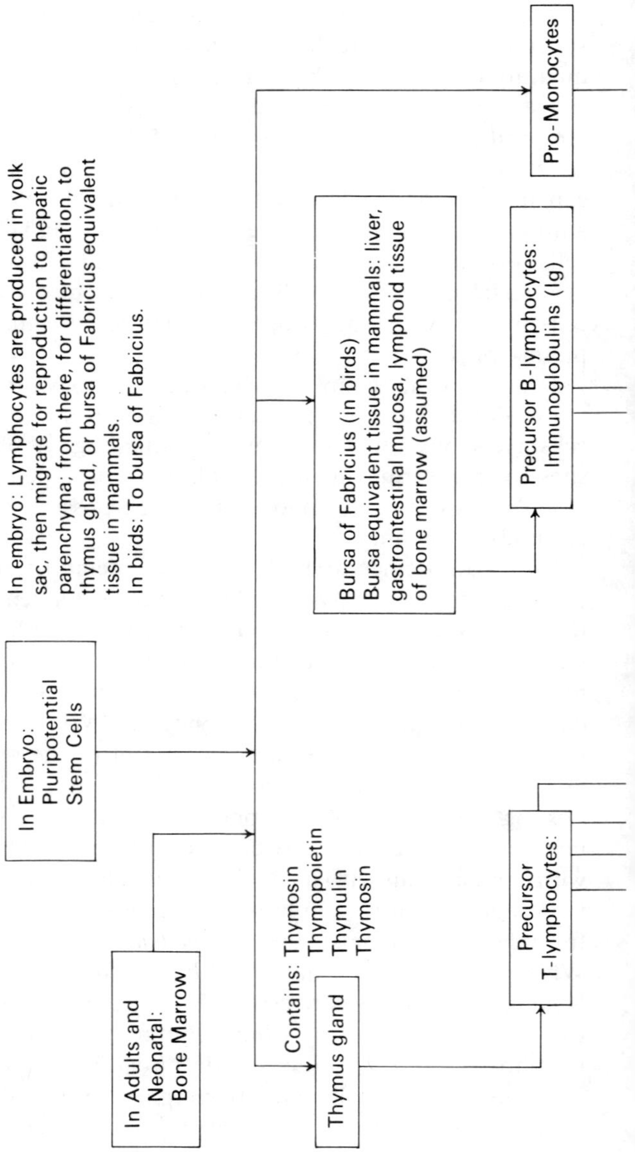
In Embryo:
Pluripotential
Stem Cells
In Adults and
Neonatal:
Bone Marrow
In embryo: Lymphocytes are produced in yolk sac, then migrate for reproduction to hepatic parenchyma; from there, for differentiation, to thymus gland, or bursa of Fabricius equivalent tissue in mammals.
In birds: To bursa of Fabricius.
Thymus gland
Contains: Thymosin
Thymopoietin
Thymulin
Thymosin
Bursa of Fabricius (in birds)
Bursa equivalent tissue in mammals: liver, gastrointestinal mucosa, lymphoid tissue of bone marrow (assumed)
Precursor
T-lymphocytes:
Precursor B-lymphocytes:
Immunoglobulins (Ig)
Pro-Monocytes

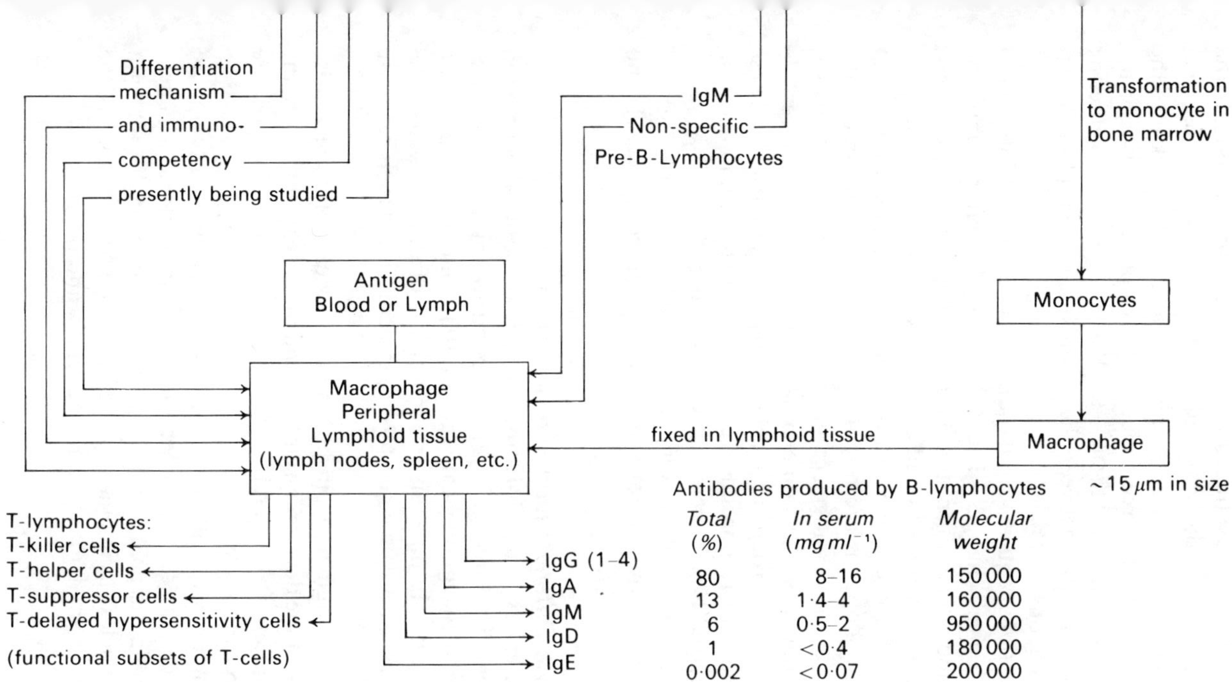

	Total (%)	In serum ($mg\,ml^{-1}$)	Molecular weight
IgG (1–4)	80	8–16	150 000
IgA	13	1·4–4	160 000
IgM	6	0·5–2	950 000
IgD	1	<0·4	180 000
IgE	0·002	<0·07	200 000

Fig. 6. Genesis and function of lymphocytes in acquired immunity.

return from the lymph nodes via the thoracic duct back to the blood. In the spleen the lymphocytes enter the white pulp to leave it a few hours later via the zone bridging channels to enter the red pulp and finally back into the venous blood circulation. The recirculation of lymphocytes also occurs through gut-associated lymphoid tissue—the Peyer's patches—by entering postcapillary venules located between the thymus-dependent (T-cells) and B-cell-dependent regions.

Both cell populations require stimulation of antigen to multiply and to undergo morphological changes. The B-lymphocyte matures into the *plasma cell series* and after antigen stimulation it engages in active synthesis and subsequent secretion of the five different types of antibodies. These five different types of antibodies are called the immunoglobulins G, M, A, D and E. The immunoglobulins are responsible for the humoral immunity.

The T-lymphocytes, after antigen contact, transform into lymphoblasts. It is these lymphoblasts that differentiate into T-helper/inducer, T-suppressor, T-killer (cytotoxic) and T-delayed hypersensitivity cells. These cells are the components responsible for the cell-mediated immunity (also called type IV, delayed-type hypersensitivity; see later).

Limited space permits only a very abbreviated outline of the very complex functions of the various immunological components and phenomena.

1.5.3. Lymphocyte Differentiation and Activation

Lymphocytes, present as T- and B-lymphocytes, take a central position in the immune system. They recognise and respond to antigen. In spite of showing very similar morphological features and coming from the same origin, they are strikingly different in their tissue of differentiation, the pathways of differentiation, their cell surface structures and their response to antigen (see Fig. 10).

Both lymphocytes originate from pluripotential stem cells which arise first in the yolk sac of the developing embryo and migrate from there to the foetal liver and spleen. In neonatal and adult life, the lymphocytes are produced in the bone marrow.

Precursor T-lymphocytes and precursor B-lymphocytes enter lymphoid tissue (lymph nodes, spleen or subepithelial accumulations of unencapsulated lymphoid tissue in lamina propria of intestinal wall with isolated solitary follicles), as in pharingeal, lingual and palatine tonsils, the Peyer's patches of the small intestine, the appendix and the genito-urinary tract.

Lymphocytes go through three stages to mediate adaptive immune responses:

(1) Induction or activation by either direct antigen exposure or, more regularly, by presentation of antigen via macrophage. (The macrophage usually entraps the antigen first, and then exhibits it to the lymphocytes on its cell surface.)
(2) After contact with antigen, the clonal proliferation of antigen-specific lymphocytes sets in.
(3) A major proportion of lymphocytes become effector cells and some quantity become memory cells which can readily be activated into effector cells upon a second exposure to antigen.

T-lymphocytes which mature in the thymus gland are responsible for the *cellular immunity*. Dentritic macrophages present the antigen, together with class II molecules of the major histocompatibility complex, to the T-lymphocyte. The T-lymphocyte provides the trigger for the B-lymphocytes to produce antigen-specific antibodies and also provides the B-lymphocyte growth and the maturation factor. B-lymphocytes, after the interaction with the activators from the T-lymphocyte, can undergo successive proliferation and maturation for production of clones of antibody-producing B-lymphocytes. The *B-lymphocytes* are responsible for the *humoral immunity* (defence against antigens in blood and also lymph).

The T-lymphocytes are activated by a combined system consisting of antigen and macrophage-derived interleukin-1. These T-cells are then cloned after interaction with interleukin-2, which is produced by a subpopulation of T-cells (the blast cells).

The cell-mediated immune response (Type IV hypersensitivity) is operated with the help of cytotoxic T-cells (killer cells). Cytotoxic T-cells are the final product of three stages. An antigen-presenting macrophage interacts with a T-helper/inducer/delayed-type hypersensitivity cell. This cell then releases lymphokines which, in turn, activate a cytotoxic precursor T-cell and matures it into a cytotoxic killer T-cell. Such T-killer cells are capable of effectively destroying, via various means, an infected cell. (Lymphokines are glycoproteins with molecular weights ranging from 10 000 to 200 000 D.)

1.5.4. Immunoglobulins

The entry of antigens into the body (bloodstream or lymph) is across the skin, e.g. after injuries; by direct entry into the bloodstream via the gums, e.g. following trauma to the gums; or via the mucosal linings of the digestive system, the airways or genito-urinary tract.

In order to prevent the entry of antigens into the body or effectively inactivate them after successful entry, the immunoglobulins are strategically produced, deposited or circulated where they can most effectively combat the attacking/invading antigens.

As discovered by Tiselius and Kabat, the immunoglobulins reside in the γ-globulin fraction of the blood serum. Five different classes of immunoglobulins were separated via electrophoresis, immunoelectrophoresis, ultracentrifugal and immunochemical separation.

The primary composition of the immunoglobulins has been determined, according to which the immunoglobulins are composed of amino acids and a smaller portion of a carbohydrate fraction.

The immunoglobulin classes are ordered according to decreasing percentage present in serum: immunoglobulin G (abbreviated IgG), IgA, IgM, IgD and IgE (see Table 8 for a summary of immunoglobulins).

Immunoglobulin G. The IgG- and IgM-producing plasma cells (B-lymphocytes) develop mainly in the lymph nodes and in the spleen. IgG is the major immunoglobulin found in the body. Approximately 80% of the immunoglobulins is IgG. It has the ability to cross the placenta, thus providing a defence against infection in the newborn during the first few weeks. Mothers' milk (especially colostrum) is also rich in IgG. Since it can cross the gut mucosa, it can distribute throughout the body to do the following. (1) It neutralises bacterial toxins. (2) It binds to invading microorganisms to prepare them for phagocytosis. (3) Complexes of IgG with bacteria activate the complement factors (also residing in the globulin fractions of the serum; see later) which, in turn, produce chemotactic factors to attract polymorphonuclear cells to the activation site. The attracted polymorphonuclear cells, stimulated by the binding of IgG to the antigen, will then actively ingest, via a phagocytotic process, the complex into the cell for subsequent deactivation and digestion.

Immunoglobulin A. Immunoglobulin A, contrary to IgG and IgM which are delivered to the circulating blood, is mainly secreted close to epithelial cells such as subepithelial sites of the gut, the upper respiratory tract, salivary, lacrimary and mammary glands, and the tonsils. IgA's prime function is to guard the mucosal linings of the body (digestive system, airways and genito-urinary tract). Relatively large quantities of IgA are secreted into the milk to provide the newborn with maternal immunity. Part of IgA enters the bloodstream. Most of these IgA molecules are taken up by liver cells and undergo transhepatic migration across the liver sinusoid into

Table 8

Immunoglobulins: physical and biological properties

Class	*Approximate concentration (%)*	*Approximate concentration in serum ($mg\,ml^{-1}$)*	*Molecular weight*	*Antigen binding sites: 'Valency'*	*Properties*	*Distribution*
IgG	80	8–16	150 000	2	Circulating antibody exists in four subclasses: IgG_1, IgG_2, IgG_3, IgG_4. Binds complement	Blood and lymph; extracellular fluid; crosses placenta
IgA	13	1·4–10·	160 000 (monomer) 320 000 (dimer)	2 for monomer 4 for dimer	Can form dimer. Secretory antibody	Extracellular and body fluids. Exists as monomer in serum, as dimer in secretions
IgM	6	0·5–2	900 000 (pentamer)	10 present as pentamer	Binds complement. First antibody to appear after immunisation	Blood and lymph
IgD	0–1	<0·4	185 000	2	Minor circulating antibody	Blood and lymph on lymphocyte surface
IgE	0·000 2	17–450	200 000	2	Contains specific mast cell binding fragment in molecule. Involved in immediate-type hypersensitivity (allergy)	Blood and lymph only

the bile. The effector functions of IgA are to impede the absorption of dietary macromolecules, toxins and allergens, and to impede the entry of viruses, bacteria and parasites into the body. For example, IgA is secreted via the sweat glands on to the skin.

Immunoglobulin A antibodies have the function of inhibiting the adsorption of coated microorganisms to the surface of mucosal cells. Immunoglobulin A aggregates and binds to polymorphic cells and thus activates the co-operation of complement (alternative pathway) and lysozyme (a hydrolysing enzyme in mucosal fluids) in killing coliform bacteria. Immunoglobulin A exists in a monomeric and also a dimeric form. The dimer consists of two IgA monomers held together by a peptide, called the J-chain.

Immunoglobulin M. Immunoglobulin M is also called the macroglobular antibody. As in IgA, where the J-chain enables the formation of a dimer, here the J-chain is instrumental in associating five IgM units, thus forming a pentamer with a molecular weight of 950 000 D. The pentamerisation of IgM also makes it the immunoglobulin with the highest antigenic valency of 10; this high antigenic valency makes it an efficient agglutinating and cytolytic agent. Immunoglobulin M is largely confined to the bloodstream; it is likely that it plays a special role against microorganisms.

Immunoglobulin D. Immunoglobulin D has the shortest half-life of all immunoglobulins. It is present in serum at relatively low levels (0–1%). It gained significance when it was found to be a major membrane immunoglobulin associated with B-lymphocytes and that it is very amenable to enzymic degradation. It was also found that most B-lymphocytes have both IgD and IgM as membrane immunoglobulins, thus demonstrating that B-lymphocytes are capable of simultaneously synthesising two different immunoglobulins. Immunoglobulin D seems to be tolerant of some antigens, leading to the assumption that it might be capable of mutually interacting with IgM to control the activation or the suppression of the lymphocyte.

Immunoglobulin E. Immunoglobulin E is the immunoglobulin present in the serum of normal people in very low concentrations (0·0002%). In 'atopic' (non-typical) allergic people, and also in parasitic disorders, this concentration is considerably increased. This immunoglobulin plays a very prominent role in the phenomenon of *immediate hypersensitivity against allergens*, and therefore is of special interest since the contact with dust associated with green coffee beans will cause immediate hypersensitivity in some individuals.

Immunoglobulin E has a specific capability to adhere strongly to mast cells and other eosinic cells. If an allergen, like certain components of green coffee bean dust, cross-links two IgE molecules adhering to the mast cell, then a mechanism is triggered which leads to the release of mediators of anaphylaxis (see Fig. 7, mast cell activation). These pharmacologic mediators can be released from preformed granules of the mast cell within less than one minute after allergen contact. A second, slower release of anaphylactic substances (about 6 h later) occurs from the mast cells. The release of the so-called 'slow-reacting substances of anaphylaxis' (SRS-A) is not from granules but from the cell wall of the mast cell.

The main role of IgE is to protect the mucosal surfaces of the body by activating plasma factors and effector cells (T-cells) for elimination of the allergen. The reaction of IgE with mast cells and antigens plays a major role in the allergies associated with the dust on green coffee beans and, therefore, is of special interest to this chapter.

1.5.5. The Complement System

In addition to immunoglobulins which are partially responsible for humoral immunity, the complement system is co-responsible for humoral immunity. The *complement system* is composed of at least some 20 distinct serum proteins (C1 through C9 and subunits). They are synthesised by macrophages, by the liver, and by the epithelium of the gut.

The complement components exist in the blood as 11 inactive precursors. However, they are activated when, for example, IgM or IgG antibodies attach to a red blood cell (formation of an immune complex) in the presence of complement. There are two pathways of complement activation: the classical complement pathway and the alternate complement pathway.

The complement system can be characterised as a multicomponent cascaded system whereby a first component, after initial activation, is hydrolysed into two components. One of the newly generated components becomes an active enzyme which, in turn, will hydrolyse the next complement component, etc. The second component resulting from the hydrolysis acts either as a chemotactic agent (contracts smooth muscle) or activates mast cells (or basophils) for release of anaphylatoxins. This type of cascade event leads from an initially small reaction to a large effect.

The classical pathway is started by antigen–antibody complexes. The alternate pathway is started in the absence of antibody by cell wall components such as bacterial polysaccharides.

Some hydrolysed fragments (1) attract polymorphonuclear leucocytes

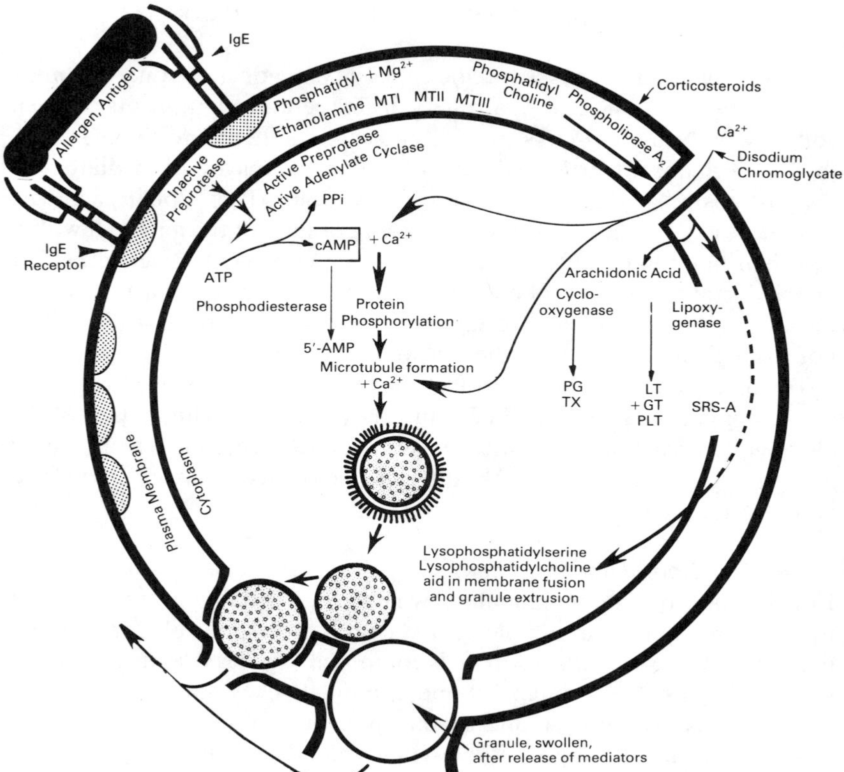

Fig. 7. Mast cell activation: simplified hypothetical mechanism.[107,108] MTI = methyltransferase I; MTII = methyltransferase II; MTIII = methyltransferase III; PG = prostaglandin; TX = thromboxanes; LT = leucotrienes; GT = glutathione; PLT = peptide leucotrienes; SRS-A = slow-reacting substance of anaphylaxis.

Agents increasing the cAMP level
Theophylline
Caffeine
Quercetin
- PGE$_1$—recept.
- Acetylcholine—recept.
- Adenosine—recept.
- Histamine, H$_2$—recept.
- Cholera toxin—recept.
- β-Adrenergic—recept.

Mediators of anaphylaxis
Histamine
(Serotonin)
Eosinophil chemotactic factor
Neutrophil chemotactic factor
Superoxide generating enzymes
Superoxide dismutase — Action in
Mast cell peroxidase — 30 s to
Heparin — 3 min
Chymotrypsin
Allergic inflammation
- Leucotrienes
- Prostaglandins — SRS-A
- Peptide leucotrienes — 6–24 h

and (2) increase vascular permeability via mediator release from mast cells (histamine, etc.) and other basophilic cells. Assembling of the terminal components of the complement system (C6–C9) causes the formation of transmembrane channels, and these cause the death of the invading cells by osmotic leaking of the cell contents.

Via these destructive processes against invading cells, etc., the complement system plays an important role in the body's defence against infection.

Finally, the complement cascade plays a role in hypersensitivity reactions, namely in antibody reactions with surfaces (e.g. in nephrotoxic nephritis) or antibody reactions with immune complexes (e.g. in farmer's lung). The complement system plays a role in the classical Type II and Type III hypersensitivities (see later).

The cascade sequence occurs in nine steps: (1) IgG or IgM attaches to cell wall membrane and binds complement component 1 (C1), actually consisting of three components complexed together by Ca^{2+} of C1q, C1r and C1s. This antigen–antibody–complement complex is now capable of acting as an *esterase* for complement component 4; more illustratively depicted as C4(ab). C4(ab) is hydrolysed into the large fragment C4b and a small fragment C4a. (2) C4b attaches to the cell wall, whereas C4a is dispersed in the serum and increases vascular permeability and contracts smooth muscle. (3) The antigen–antibody–complement complex repeats the same hydrolytic step for complement 2(ab) as was done for C4(ab). C2b attaches to C4b while C2a exhibits kinin-like activity and contracts smooth muscle. This newly formed complex exhibits enzymatic activity (C3-convertase) and (4) is capable of hydrolysing C3(ab) into C3a and C3b. C3b attaches to the complex consisting of C4b and C2b, and the newly formed complex, in turn, (5) is capable of hydrolysing C5(ab) (C5-convertase). (6–9) C5b attaches to the cell membrane and assembles the terminal complement components C6–C9 to produce a complex that is capable of cytolytically disrupting the cell membrane. The disrupted cell membrane no longer is capable of retaining the cell contents and therefore the cell terminates.

In summary, the complement system fulfils at least six different functions:

(1) Components of the complement system attach to the surface of bacteria and other antigens, while
(2) simultaneously generating chemotactic factors for neutrophils and macrophages. The neutrophils and macrophages then neutralise/remove, via phagocytosis, the invading antigen.

(3) The complement factors, after attachment to some antigens, cause them to agglutinate and thus render them inactive.
(4) The terminal components of complement form tube-like structures. These structures are capable of lysing the cell walls of invading microorganisms. This, in turn, causes leakage of the cytoplasm and thus cell death.
(5) The molecular structures of viruses can be attacked by the complement system and it thus inactivates them to a non-virulent state.
(6) Finally, the inflammatory effect of the activated complement system leads to hyperaemia and precipitation of proteins, thus arresting the invading agent and preventing it from further movement to other tissues (wall effect).

1.5.6. Mechanism of Hypersensitivity Reactions

A system of classification exists which differentiates the diseases of immunity on the basis of the immunological mechanism.[103–105] This approach describes the means by which immunological response in defence against the extrinsic (or intrinsic) antigens also causes tissue injury. This system has five divisions:

(1) *anaphylactic type (Type I)* (antigen reacts with antibodies present on mast cells, degranulation of mast cells, release of vasoactive amines and other mediators);
(2) *cytotoxic type (Type II)* (mediated by antibodies against antigens present on cells or tissue components, complement-mediated cytotoxic hypersensitivity);
(3) *immune complex-mediated disease (Type III)* (induced by antibody–antigen–complement interaction);
(4) *cell-mediated, delayed hypersensitivity (Type IV)* (sensitised lymphocyte-mediated); and
(5) *stimulatory hypersensitivity.*

These types were originally classified by Coombs and Gell in 1963.[106]

Newer approaches no longer differentiate between Type II and Type III hypersensitivity since the mechanistic details of tissue injury elucidate that both reactions have a common basis of involvement of the complement system.

Type III hypersensitivity is depicted in Fig. 8. The figure shows that the Type III hypersensitivity also contains the elements of Type I and also of Type II hypersensitivity. Therefore it is difficult to maintain a strict

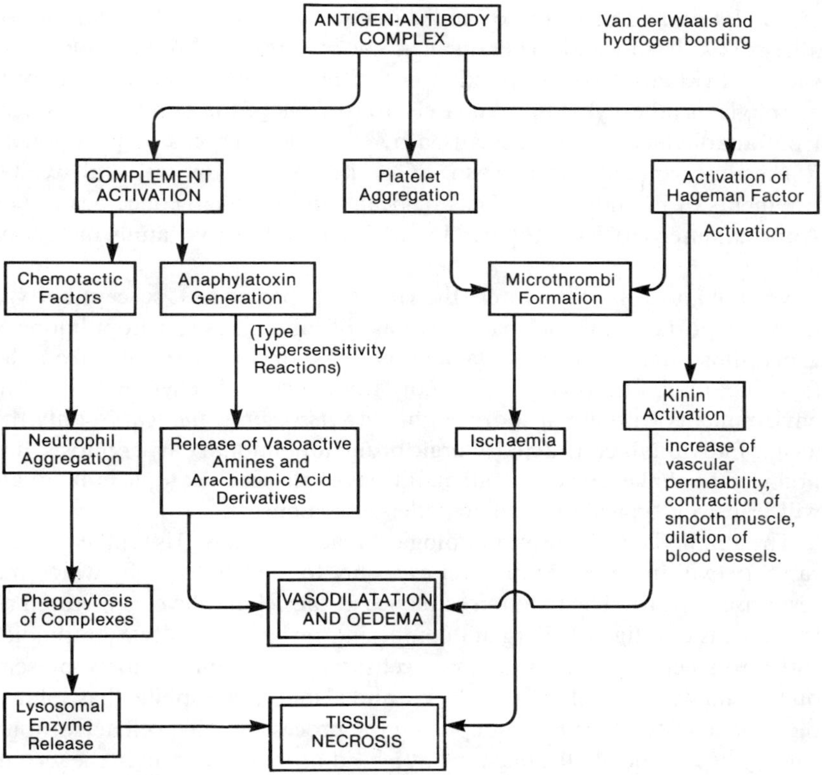

Fig. 8. Type III hypersensitivity (immune complex mediated).

separation of the three classifications since all three types have both actors of humoral immunity included, namely antibodies and complement system.

The cell-mediated, delayed-type hypersensitivity, Type IV, acts independently of antibodies. The term 'cell mediated' is somewhat unfortunate, and the term 'T-cell (or T-lymphocyte) mediated' appears less ambiguous, more direct and descriptive.

1.5.7. Allergic Reactions in the Atopic (Allergic) Individual (Anaphylactic Type I)

This type of allergy occurs in people who are predisposed (e.g. hereditarily) to allergic reactions. In this case an abnormally large quantity of immunoglobulin E (IgE) antibodies is present in the serum.

Inhalation and ingestion of and skin contact with minute quantities of allergens will cause a local allergy in atopic individuals. Atopic implies that such individuals have a specific genetic propensity to develop a local anaphylaxis (allergy). Approximately 10 to 20% of the population suffers from anaphylactic reactions caused by extrinsic allergens such as pollen, house dust, animal dander, fish, SO_2, the conidia (spores) and hyphal fragments of moulds, dust of castor beans and/or green coffee beans, etc. Approximately 50% of atopic individuals have a positive family history of allergy.

Vernon and Allan[102] express the view that specific host susceptibility is more important in the pathogenesis of allergic bronchopulmonary aspergillosis than environmental factors. This conclusion resulted from the fact that, of eight atopic and five non-atopic urban dwellers who lived in environments with equal spore counts for *Aspergillus fumigatus*, only the atopic individuals contracted allergic bronchopulmonary aspergillosis. It is noteworthy that even very small quantities of allergen in atopic individuals will lead to a typical, immediate, allergic response.

The anaphylactic-type immunological reaction (allergy) is typified by the rapidity (within 30 s to 3 min after exposure to the 'allergen') by which the response appears. The reaction sequence at the cellular level is pictured as the extrinsic antigen ('allergen') combining/reacting with two antibodies (IgE) attached specifically to mast cells, which are immediately present outside many of the blood capillaries, and also to eosinophilic (basophilic) blood leucocytes. (See Fig. 7 depicting the process of mast cell activation.) The IgE is attached to the mast cell by a binding factor present in the serum. The IgE binding factor from three of four patients with hyper-IgE syndrome enhanced the production of IgE in B-lymphocytes derived from patients with perennial allergic rhinitis or from patients with seasonal allergic rhinitis after pollen exposure. However, it enhanced neither the IgE production by B-lymphocytes from non-atopic donors nor that from patients with seasonal allergic rhinitis without previous pollen exposure. The binding factor from a fourth patient with perennial allergic rhinitis could be separated by affinity chromatography into a fraction showing IgE production enhancement and a second fraction showing suppression of IgE production in B-lymphocytes. This suggests that the IgE binding factors from different individuals can either suppress or stimulate IgE production in B-lymphocytes.[109] Figure 7 shows a mast cell to which two molecules of IgE are attached. An allergen is bound by two IgE molecules, which are attached via an attachment factor to the cell. IgE production is normally suppressed by T-suppressor cells. However, in response to the presence of specific allergens, IgE is produced in B-lymphocytes and plasma cells. (The synthesis of IgE also requires T-helper cells.) IgE, once formed, has a strong affinity for mast cells or basophil where the crystalline fragment portion (Fc) of IgE attaches to cell surface receptors (binding factors). The antigen then binds in crosslinking fashion to two IgEs, which leads to perturbations via the IgE–Fc receptors in the cell wall of the mast cell. These perturbations in turn initiate two parallel and interdependent reactions. One reaction occurs within the cell and a second process occurs in the cell wall.

(a) Reactions within the cell. As shown in Fig. 7, after the antigen crosslinks the two IgE molecules, pre-protease is activated into protease. Protease in turn will transform the inactive adenylate cyclase to the active state. The thus activated adenylate cyclase will transform adenosine triphosphate into cyclic adenosine monophosphate (c-AMP). The c-AMP is then transformed by phosphodiesterase into 5′-adenosine monophosphate (5′-AMP). It appears that, with the initially transient increase of c-AMP, the release of the mediators of anaphylaxis is activated as depicted in Fig. 8.

The level of c-AMP in a mast cell can be durably increased by:

(a) addition of exogenous c-AMP or by addition of dibutyryl-c-AMP which is more stable and penetrates cell membranes more readily than c-AMP; or
(b) by blocking the activity of the phosphodiesterase: this is achieved by addition of theophylline, caffeine and/or quercetine. The beneficial influence of caffeine on the blocking of the release of anaphylactic mediators is, of course, of great interest.
(c) Additional blocking agents for phosphodiesterase, acting via cell surface receptors, have been identified, namely β-adrenergic agents (acetylcholine), adenosine, some prostaglandins (PGD_1), cholera toxin and others.

As stated above, the transient increase of c-AMP concentration is essential for mediator release. However, prolonged increase of c-AMP by reactions (a) to (c) results in a decrease in mediator release. The normal sequence of mediator release, without the interference of the inhibition reaction, continues as follows: the transient increase of c-AMP activates a protein kinase. This protein kinase causes the phosphorylation of perigranular proteinaceous membranes, thus causing them to swell with water, and also permits the influx of Ca^{2+} into the mast cell. The Ca^{2+} activates enzymes which release energy required to form microtubules and filaments around the membrane structures. After formation of filaments, the vesicle content is released outright, or the vesicle moves towards the cell wall of the mast cell and fuses; eventually the cell wall opens (helped by lysophosphatides), releasing the mediators of anaphylaxis contained in the vesicle. What are the contents of the vesicle, and what are the physiological consequences to the tissue structures in the vicinity of granule release? The granules contain histamine, serotonin and bradykinin. Bradykinin forces the opening of small gaps in the venules which are caused by the contraction of endothelial cells of the latter, which are about 100 μm or less in diameter. These small gaps permit the leakage of serum through the venule wall. The specific action of histamine on venules occurs because they seem to have specific high-affinity binding sites for histamine. Such binding sites are not present in arteriolar or capillary endothelium,[110] and therefore no effect is caused to these endothelial walls by histamine. Histamine is also responsible for the intense contraction of the bronchial smooth muscle. Further, increased vascular permeability, and increased mucous secretion by nasal, bronchial and gastric glands, which leads to an immediate allergy attack, are caused by the action of histamine. Chemotactic anaphylactic factors, which specifically attract eosinophils and neutrophils to the site of granule release, are also contained in the granules. In addition, the granules contain large quantities of neutral proteases.

(b) Reactions in the cell wall (formation of prostaglandins, thromboxanes and leucotrienes). The perturbations in the cell wall, caused by the binding of antigen to a pair of antibodies (IgE), activate phospholipases and methyltransferases (I, II and III) in the cell wall (see Fig. 7). These enzymes hydrolyse phospholipids and cause the release of arachidonic acid or other unsaturated fatty acids. The arachidonic acid, for example, is then transformed via two different pathways into biochemically active compounds that play a very important role in the production of some of the allergy symptoms, namely through (i) the lipoxygenase pathway and (ii) the cyclo-oxygenase pathway.

(i) *The lipoxygenase pathway* transforms arachidonic acid into vaso- and broncho-constrictors 1000 times more active than histamine, such as leucotriene $C_4(LCT_4)$ and leucotriene $D_4(LTD_4)$. The interesting feature of these two compounds is that they are derivatives of glutathione. Leucotriene B_4, a dihydroxy derivative of arachidonic acid, is the most potent chemotactic factor known. Oxidation of the alkyl end of the molecule to the alcohol or carboxy stages leads to biochemical inactivation of the chemotactic compounds (2O—OH—LTB_4, 2O—COOH—LTB_4).

(ii) *The cyclo-oxygenase pathway* transforms the various unsaturated free fatty acids obtained by hydrolysis from the phosphoglycerides in the cell wall, e.g. linoleic acid, arachidonic acid and eicosapentaenoic acid, into prostaglandins (PG) and thromboxanes (TB).

(c) Management of prevention of anaphylactic mediator release

(a) Removal of antigen (allergen) from sensitive (atopic) individual. Antigen may consist of dust (e.g. house dust, green coffee bean dust), animal dander, pollen, mould mycelia (fragments), mould spores and various foods (fish, flour products, etc.).

(b) Hyposensitisation therapy. This is done by continued injections of dilute allergen-containing extracts.

(c) Application (insufflation) of disodium chromoglycate (see formula). This agent prevents not only the immediate release of mediators but also the formation of the slow-reacting substance A (SRS-A). Since disodium chromoglycate affects extracellular Ca^{2+} transport, the inhibition of mediator release indicates not only that the intracellular release mechanism of mediators is tied to the presence of Ca^{2+} ions, but also that the release of SRS-A (occurring about 6 h after contact with the allergen) depends on the presence of free Ca^{2+} ions. It has been further shown that disodium chromoglycate increases the c-AMP level and also inhibits exocytosis of the granules of mast cells.

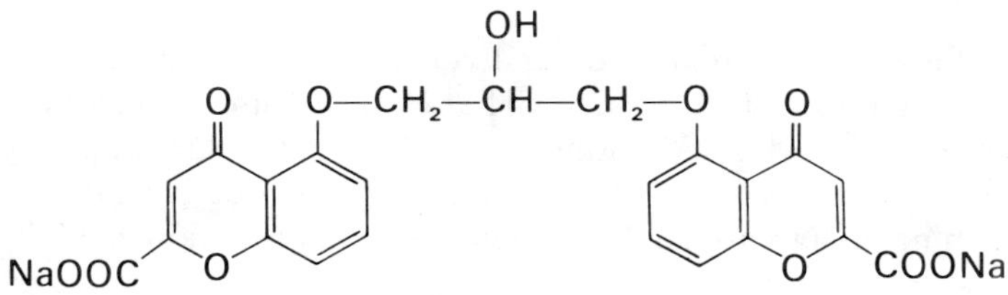

Disodium chromoglycate (Inhal)

(d) The application (by inhalation, per os, or injection) of corticosteroids (cortisone), hydrocortisone or adrenocorticotropic hormone (ACTH) will deplete the number of basophilic cells (including mast cells) in the organism and thus decrease the hyper-reactivity towards allergens.

(e) The application (by inhalation, per os, or injection) of catecholamines (epinephrine, isoproterenol, salbutamol, etc.) acutely inhibits an anaphylactic reaction. Catecholamines are β-adrenergic stimulators and act to relax the smooth muscle system in the lung to alleviate asthmatic attacks. They also apparently act on the venules of the blood vessel system and keep the

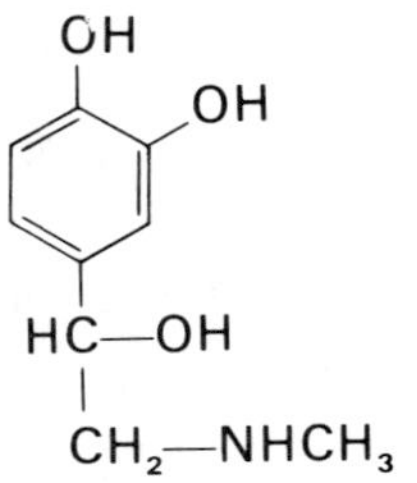

Epinephrine (adrenaline)

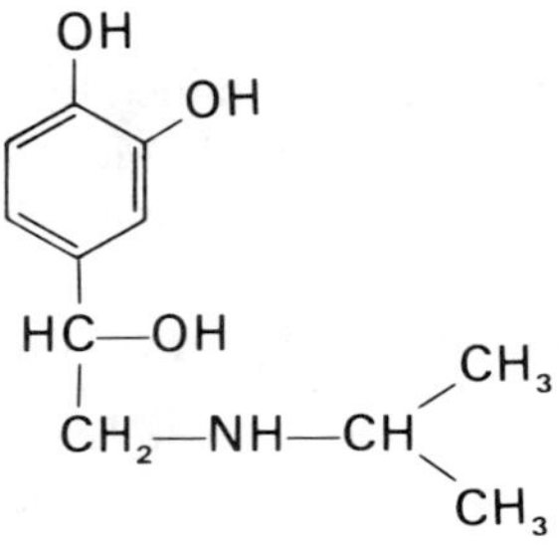

Isoproterenol

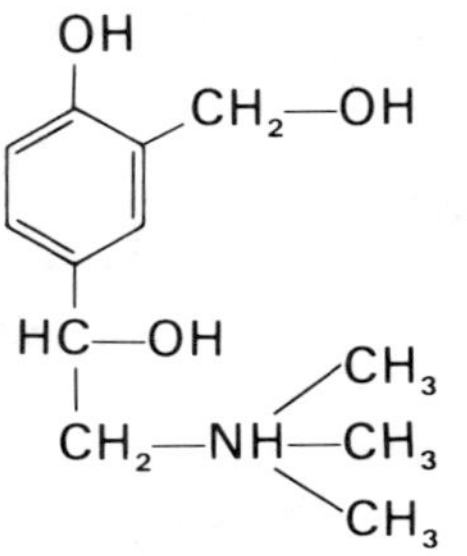

Salbutamol

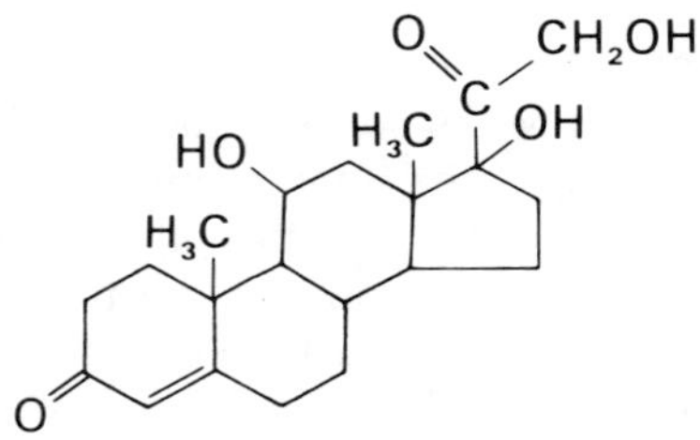

Hydrocortisone (cortisol)

endothelial cells from contracting; they will not permit formation of interendothelial pores (gaps) and thus will prevent plasma leakage (oedema formation).

(f) Antihistamines are to be subdivided into H_1 and H_2 receptor-dependent actions of histamine. The H_1-histamine antagonists are represented by the classical antihistamines like ethanolamines, ethylenediamines, alkylamines, piperazines and phenothiazines, etc. Some of the H_2-histamine antagonists are impromidine, dimaprid and cimetidine.

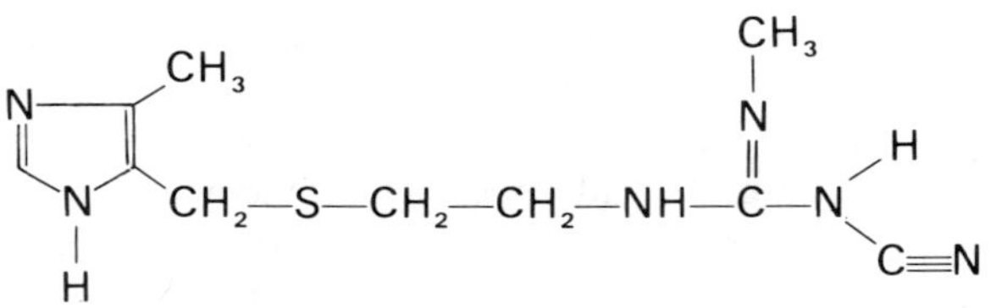

Cimetidine

(g) Methylated xanthines and also certain flavonols act as inhibitors of cyclic adenosine monophosphate phosphodiesterase, thus leading to elevated concentrations of c-AMP. As mentioned before, prolonged increase of c-AMP will decrease the release of anaphylactic mediators from the mast cells (and eosinophilic cells in general), most likely via a decoupling mode.

It is of interest to point out that there exists a long medical history of xanthine derivatives as a remedy for asthma, with theophylline now found to be more effective than caffeine.

In a 1986 overview of effects of theophylline in asthma therapy, G. A. Persson[208] summarises that theophylline

(1) relaxes the smooth muscle of large and small airways in humans and animals;
(2) is effective and is relatively independent of type of mediator causing the constriction of airways;
(3) shows anti-inflammatory effects by deactivating inflammatory mast cells, basophils, neutrophils, macrophages, thrombocytes, etc., at 10^{-5} to 10^{-4} mol litre^{-1} therapeutically relevant low concentrations; and
(4) may be a potent adenosine receptor antagonist.

1.5.8. Immune Complex Mediated Type (Type III)

In Type III hypersensitivity, immunoglobulins (other than IgE) are reacting directly with the antigen (see Fig. 8) to form an antigen–antibody complex. IgG and IgM are the main reactants that form a complex with the antigen. It is this complex that activates a variety of serum mediators. However, activation of the so-called complement cascade systems is viewed as the central event of this hypersensitivity reaction that leads to tissue damage. The complement system is activated via (i) the classic pathway (by the complex formation of antigen with antibody), or (ii) the alternate pathway (activated by bacterial endotoxins, cobra venom or aggregated globulins). The complement system is found in its highest concentration in the plasma. The system consists of initially 11 proteinaceous (serum) components. They are capable of mediating increased vascular permeability, chemotaxis, opsonisation of the antigen prior to phagocytosis and lysis of the target organisms, and inflammation.[111,112] Again the antigen may also be from an endogenous source, e.g. trace components in the blood or antigenic components contained in tissues or cells. The latter antigen–antibody reaction can involve many vessels in many organs (serum sickness) or, in the case of glomerulonephritis, may engage only the kidney, or in arthritis the joints, or only the small blood vessels in the skin, as in the Arthus reaction.

T-Cell-mediated, delayed hypersensitivity (Type IV). Contrary to

Type I hypersensitivity (anaphylactic type), where hypersensitivity reactions occur within minutes after allergen exposure, the Type IV hypersensitivity requires several hours (faster in insect stings, though) before a reaction is noticed, and mostly the reaction reaches a maximum after 24–72 hours before it subsides. Many different stimuli can provoke Type IV hypersensitivity:

(1) Fungal diseases: candidiasis, dermatomycosis, coccidiomycosis, histoplasmosis;
(2) Bacterial diseases: tuberculosis, leprosy, anthrax; parasitic diseases: leishmaniasis, schistosomiasis.
(3) Contact dermatitis (skin sensitivity): *chemicals*, e.g. chromates, picryl chloride, *p*-phenylenediamine (in hair dyes), neomycine (in topically applied ointments), nickel salts (in thimbles and jewellery clasps), etc.; *plant allergens*, e.g. urushiol of poison ivy, poison oak.
(4) Insect bites.
(5) Graft rejections (circulating antibodies are engaged).
(6) Virus-infected cells, measles, smallpox, herpes virus, AIDS; tumour cells.

Delayed-type hypersensitivity reactions are initiated by sensitised T-lymphocytes which are produced during the initial or, many times repeated, contact with the antigen. T-memory cells, after initial sensitisation, often remain for several years in circulation. The inflammatory reaction is activated by memory T-lymphocytes on re-exposure to sensitised/reactive antigen. The memory T-lymphocytes divide rapidly and release biologically active lymphokines. The lymphokines recruit, in turn, inflammatory cells, initially mononuclear cells, followed by exudation of mononuclear and polymorphonuclear types. When these cells later migrate out of the lesion, a lymphocyte and monocyte–macrophage-type cell infiltrate is left behind.

Tissue damage results from lesions caused by bacterial allergy such as cavitation, drying out of tissue (caseation), and general toxaemia as observed in tuberculosis and granulomatous skin lesions in borderline leprosy.

With this abbreviated background information on hypersensitivity, we can now turn, with some appreciation of the complexity of this subject, to the consequences suffered by some predisposed people when they come in contact, on an occupational basis, with allergens that accompany green coffee beans or many other occupational sources of antigens.

2. ALLERGENS

This section presents the ongoing studies of allergies caused by green coffee bean dust which were first reported by Figley and Rawling in 1950.[101] A detailed report on these ongoing studies is presented in Section 2.4. However, no clear picture has been obtained to date on the causative agent(s) of these allergies or whether the symptoms encountered are totally allergenic in character. Microscopic examination of green coffee bean dust revealed a multiplicity of components ranging from cellulosic fibres, pollen, mould spores, mould mycelia, bacteria and, of course, fragments derived from coffee beans, including cuticle wax (see Figs 9a–d). Electron microscopic pictures of green coffee bean dust are presented in this chapter for the first time. These pictures are presented to shed some light on the probable actors of the green bean dust allergy.

2.1. Respiratory System

Many of the allergic reactions to be described in Section 2 and some of the mycoses resulting from inhalation of mould conidia or mould mycelium are related to the respiratory or breathing system, therefore some basic description of the latter is necessary. Figure 10 shows the elements of the system in which the air passage, starting from the nasal cavity via the turbinates through the pharynx and larynx, leads to the trachea (windpipe), after which the bronchial tree, consisting of the two bronchi and their subdivisions, and the bronchioles from which lead out around 300 million alveoli (blind-ended sacs of microscopic size; see inset of Fig. 10), leads to the lungs. The bronchi enter the lungs (about 5·5 litres average capacity), and the alveoli are the sites of gaseous exchange from the blood of the pulmonary capillaries. The respiratory tract is lined with mucous membrane, but only the epithelium of the upper tract is ciliated.

The ciliated mucous-covered linings of the upper tract (nasal walls, primary, secondary and terminal bronchial tubes) have the function of trapping inflowing airborne particular pollutants in the mucous layer and then transporting it in co-ordinated ciliary action (called the mucociliary escalator) to the throat where the mass is either usually swallowed or expectorated.

Large conidia, pollen or dust particles (ranging from over 60 μm down to about 6 μm) are trapped mainly in the upper respiratory tract. Smaller particles ($<5\ \mu$m) may reach the respiratory bronchioles, alveolar ducts and the alveoli (see Fig. 10). Deep-penetrating conidia have been known for typically causing delayed allergy symptoms.[28] Secretion of IgA from

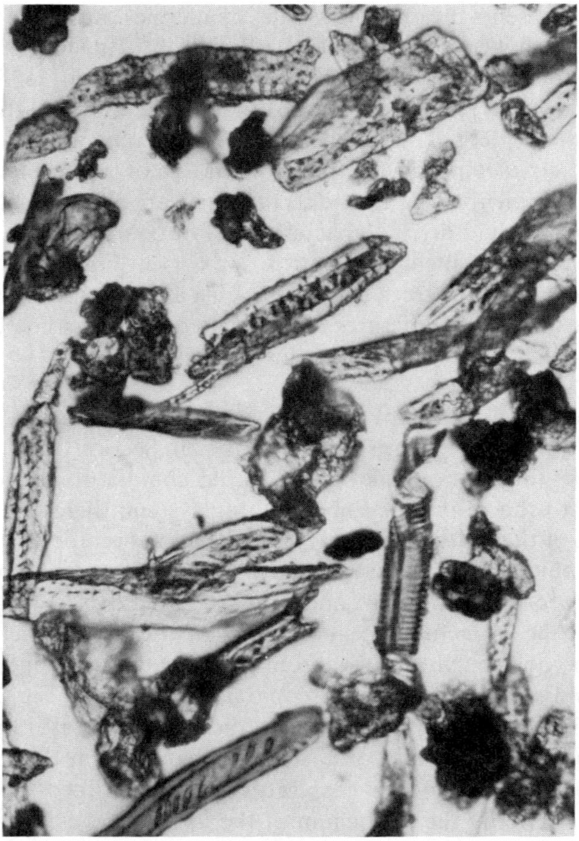

Fig. 9a. Photomicrograph of green coffee bean dust (original magnification ×200). Most of the particles are fragments of fibres from the green bean coffee sacks with an average length of about 1–2 mm and an average thickness of about 30–40 μm. The smaller particles (10–30 μm thick) are non-fibrous and constitute, for the most part, coffee bean fragments. All pictures concerning green coffee bean dust were taken and interpreted by Dr D. Montezinos. Green coffee bean dust was supplied by Ms E. Chartier and Mr M. Noyes. The dust was removed from the filter of the exhaust ventilator of the green bean unloading station in one of the coffee plants.

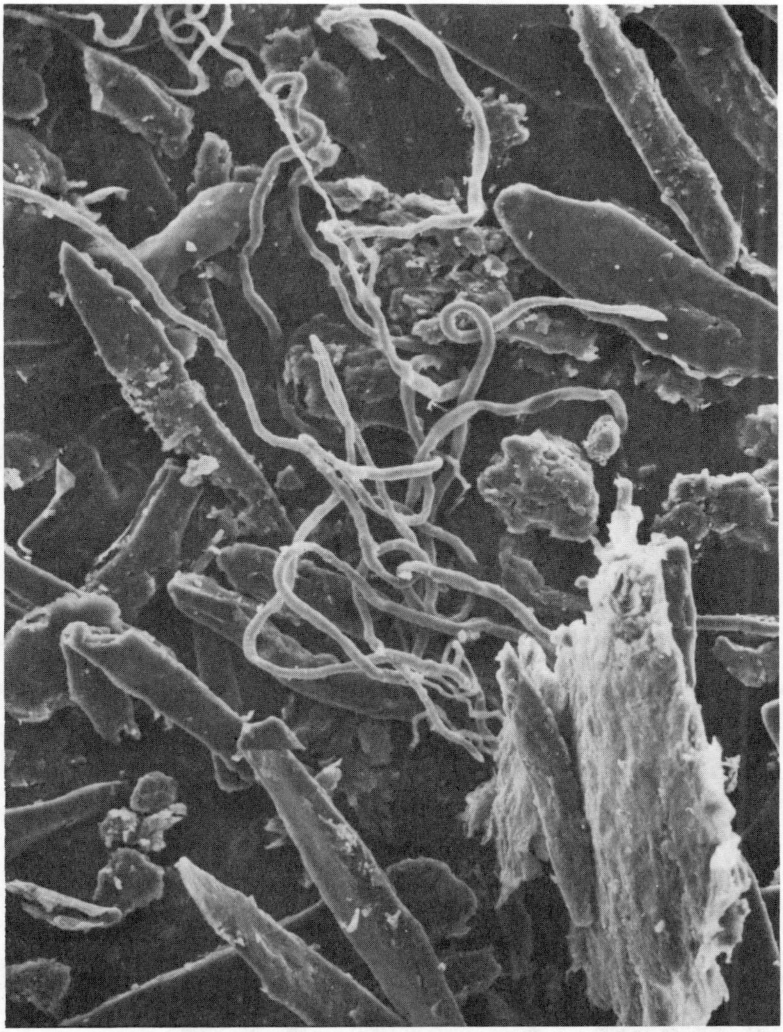

Fig. 9b. Electron scanning micrograph of coffee dust (original magnification $\times 375$). Note the long intertwining mould mycelium in the centre of the picture. The average thickness of the mycelium ranges from 4–6 μm. The scale-like fragments are tentatively identified as coffee wax with a thickness of about 20–40 μm and a length of 50–150 μm. The larger particle in the lower right-hand corner is a coffee bean fragment.

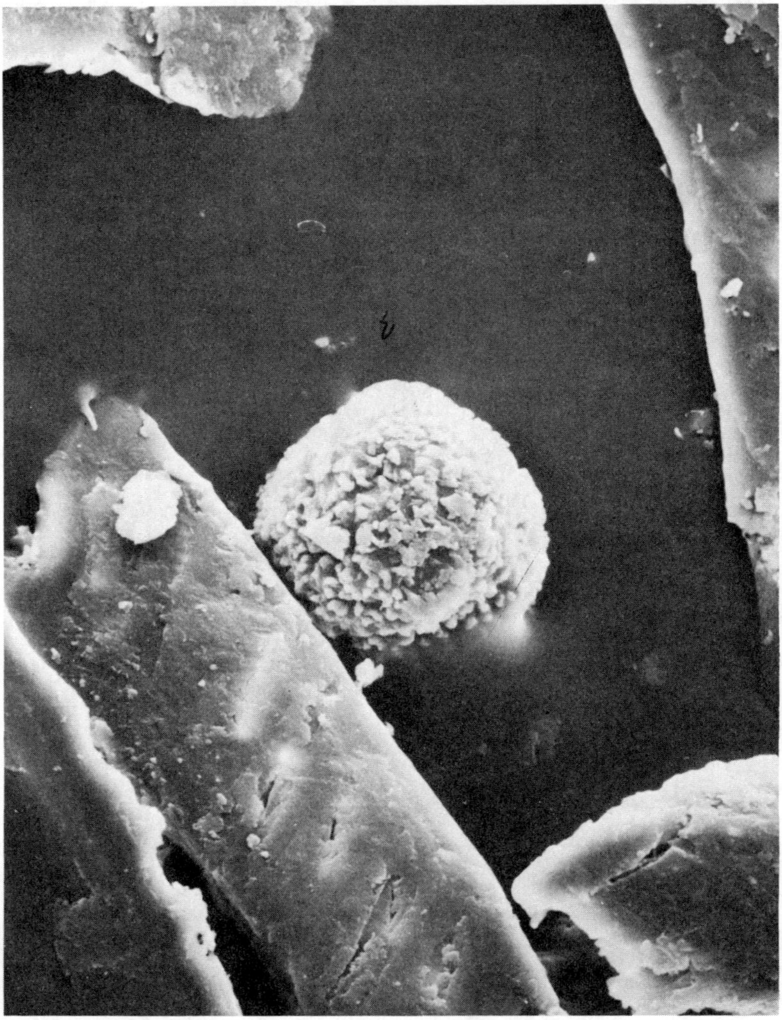

Fig. 9c. Electron scanning micrograph of coffee dust (original magnification ×1500). The spherical particle in the centre of the picture is tentatively identified as a pollen grain with a diameter of 28 μm. The adjacent elongated particle is a coffee wax particle with an approximate thickness of 23 μm.

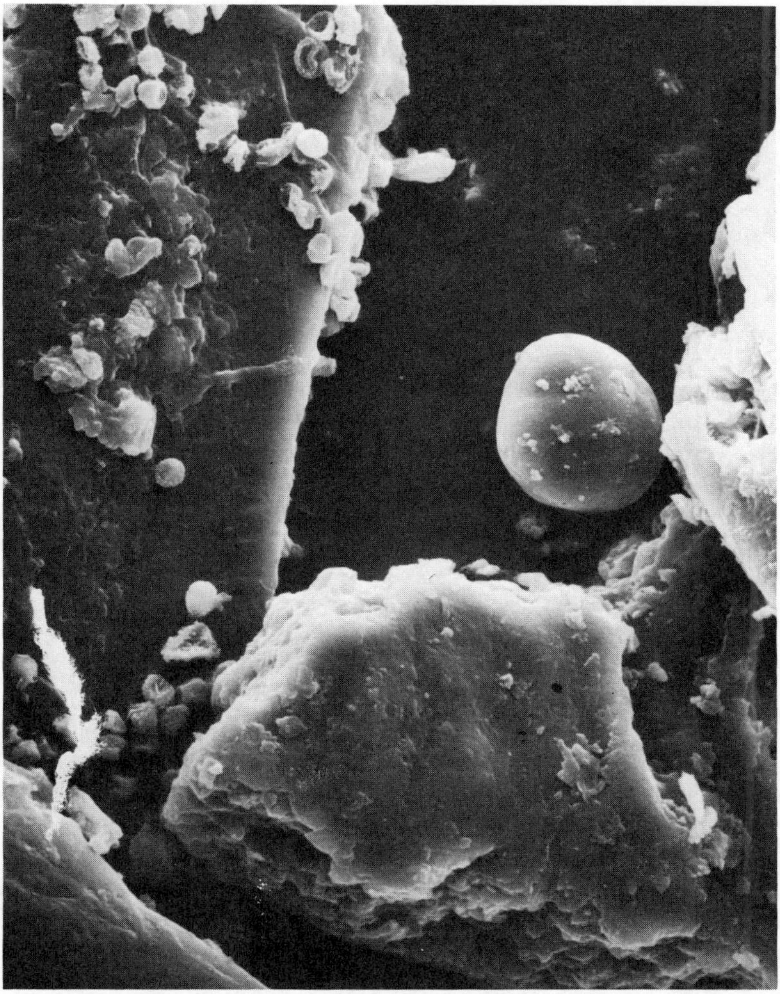

Fig. 9d. Electron scanning micrograph of green coffee bean dust (original magnification ×2250). The spherical body in the upper centre of the picture is tentatively identified as a mould spore with a diameter of 10·6 μm. The smaller spherical bodies clinging to a coffee wax fragment located on the upper left-hand side of the picture are tentatively identified as smaller mould spores with diameters ranging from about 1·7 to 2·6 μm.

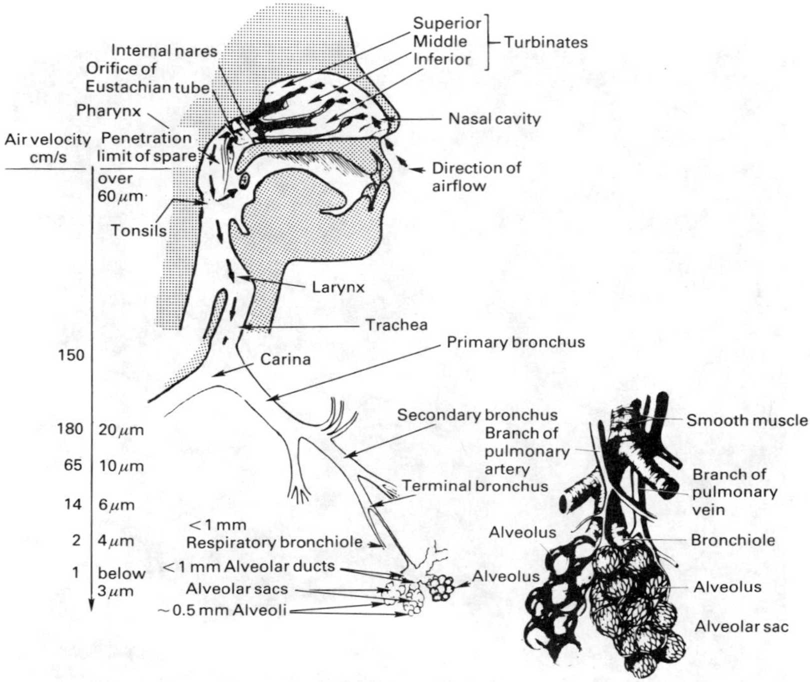

Fig. 10. Human respiratory tract figure (modified from refs 97–100).

laminal glands distributed throughout the respiratory tree constitutes an important defence against foreign chemicals. Furthermore, alveolar macrophages are found in the alveoli. These cells, in phagocytic fashion, engulf and destroy airborne particles reaching the alveolar region.

2.2. Microscopic Analysis of Green Coffee Bean Factory Dust

The case and clinical studies on coffee workers handling green coffee beans described in forthcoming paragraphs provoke a number of questions.

(a) Why are symptomatic green bean coffee workers more sensitised to castor bean allergen than to actual green coffee bean extract?[127,128,138]

(b) Why is the percentage of coffee workers with allergic symptoms similar to the percentage of atopic people who do not come in contact with green coffee dust?[28]

(c) Why do burlap sack extracts (not containing green coffee beans) also cause allergy reactions?[128]

All these and more questions are begging for answers. However, one must not overlook the main question, which is whether green coffee bean allergy is an allergy caused directly by green beans or by other components, such as contaminants, also present at the green bean coffee unloading station. The question 'Are chlorogenic acids (one of the major components of the green coffee bean) causing the allergy reactions seen in some coffee workers who come in contact with green coffee beans?' has been affirmatively answered. Numerous investigators tested chlorogenic acids and found that neither chlorogenic acids directly isolated from coffee beans nor any of those obtained from other sources caused any allergy.

Green coffee bean unloading from jute sacks into hoppers feeding the coffee roasters is accompanied by the stirring up of dust. It is well known that atopic individuals, consisting of about 20% of the population, will show symptoms of allergy sooner or later as a result of coming into contact with such dust.[28] Even non-atopic workers when subjected to chronic exposure to heavy dust loads—any dust, not just green coffee bean dust—sooner or later may become sensitised to some of the dust ingredients.[28] Atopic individuals are easily sensitised by air spora at concentrations of 10^6 spora per cubic metre. Non-atopic individuals require loads of spora ranging from 10^6–10^9 m^{-3} before sensitisation sets in.[28]

To provide perspective on the green bean coffee dust question it should be pointed out that some industrial dusts, ranging from strictly inorganic to mainly/purely organic in composition (carbon, clay, asbestos, flour-mill dust, cotton gin dust, purified cotton linters, etc.), can present serious health problems. Fortunately no such threatening health problems are revealed in the case studies of coffee dust.

Microscopic examination of the composition of dust deposited on ventilator filters installed in the green bean coffee unloading stations revealed some interesting results. The dust particles were composed of (in descending order of estimated quantity):

(1) cellulosic fibres (jute sack fibres),
(2) fragments of jute sack fibres,
(3) fragments of green coffee beans,
(4) mould mycelia,
(5) mould spores, and
(6) pollen.

The pictures of green bean coffee dust obtained with light microscopy and electron scanning microscopy are shown in Figs 9a–d. These figures demonstrate that only one component from a total of six tentatively identified actually originates from green coffee beans. All other components are contaminants either from the packaging material (jute sacks) or from general environmental materials such as mould spores, mycelia and pollen. Bacteria and inorganic dust particles (not shown in the micrographs) are also part of the dust.

Other than the known allergens, such as pollen, mould spores, fragments of mould mycelia, cellulosic fibre fragments and bacteria, there is one potential coffee bean derived allergen, namely cuticular coffee bean wax particles containing 5-hydroxytryptamide. Qualitative and quantitative studies are under way to determine whether or not 5-hydroxytryptamide is a component of coffee dust, and if yes to what extent.

In order to alleviate the occupational discomfort potentially connected with the dust generation, efforts can be made to remove most of the dust via ventilators and air suction devices. This is already the case in more modern coffee-roasting installations. Replacement of jute sack transport with bulk containers will eventually eliminate the main source of dust particles, namely jute fibres and jute fibre fragments. Wearing of face masks during emptying of coffee sacks is one practice that can be exercised immediately and readily by those individuals who experience symptoms when exposed to green bean coffee dust, as either a precautionary or a preventive means. In both hypersensitivity reactions, whether via IgE–allergen–mast cell interaction (anaphylactic immunological response, Type I) or IgG/IgM–allergen interaction (immune complex-mediated, Type II/III), the removal of the allergen will successfully limit any hypersensitivity reaction.

2.3. Chemical Analysis of Green Bean Coffee Dust

In order to obtain an estimate on the actual contribution of green bean components to the composition of green bean coffee dust, the following quantitative analyses were carried out:

(1) chlorogenic acids;
(2) caffeine; analytical values (1) and (2) should give an indication of how many coffee bean fragments are part of the dust associated with green coffee beans;
(3) 5-hydroxytryptamides.

Cuticle wax fragments were tentatively identified by means of electron microscopy. Since cuticle wax is the only morphological component of the coffee bean that contains 5-hydroxytryptamide, a determination of 5-hydroxytryptamides will reveal the approximate ratio of actual endosperm-derived green bean fragments vs cuticle wax fragments.

All analytical data are reported on a weight basis. One has to caution that the vastly different densities of the different dust components will exaggerate those components with high densities and will diminish those with low densities. Visual observation and estimation suggests that sack fibres and fragments thereof, from a volume point of view, represent the prevailing dust component.

Chlorogenic Acids and Caffeine

Results of analysis of caffeoylquinic acids, feruloylquinic acids and dicaffeoylquinic acids for dust, two arabica bean samples, a robusta bean sample and a typical coffee blend are shown in Table 9. The content of chlorogenic acids in green coffee beans shown in the table ranges from 7·3 to 12·0%.

Table 9
Percentage of chlorogenic acids and caffeine in green coffee bean dust

	Green coffee bean dust	*Arabica I*	*Arabica II*	*Robusta*	*Coffee blend*	*CA and caffeine in coffee dust/ coffee blend*
Chlorogenic acids	1·902	7·900	7·388	12·008	9·099	0·210
Caffeine	0·664	0·847	0·887	1·659	1·108	0·600
CA/Caffeine ratio	2·86	9·32	8·33	7·23	8·21	

In accordance with literature values, the arabica beans show a lower chlorogenic acids content than the robusta beans.

The caffeine contents in green arabica bean samples vs robusta beans analysed range from 0·85 to 1·66%. These values also fall into the ranges found in the literature.

Ratios between the chlorogenic acids content vs caffeine content were calculated. These ratio values for arabica and robusta beans range between 9·3:1 and 7·2:1. The coffee blend shows a ratio of 8·2:1.

The coffee dust samples show a strikingly different ratio, namely only 2·9:1. The caffeine content in the dust comes only to about 60% of the caffeine in the coffee blend. The content of chlorogenic acids in contrast amounts to only about 21% of chlorogenic acids contained in the coffee blend. One explanation of this discrepancy is to assume that the coffee bean dust fragments are derived mainly from the outer bean region where the content of chlorogenic acids is lower than in the centre of the coffee bean.

Therefore using the chlorogenic acids as an indicator for green bean coffee particle concentration may result in a misleadingly low estimate. Using the caffeine content as a more reliable indicator also is not entirely justified since no or very little caffeine is present in the non-cellular centre of the coffee bean. The estimate for green bean components in the coffee dust resulting from the determinations of chlorogenic acids and caffeine can be given at best at this time as a range value, namely from about more than 20% to less than 60% or an average estimated rough value of about 40%.

5-Hydroxytryptamide

The analysis of 5-hydroxytryptamide by determining the main components—arachinoyl, behenoyl and lignoceroyl-5-hydroxytryptamide—reveals a relatively high content of the coffee cuticle wax in the green bean coffee dust (see Table 10).

Table 10

Content of 5-OH-tryptamides (mg kg^{-1}) in green bean coffee dust and commercial blend of green beans

Component	*Coffee dust*	*Commercial bean blend*	*Ratio of 5-OH-tryptamide in dust vs bean blend*
Arachinoyl-5-OH-T	1 003·0	229·85	4·36
Behenoyl-5-OH-T	1 423·2	395·00	3·6
Lingnoceroyl-5-OH-T	221·4	43·00	5·15
Total	2 647·6	667·85	Av. 4·37

It is of interest to point out that the ratios of chlorogenic acids and caffeine in the green bean dust vs a commercial green bean coffee blend yielded numbers smaller than one, whereas the ratio of 5-OH-tryptamides in the coffee dust vs a commercial green bean coffee blend showed numbers greater than one. This indicates that the cuticle wax is by far the prevailing coffee bean component of the green bean dust.

2.4. Case and Clinical Studies on Coffee Workers Handling Green and Roasted Coffee

The occurrence of asthmatic symptoms in workers handling green coffee beans was first reported in 1950 by Figley and Rawling.[101] The authors suggested that residues of castor beans, transported in the coffee sacks before using them for green beans, were the factors responsible for the asthmatic symptoms. Bronchial asthma is a well known allergic condition for people in many walks of life. Allergies, especially from inhalants, are studied through a number of clinical tests, e.g. skin prick (or patch) tests, respiratory functioning, bronchial provocation, and serological analyses for IgE and IgG antibodies by means of the radio-allergo-absorbent test (RAST). Since this limited early work, a large number of other studies have been carried out with actual coffee bean dust and with aqueous extracts prepared from green and roasted coffee. These studies are described in Section 2.4.1.

The electron microscopic pictures of green bean coffee dust indicate the presence of:

(1) Pollen
(2) Mould spores
(3) Mould mycelia
(4) Bacteria (shown by preparation of cultures)
(5) Sack fibres and fragments thereof
(6) Coffee bean particles
    (a) Endosperm (high in caffeine, low in chlorogenic acids)
    (b) Cuticle wax scales from coffee bean surface (high in 5-OH-tryptamides)

The components (1) to (5) all have been shown to provoke allergic responses to higher or lesser degrees in atopic people. Green coffee bean particles [(6), (a) or (b)], as such, are not known to cause allergies. Based on the allergy-suppressing capability of methyl xanthines, one would expect that coffee bean particles containing relatively high quantities of caffeine actually should help to suppress the provocation of allergies and allergic inflammatory subsequent reactions. Some authors of the present studies actually express that green coffee bean allergens are of moderate potency and that high concentrations are required to provoke sensitisation. This is in contrast to castor bean allergen where very small quantities suffice to provoke sensitisation and subsequent allergic response.

2.4.1. Studies from 1961 to 1978

In 1957 Bruun [113] showed that coffee workers showing hypersensitivity symptoms gave positive skin tests to green coffee bean extracts. Investigations by Freedman *et al.*[114–117] in 1961 to 1964 led these authors to propose chlorogenic acids as the active ingredient in green coffee beans, citrus fruits and castor beans that caused the allergic symptoms. Layton and co-workers,[118,119] challenging the results by Freedman *et al.*, suggested that the chlorogenic acids preparations were contaminated with proteins of high molecular weight, and proposed alternatively that the proteinaceous components were the cause of the allergic symptoms. These authors[118,119] found no allergic activity in other low molecular weight, proteinaceous, green coffee components, in chlorogenic acids or in extracts derived from roasted coffee beans. After spirited debate,[120–123] the findings of Layton *et al.*[123] prevailed when human non-atopic volunteers and macaque monkeys, on treatment with sera from allergic coffee workers, showed a positive cutaneous hypersensitivity for green coffee bean extracts but not for chlorogenic acids. Turula *et al.*[124] reported on allergies in workers handling green coffee beans; and there were two further contributions, in 1970[125] and in 1975.[126] There was never any debate, though, concerning the course of the hypersensitivity reactions. Typical, immediate-type weal and flare skin reaction suggested a reagin-mediated hypersensitivity reaction (Type I hypersensitivity; see Section 2.2). Such reactions were brought about by green coffee bean, castor bean and burlap sack extracts.[127] However, no direct 'cause and effect' relationship between the allergens mentioned above and the asthma of coffee workers was established. In 1978 Karr *et al.*[128] answered this question by exposing eight coffee workers who had job-related respiratory symptoms (asthma, rhinitis, conjunctivitis, urticaria or pruritus) to phosphate-buffered saline extracts of green coffee beans, castor beans and three different sack samples. Three different test methods were applied to determine the allergenic activity: the serum radio-allergosorbent test (RAST), provocative inhalation challenge, and intradermal skin testing. The results indicated that coffee workers with occupational allergic disease demonstrated the presence of serum IgE antibodies specific for antigens which stemmed from green coffee bean dust, castor bean and one sack sample. These antigens were distinct and unrelated to chlorogenic acids and were capable of producing asthma in sensitised atopic subjects. Lehrer *et al.*,[129] in a second study, determined, via the mouse model, the reaginic antibody response as produced by the passive cutaneous anaphylaxis reaction. Mice (and rabbits) were immunised with phosphate-buffered saline extracts of (a) green coffee bean dust, (b) green coffee beans, (c) roasted coffee beans, (d) coffee cleaner debris from green beans, (e) chaff, and (f) raw castor beans. Cross-reacting allergens* were found in cleaner debris, in green beans, but not in chaff or roasted coffee beans. None of the

* Cross-reacting allergens may be from different sources, but they are of close similarity in molecular shape and charge distribution, and therefore react with the same antibodies.

coffee bean-derived allergens cross-reacted with castor beans, hence indicating the different chemical nature of the two allergens. Column separation (Bio-Gel A and Sephadex G-100) of the green bean allergen indicated heterogeneity in molecular size, but all of the fractions isolated showed immunological activity, indicating that allergenicity was distributed over all the fractions—regardless of molecular size. Testing with chlorogenic acids again showed no immunochemical activity. In 1978 Osterman *et al.*[130] used the skin prick test, the radio-allergo-sorbent test, various allergic symptoms (such as itching in the nose, rhinitis and breathing discomfort) and IgE determination to study the effect of green coffee dust and of chlorogenic acids on 48 coffee workers. Twenty-three workers complained about positive allergic symptoms, ten showed a positive prick test, four had a marginally positive prick test and eight had elevated IgE. None of the workers showed positive allergic reactions to chlorogenic acids.

2.4.2. Studies from 1979 to 1982

A different approach to the determination of respiratory effects on coffee workers of green and roasted coffee bean dust was taken by Zuskin *et al.*[131] The respiratory function of the coffee workers was determined by measuring changes of the vital lung capacity and in the forced expiratory volume. Measurements of forced expiratory volume on Mondays before and after work showed only very small decreases in the vital lung capacity—ranging from 1·3 to 2·8%. Measurements of maximum expiratory flow rate at 50% vital capacity, however, decreased from 4·0 to 8·7% (for roasted and green coffee dust) and at 25% vital capacity from 6·0 to 18·5% (for roasted and green coffee dust). On Thursdays the ventilatory changes before and after work were less than those observed on Mondays. Decreases in flow rates at low lung volumes suggest that the bronchoconstrictor effect of dust acts mostly on smaller airways. Administration of Inhal (disodium chromoglycate) before work considerably diminished marked reductions in flow rates. Of 33 workers, 16 were completely protected by administration of Inhal. Dust measurements in the air of the green and roasted coffee working areas revealed an average of 11·2 and 4·3 $mg\,m^{-3}$, respectively.

Previous studies in 1976 by Zuskin and Bouhuys[132] and Zuskin *et al.*[133] had shown that Inhal protects against the airway constrictor response to textile dust or dust extracts. According to Cox,[134] Inhal prevents the release of chemical mediators from mast cells. In a follow-up study in 1981, Zuskin *et al.*[135] determined the immunological status and respiratory function of

45 female coffee workers. Aqueous extracts of dust collected from emptying green coffee sacks, from green coffee and from roasted coffee were used for intradermal skin testing. Serum levels of IgE were determined by the radioimmunological sandwich technique; IgA, IgG and IgM were determined by single radial diffusion on partigen plates. Histamine and aqueous extracts of house dust, bacterial antigen (obtained from six different bacteria), mould antigen (from six different moulds) and *Dermatophagoides pteronyssinus*, for comparison, were also used in the skin testing. Histamine gave a positive reaction in 100% of the workers, whereas only 40% of the coffee workers and 14·7% of the non-coffee workers in the control group showed a positive reaction to dust collected from emptying coffee sacks. Twenty per cent of all the workers, however, also showed a positive skin reaction to house dust. About 10% of the coffee workers showed a positive skin test to the dust of green coffee beans and roughly 5% of the coffee workers showed positive skin tests to roasted coffee bean dust. None of the workers in the control group showed a positive skin reaction to the latter two dusts. However, 14·7% of the control group were sensitive to the green coffee sack dust. IgE was found at above normal levels in 24·4% of the coffee workers and in 5·9% of the control group. IgM was increased in 18·2% of the coffee workers and no increase was found in the control group. IgA and IgG were not increased in either the coffee workers or in the control group. The increase of IgE in some of the coffee workers indicated a Type I immunological mechanism mediating some forms of bronchial asthmatic or other atopic symptoms. Of 11 coffee workers with increased IgE, seven (12·9% of all coffee workers) complained of rhinitis and conjunctivitis. Only two of the four coffee workers with asthma had increased IgE levels. Among those coffee workers with increased IgM, only one showed a positive reaction to allergens associated with green coffee beans. Lung function measurements (recording the maximum expiratory flow volume) were also carried out. Measurements on Mondays before and after work were taken on a portable flow volume spirometer,[136] and flow rates at 50% and 25% of the control vital capacity were taken from these curves. The mean of the two highest maximum expiratory flow volumes was used as the test result. The workers in the control group showed 5·18 litres s^{-1} at 50% maximum expiratory flow rate, whereas the coffee workers had 4·80 litres s^{-1} before work. After work the coffee workers showed a reduction in flow rate of 380 ml s^{-1} or 7·9%. Unfortunately no measurements of the control group were taken after work. At 25% of vital capacity, the coffee workers showed 2·25 litres s^{-1} while the control group showed 2·85 litres s^{-1}. After work the coffee workers showed a decrease to

1·85 litres s^{-1}, which amounts to a reduction in flow volume of 17·8%. Again no measurement was recorded for the control group. In 1974 the National Institute for Occupational Safety and Health[137] recommended (for cotton dust) that reductions in ventilatory capacity by more than 10% should be classified as an effect of dust.

Osterman *et al.*,[138] reporting in 1982, studied two groups of coffee workers at two-year time intervals with 170 (study I) and 129 (study II) workers, respectively. They looked at the allergies of 25 sensitive coffee workers and at 25 workers showing hypersensitivity symptoms to determine:

(a) the type of symptoms and their prevalence;
(b) the occurrence and importance of IgE-mediated sensitisation and the relative importance of allergens from coffee beans and castor beans; and
(c) the factors contributing to allergy symptoms such as atopy, degree of exposure and smoking habits.

The symptoms ranged from coughing, wheezing in the chest, sneezing, itching of the eyes and nose, nasal congestion, lacrimation, itching in combination with nasal symptoms, eczema, itching red (skin) spots or itching alone. The most prevalent symptoms were allergic symptoms of the nose and eye (study II). Most of the workers reported that the symptoms were aggravated by the increase in green coffee bean dust. In study II, 28 of the 129 workers (22%) showed sensitisation to green coffee dust, as evidenced by the skin-prick test. Of the 20 workers involved in handling the green coffee beans with very high dust occurrence, 10 showed a positive skin-prick test. Of the 109 workers not working in the green coffee unpacking area, 18 showed a positive skin-prick test. The authors concluded that the green coffee bean allergen was of moderate potency and that high concentrations of dust were needed to provoke sensitisation. In contrast, castor bean had one of the most potent allergens ever experienced and even very small amounts sufficed to cause sensitisation. Castor bean is only a contaminant in the green bean unpacking area; nevertheless, more workers were found to be allergic to castor bean extract than to coffee bean extract. Smoking was found to contribute greatly to the risk of sensitisation to green coffee dust. The geometric mean value of log IgE for smokers was 88 and for non-smokers 30·7 ku $litre^{-1}$. In 1985, in a more refined study, Osterman *et al.*[139] described tests with a partially purified, water-soluble extract from green coffee beans on 22 coffee workers afflicted with work-related symptoms of varying degrees from eyes, nose or bronchi. The study was undertaken in order to find the most valuable diagnostic test to reveal sensitisation. The workers themselves measured their lung function with a spirometer three times a day for one week. Skin prick, intradermal and bronchial (via inhalation) provocation testing (of 1 ml of nebulised extract) was carried out. The partially purified extract was diluted to a 1:10 000 ratio, and this concentration was increased, in 10-fold increments, up to 1:10 dilution (w/v) or until a reaction became visible. A positive reaction was determined by the peak expiratory flow rate measurement with a Wright flow meter. A decrease in the flow rate by 15% was considered evidence for positive bronchial provocation response. The nasal provocation test consisted of instilling 0·1 ml of an extract into a nostril via a syringe. Extract dilutions of 1:1000 were applied in 10-fold increments until a positive reaction occurred or until the dilution of 1:10 was reached. Occurrence of sneezing, nasal blockage and nasal discharge was considered a positive response. Serum IgE levels were measured and IgE antibodies were determined. The most common symptoms were nasal congestion, sneezing and itching in the nose (20 workers) and cough and wheezing in the chest (16 workers). Three workers complained of skin itching, one of sweating and one of burning sensations in the throat.

Some of these workers had symptoms when they came into contact with the burlap sacks used for green coffee transportation. The 22 symptomatic workers were subdivided into three different groups according to the severity of their reactions to the various tests. Eleven workers had mild reactions; they were not responsive to either nasal or bronchial provocation. Four of them had a decrease of 15% or more of peak expiratory flow rate after challenge with methocholine. Although none of the workers showed detectable IgE antibodies via the radio-allergo-sorbent test with coffee bean extract, six of the 11 workers did show detectable levels with castor bean extract. Serum IgE in this group occurred at relatively low levels—from 3 to 180 ku litre^{-1}. Five workers had reactions of moderate severity and six workers showed reactions of greater severity. Seven workers of the moderate/severe groups were considered atopic. Six of the 11 workers with mild reactions, all five workers with moderate reactions and also three of six workers with more severe reactions were smokers. Smoking is often associated with increased reactivity since it affects the bronchial mucosa.[140]

2.4.3. *Studies on the Associated Effect of Smoking*

Becker *et al.*,[141] consistent with the finding of Osterman *et al.*[138] that smoking enhances the risk of sensitisation to green coffee bean dust, showed that tobacco leaf extract and tobacco smoke condensate (and also cocoa, coffee and ragweed extracts) contain polyphenol haptens that are capable of activating pathways dependent on factor XII (Hageman factor—named after the patient in whom it was first discovered) in normal human plasma. Activation of factor XII, among other effects (e.g. activation of complement system and release of vasoactive nonapeptide, bradykinin, C3), leads to blood clotting within intact vessels and thus causes pathological conditions. Approximately one-third of human volunteers, smokers and non-smokers alike, exhibited immediate cutaneous hypersensitivity to the purified phenolic glycoprotein isolated from tobacco leaves and tobacco smoke. The purified fractions, after polyacrylamide chromatography, showed molecular weights ranging from 20 000 to 40 000 daltons. In 1976 and 1977 Becker *et al.*[142,143] showed that tobacco leaf extracts and tobacco smoke contained rutin or quercetin-like polyphenols capable of activating the factor XII dependent pathways.

In 1975 Malley *et al.*[144] showed that an important allergen of Timothy grass contained quercetin, the compound that was potentially capable of factor XII activation. Becker *et al.*[141] determined the inhibition efficiency of tobacco leaf glycoprotein, as well as tobacco smoke, cocoa, coffee and ragweed extracts for agglutination of erythrocytes coated with tobacco glycoprotein. This was done by using rabbit antibodies against tobacco glycoprotein or against rutin–bovine serum albumin conjugate. The results indicated that all the antigens tested were cross-reactive. However, tobacco leaf glycoprotein was found to be the most active agglutination inhibitor. Additional testing showed that the agglutination efficiency decreased in the following order: tobacco leaf glycoprotein, rutin, quercetin, chlorogenic acids, naringin, ellagic acid and, finally, hesperitin. Becker *et al.*[141] concluded that the cross-reactivity of the various materials was in part due to the similarity of the various polyphenol haptens resembling quercetin or rutin. Using the neonatal sensitisation technique of Pinckard *et al.*,[145] they further demonstrated that rabbits, when sensitised neonatally with tobacco glycoprotein, preferentially synthesise IgE antibodies which are capable of mediating passive cutaneous anaphylaxis[146] and pulmonary/cardiac anaphylaxis after challenge with tobacco glycoprotein.[147] The experimental data of Becker *et al.*[141] showed that rabbit IgE antibodies to tobacco glycoprotein could mediate a passive cutaneous anaphylaxis response to glycoprotein isolated from tobacco leaves and from tobacco smoke, and to similar allergens obtained after chromatographic separation from cocoa, coffee and ragweed. They also concluded that maternal cigarette smoking may result in sensitisation to tobacco glycoprotein of the offspring during the foetal/neonatal period, thus making it prone to *multiple allergies.*

In 1980 it was demonstrated by Becker *et al.*[148] that *in vivo* activation of the factor XII dependent pathway in dogs and African green monkeys by intravenous injection of ellagic acid or rutin leads to the development of acute inflammation of the lung and gall bladder. Similar lesions were produced in dogs after inhalation of the water-soluble, non-dialysable fraction of cigarette smoke condensate.

2.4.4. Studies from 1983 Onwards

Zuskin *et al.*[149] studied the effects of green and roasted coffee bean extracts on isolated guinea pig tracheal spirals. The extract of green coffee beans induced a concentration-dependent contraction of the tracheal spirals. However, 1000 mg of green coffee dust was needed to produce the biological equivalent of a reaction to 1·23 mg of histamine. Whereas potency of histamine was reduced 500-fold when applying an antihistamine, the potency of green coffee bean extract remained unaffected, thus suggesting that the effective contractile agent may be of a different chemical/physiological nature than a cationic organic amine such as histamine. In contrast to green coffee bean extract, the extract from roasted coffee beans produced a concentration-dependent *relaxation* of uncontracted, histamine contracted and green coffee extract contracted tissue. The relaxant activity of 1000 mg of roasted coffee extract was equivalent to 1·95 mg of theophylline (theophylline and caffeine are inhibitors of the phosphodiesterase reaction in mast cells; see Fig. 7). These authors concluded that the disappearance of contractile activity after roasting was due to the formation of potent bronchodilator compounds which counteracted the contractile agent of green coffee beans. They saw this conclusion supported by the finding that regular *and* decaffeinated roasted coffees contained a parasympathomimetic substance.[150] In addition, they also pointed out that reactions of workers to green bean coffee dust has always been considered to be of an exclusive allergic nature. Considering the data presented, there is the possibility that green coffee dust may induce acute bronchoconstriction without allergic hypersensitivity. This latter possibility found some support in a subsequent study by Zuskin *et al.*[151] Of nine coffee workers with job-related respiratory symptoms, six described typical symptoms for occupational asthma. Bronchoprovocation with green coffee bean allergen provoked immediate asthmatic reaction with acute reductions of ventilatory capacity in four workers. Eight of the nine workers had an increased total IgE serum level. Interestingly, the one worker without the IgE increase had been exposed for 17 years (the longest exposure of all workers) to green coffee bean dust. He did not have asthma but suffered from dyspnoea (severity grade 3, of a possible 4). He also had chronic cough, chronic phlegm, chronic bronchitis and rhinitis. He was a non-smoker. He was the only worker who showed a positive intradermal skin test to roasted coffee extract.

In the 1985 study by Osterman *et al.*[139] seven of 11 workers with IgE values of less than 50 ku $litre^{-1}$ had mild reactions towards green coffee bean dust. In the group of six workers with severe reactions towards green coffee bean dust, one had an IgE value of only 35 ku $litre^{-1}$. These results indicate that an increase in IgE values is not always present when workers show a hyper-responsiveness of the bronchial tree towards green coffee bean dust, and suggest that green coffee bean components can cause direct bronchial effects rather than acting via the allergy route.

Van den Bosch *et al.*[152] give a follow-up report on a coffee worker who had worked in a coffee roasting factory for 20 years[125] and had been tentatively diagnosed as having extrinsic allergic alveolitis (coffee worker's lung). The clinical evolution of this patient 13 years after the first diagnosis showed increased respiratory symptoms. Also the radiological results and pulmonary function abnormalities had progressed in spite of avoidance of green coffee bean dust and treatment with prednisone. 'Due to the involvement of the pleura and the hyperplasia of bronchus associated lymph tissue, suggest that the patient suffers from cryptogenic fibrosing alveolitis associated with rheumatoid arthritis rather than "coffee worker's lung".'[152] This new diagnosis thus eliminated the only case of 'coffee worker's lung' described in the literature.

Weill *et al.*[153] point out that it is becoming increasingly evident that chronic airways obstruction is a multifactorial disorder. Smoking, in their opinion, is the primary cause thereof. However, workplace inhalants of mineral or organic composition and of vapours or gases affect the small alveolar and interstitial airways. Coal dust, cotton dust and green coffee bean dust may be associated with long-term effects on airways without clinically apparent asthma. Workers exposed to isocyanates and susceptible to them develop dose-related acute airways responses. The role of lower level intermittent exposures to, for example, industrial gases such as chlorine is being investigated.

Sixty-nine coffee factory workers in a remote valley in Papua New Guinea were found by Smith *et al.*[154] to have more chronic symptomatic respiratory impairment than a group matched for age, height, ethnic origin, housing and smoking drawn at random from the Asaro Valley village (31 people) census book and a group from a neighbouring soft-drink factory (22 people). The green coffee beans, before receipt at the coffee factory, are removed from the coffee cherries and dried in the sun. No dust is produced in this process. At the factory the parchment skin surrounding the beans is mechanically removed by a dehuller/polisher. (See Volume 2, Chapter 1, for details of this process.) This machine also removes the silverskin. The green coffee beans are graded for size and colour and then bagged. Considerable dust is produced during the dehulling/polishing step. Dust levels in the air were measured with a Sartorius MB2 sampler with 5·0 μm Millipore SM filters, and values ranged from 0·7 to 10·0 mg m^{-3}.

Light microscopic pictures of dust filtered from the air of the coffee factory showed large bean husk particles measuring about 300 to 800 μm in length and 20 to 40 μm in width. Many small circular particles with a diameter of about 5 μm, 'whose origin was less clear', were also seen. No skin hypersensitivity to coffee bean husk extract was found in the six workers with hypersensitivity symptoms nor in the 12 workers without.

In summary, the green bean coffee dust as shown by light microscopic and electron microscopic pictures consists of many different components. Only several of these are of actual coffee origin. These are coffee bean husks, cuticle wax particles and coffee bean fragments derived, according to their analytical composition, from the outer regions of the coffee bean.

Coffee bean husks apparently do not contribute to the hypersensitivity symptoms as established by skin testing.[129,154] Extracts of green coffee beans showed some reactivity.[123,127,128,133] Cuticle wax flakes present in green coffee dust, as indicated by electron microscopic examination, have neither been isolated nor tested as a single entity for hypersensitivity reactions. Therefore no statement about its hypersensitivity activity can be made at this time.

Non-coffee-derived 'green bean coffee dust' constituents such as (1) sack fibres and fragments thereof,[128] (2) bacteria,[133] (3) mould mycelia and fragments thereof, (4) mould conidia spores,[133] (5) castor bean allergen (from transporting castor beans in sacks used later for coffee bean transporting)[128,129,138] and (6) pollen have been identified and have been tested for provoking allergic reactions. All six components were found active. However, the castor bean allergen was found by far the most active allergen of all the allergens tested.[101,138]

According to the different symptoms and different hypersensitivity reactions described by the various authors, it becomes clear that the dust associated with green coffee beans affects people showing hypersensitivity symptoms in a non-uniform fashion.

Most of these individuals react to extracts of green coffee bean, castor bean and burlap sacks in an immediate-type weal and flare skin reaction. This suggests a reagin-mediated Type I hypersensitivity reaction with the involvement of IgE and subsequent mast cell degranulation.[127–130]

The increased quantities of IgE in the serum of about 20 to 40% of symptomatic individuals supports the Type I hypersensitivity directly. Release of the symptoms is achieved by administering Inhal (disodium chromoglycate).[131–133] Its known inhibitory effect of mast cell degranulation via increase of c-AMP and—most importantly—the arresting of extracellular influx of Ca^{2+} into the mast cell[134] give circumstantial evidence for IgE mast cell-mediated (Type I) hypersensitivity.

It has to be pointed out that the causative agents in the green bean coffee extracts used for testing hypersensitivity are not necessarily coffee bean derived agents. Since more than 90% of the coffee beans have adhering mould spores[193] (see Table 11) and also have adhering bacteria (L. Judge, pers. comm.), the allergenic activity of green bean water extracts may actually be due partially or entirely to extractable microbial endotoxins. This question is important and should be answered.

The most prevailing symptoms caused by green bean coffee dust were those affecting the nose and the eye (rhinitis, conjunctivitis) and asthmatic symptoms (wheezing, shortness of breath). While rhinitis and conjunctivitis are reaginic (IgE) symptoms, the asthmatic symptoms in some workers were not related to increased IgE levels in the serum (Type I hypersensitivity). The increase of IgM in nearly 20% of symptomatic subjects tested,[133] without IgE involvement, indicates that hypersensitivity mechanism Type II complement-mediated type plays a role.

In those individuals showing hypersensitivity symptoms, where neither IgE, IgM, IgG and IgA was elevated in the serum, a T-cell mediated-type hypersensitivity (Type IV) has to be assumed as the cause of the asthmatic symptoms.

It is of importance to point out that the green bean coffee dust allergens are of moderate potency and that high concentrations are required to provoke sensitisation.[138] This is in contrast to many other occupational airway diseases.

2.5. Other Occupational Pulmonary Diseases

Since the effects of green coffee dust are caused not only by the actual green bean derived substances (protein, polysaccharides and other surface matter) but also, and perhaps more so, by adhering microbial organisms (bacterial and fungal) and other biological contaminants, it is necessary to relate the green coffee bean studies concerning pulmonary effects and allergies to those of other occupational activities, where other similar organic substances, dusts and microbial organisms are the causative agents. Various mycoses have already been described, some with vernacular names such as 'farmer's lung' (arising specifically from breathing certain fungal conidia from improperly dried hay) and 'cheeseworker's lung', and also byssinosis from the pericarp of cotton seeds. Fungal toxins, when inhaled, produce ill-effects in the mucous membrane of the respiratory tract. Kovats and Bugyi[155] extended mucous membrane reactions to the alveolar regions of the lung and named the symptoms hypersensitivity pneumonitis (North American term) or extrinsic allergic alveolitis (British term). Schuyler and Salvaggio[156] identified various microorganisms, plant protein (glycoprotein), animal protein, chemicals of low molecular weight and drugs as the antigens. Hypersensitivity pneumonitis represents a strictly pulmonary granulomatous disease which affects the interstitial as well as the alveolar region of the lung (see Fig. 10).

Extrinsic allergic alveolitis has been described as being caused by repeated severe insult with mould conidia and mycelial debris in the normal individual.[157] Symptoms appear about 6 h after exposure. They normally include cough and dyspnoea ('air hunger'), and sometimes fever and chills. No peripheral or pulmonary eosinophilia is found in lung biopsy. In general, the patients have an elevated IgG level but not an elevated IgE level. Continued exposure to the causative agent when derived from aspergilli conidia in the presensitised or atopic individual can lead to the more severe *allergic bronchopulmonary aspergillosis*, also called mucomembranous aspergillus bronchitis in so far as aspergilli conidia are involved. With this bronchial plugging occurs. Bronchoscopy reveals active growth of fungus colonies perpendicularly to the bronchi. The underlying mucous membrane is red and congested but not invaded. In more than 200 patients positive sputum cultures of aspergilli were consistently demonstrated.[158,159] Immediate reactions *and* delayed skin reactions are demonstrated by such patients. The pathology of this disease is thought to be due to both the immediate and delayed types of hypersensitivity. Most of the patients with alveolar bronchopulmonary aspergillosis are adults; however, numerous pediatric cases have also been described.[158]

In general, the antigenic material, which is usually organic in character (but which can also be an inorganic material), will include respirable particle sizes in the range from 6 μm down to less than 1 μm. According to Cole and Samson,[97] 'retention of larger mould conidia in the lower airways, like bronchioles and alveoli, is usually the result of bypassing the nasal filtration system during inhalation through the mouth'. The air velocity in the respiratory tract steadily decreases deeper within the system of airways (see Fig. 10). Sedimentation by gravity is also responsible for deposition of conidia. Surface topography of the conidium, in addition to its size and density, plays a pivotal role in determining sites of deposition. The larger echinulate conidia of *Aspergillus glaucus* and *A. niger* would be expected to impact higher in the respiratory tract than the smaller, smooth- to slightly rough-surfaced conidia of *A. terreus* and *A. fumigatus*. In fact these latter conidia ($<5\,\mu m$) are capable of reaching the respiratory bronchioles (<1 mm in diameter) and even the alveoli ($\sim 0{\cdot}5$ mm in diameter). 'These deep penetrating conidia are typically responsible for delayed allergy.'[28]

It must be pointed out that the walls of alveoli are not ciliated, unlike those of the bronchi and trachea. Therefore no removal of mould particles or other allergens by means of ciliary movement is possible. The mechanism of removal in the alveoli is by means of phagocytic cells. Godard *et al.*[160] have found that alveolar macrophages from asthmatics (obtained by bronchoalveolar lavage) are impaired in both viability and functional ability, indicating that the asthmatic patient harbours conidia longer in the alveoli than the normal person. This situation may also contribute to a greater risk of disease. Asthma itself is an allergy affecting atopic rather than normal individuals and is a well established disease. The symptoms (cough, wheezing, chills, etc.) are similar to those observed in allergies caused by pollen, animal dander and mould conidia in atopic individuals. Reaginic antibodies, immunoglobulin E (IgE), are present (see Section 2.2), and peripheral blood and sputum eosinophilia (an excess of a variety of white blood cells) have been observed.[161] Symptoms appear within minutes after exposure to the causative agent.

Hypersensitivity pneumonitis was considered to be one of the most typical examples of a lung disease mediated by a Type III mechanism (see Section 2.2). This was based on the similarity of symptoms to serum sickness, the presence of precipitating antibody in the serum against the offending antigen and the delayed time span until the allergy symptoms began to appear. More recently, however, Stankus and Salvaggio[162] suggested that the Type III mechanism may be most important in the acute mode of the illness. However, in subacute and chronic cases cell-mediated Type IV immune mechanisms, where sensitised T-cells mediate the immunological reactions after several days of exposure to the antigen without involvement of either circulating antibodies or complement, must be considered. According to Rippon,[163] allergy to the metabolites of *Candida* is a well established phenomenon. The clinical condition is called candidosis. Less well documented is the role of *Candida* in other syndromes involving hypersensitivity, e.g. eczemas, asthma, gastritis and an eye condition similar to histoplasma uveitis. A *Candida* gastritis syndrome has been the subject of conflicting reports.[164,165] Extracellular cellular culture filtrates ('Oidiomycin') of *Candida albicans* can induce both immediate and delayed types of hypersensitivity. Most adults without clinical symptoms of infection or allergy react to intradermal injection of such culture filtrates. This suggests that antigenic material can pass through the mucosal wall to evoke an immunological response. *Candida* allergies, in the form of gastritis, urethritis, balanitis, rhinosinusitis, headache and irritable colon syndrome, have been reported. A typical example of Type II reaction is Goodpasture's syndrome.[166] This is an *immunological disease* that originates in the lung as a necrotising haemorrhagic interstitial pneumonitis and as a mostly rapidly progressing glomerulonephritis. The pulmonary and renal lesions in Goodpasture's syndrome are the consequence of antibody interaction with antigens common to both the pulmonary and glomerular basement membranes.

According to Pepys,[167] production of allergic diseases of the respiratory system by inhaled organic or inorganic allergens depends on three main factors: (a) the immunological reactivity of the subject (whether atopic or normal), (b) the nature of the inhaled material (the specific activity of the allergen), and (c) the circumstances of the exposure (quantity of allergen). Occupational airway diseases include a large number of different named diseases according to the industry in which the people work, e.g. farmer's lung and cheeseworker's lung.

3. MYCOTOXINS FOUND IN COFFEE BEANS

As an introduction, it should be stated that coffee beans, like all other agricultural commodities, if planted, cultivated, harvested, processed, stored and transported under proper conditions, are not contaminated with mould growth. However, because of the ubiquitous presence of mould spores, one has to consider that all agricultural commodities are harbouring mould spores. Therefore, if green coffee beans are exposed to high humidity or even wetness and at the same time to moderate to warm temperatures, the growth of mould is unavoidable. In the beginning this growth of mould, from a health and mycotoxins contamination point of view, is of little concern. During the first two weeks of mould spore germination and mould growth no mycotoxins are usually formed.

However, mere mould growth on green coffee beans is of great economic concern to the coffee manufacturer since coffee beans, when infested with mould growth, develop strong off-flavours which taint/spoil the coffee beverage resulting from such mould-infested beans. Therefore the coffee manufacturer is interested in determining, at the port of entry, whether the coffee lots coming from many different countries are clean and unspoiled. Safeguards are set up at the port of entry (especially in the USA) to discover mould spoilage. Mould spoilage is relatively easy to detect because of the malodorous emanations and the relatively easily visible growth of mould mycelia.

Heim[18] in 1946 and Nicot[21] in 1953 reported on the contamination of green coffee beans with moulds (*Aspergillus tamarii*). Natarajan *et al.*[19] and Majumder *et al.*[20] analysed the mould organisms in contaminated coffee beans and found *Cunninghamella*, *Trichoderma*, *Aspergillus*, *Penicillin* and *Fusarium* species. Immediately after the isolation of the aflatoxins from turkey and duck feed by British and Dutch scientists[168–170] and the elucidation of the structure of aflatoxins by US scientists[171–173] in conjunction with several coffee companies in the US and Europe (General Foods Corporation, US, Procter & Gamble, US, the US Food and Drug Mycotoxin Analytical Laboratory located in New Orleans, Illicaffe SpA of Italy and Nestlé's of Switzerland), Levi and Borker[17] at General Foods started to analyse green coffee beans on a routine basis. These studies are described by Levi in publications.[17,22,174,175] The results of these surveys are summarised in Table 12 (see Section 3.4). In 1968 Scott[176] analysed several green coffee bean samples. No aflatoxin was found. Boutibonnes and Jacquet[177] analysed human foodstuffs and animal feed for aflatoxin and mycotoxin-producing mould. In this study 18 coffee samples were included and no aflatoxin was found. In 1971 the same group[178] again looked for aflatoxin in coffee beans during the development of a reliable thin-layer chromatographic procedure. Beebe[179] also included coffee samples when analysing for mycotoxins in foodstuffs after he developed an analytical procedure via high-pressure liquid chromatography, and Kiermeyer[180] looked at coffee beans when analysing a wide variety of foodstuffs for mycotoxins. He too did not find any mycotoxins.

3.1. Mould and Other Microbial Growth on Coffee Beans at Elevated Relative Humidities

In 1965 Hiscocks[2] and also Majumder *et al.*[183] demonstrated that *Aspergillus* spp. would grow on green coffee beans when the relative humidity of the air exceeds 75%. In 1971 Corte de Santos *et al.*[182] studied the natural bacterial and fungal populations on coffee beans at five different relative humidities; these, fixed at 25°C, ranged from 51 to 92%. Mould growth ceased at 63% relative humidity (13·4% db moisture) and bacterial growth ceased much earlier at 92% relative humidity (29·3% db moisture).

In 1973 Multon *et al.*[183] correlated the changes in the water content, colour and taste, and in the microflora, of coffee stored for 250 days at five different relative humidities (ranging from 60 to 90%) and four different temperatures (ranging from 20 to 36°C).

No mould growth was observed at 60% relative humidity. Slow mould growth was noticed at 75% relative humidity. However, rapid mould growth set in at 85–95% relative humidity. *Aspergillus* and *Penicillin* spp. were the main species found.

In 1975 Poisson *et al.*[184] developed a standard procedure for studying the microflora for the purpose of obtaining results that could be compared from one experiment and one experimenter to the next. Flavour changes were followed over a period of 53 days at 95% relative humidity and 20°C incubation temperature. The experiment was carried out under aerated and non-aerated modes. In both modes the moisture content increased from the usual 12% db to an elevated content of about 25% within 53 h. Cup flavour began to deteriorate noticeably at about 16%. At above 20% moisture content (after about 40 h incubation) the odour became very disagreeable; the flavour was described as mouldy. The aerated samples held up only slightly better than the non-aerated samples. At moisture contents above 20% the coffee was considered unfit for roasting. There was a linear relationship between increase in flavour deterioration and increase in microbiological counts. The increasing growth of moulds with increasing moisture content caused the suppression of bacterial growth.

Rao *et al.*,[185] who studied Indian coffees damaged by monsoon rains, observed in 1971 that mould growth ceases when the moisture level in the coffee beans was less than 14·5% db. This moisture content is 1·1% higher than that found critical by Corte de Santos *et al.*[182] as being too low to support mould growth. However, it is in excellent agreement with the critical moisture value found by Levi *et al.*[186] in 1974.

3.2. Ochratoxin in Mould-contaminated Coffee Beans

Corte de Santos *et al.*[182] reported in 1971 that when the green coffee beans were stored for 3 months at 29·3% db moisture (26°C, 92% relative humidity), then *Aspergillus ochraceus* accounted for more than 60% of all *Aspergillus* spp. *Aspergillus versicolor* (20%), *A. flavus* (10%) and *A. wentii* (10%) were the main inhabitants. Smaller numbers of *A. niger. A. candidus,*

A. restrictus, *A. ruber* and *A. chevalieri* were also found. In the same coffee samples, although less frequently, the following *Penicillium* spp. were found: *P. brevicompactum*, *P. chrysogenum*, *P. purpurogenum* and *P. conescens*. The genera *Chaetomium*, *Mucor*, *Sporendenoma* and *Scopuliaropsis* were occasionally found.

It was observed at Illicaffe that one wet coffee sample which had been stored for over two years showed, according to Levi,[174] estimated ochratoxin levels up to 12 000 $\mu g\,kg^{-1}$. Levi *et al.*,[186] when analysing mouldy green coffee samples 'that had become wet and mouldy due to improper handling and abnormally long storage', also found ochratoxin A. Subsequent analysis of 335 green coffee samples (68 samples of commercial grade; 267 samples hand cleaned) showed that 22 samples had detectable levels of ochratoxin A. Only one sample of mould-contaminated green beans showed a relatively high content of 360 μg of ochratoxin A per kilogram of beans (see Table 12).

Levi *et al.*,[186] in agreement with Corte de Santos *et al.*,[182] determined in mould-damaged coffee beans *Aspergillus ochraceus*, *A. flavus*, *A. niger*, *A. glaucus*, *A. candidus* and *A. wentii*.

In 1984 Tsubouchi *et al.*[187] found ochratoxin A in 22 samples of Brazilian, Indonesian and Mexican green coffee beans that had been shipped to Japan. The mycotoxin was found in only four samples at levels ranging from 9·9 to 46 $\mu g\,kg^{-1}$ (Table 12).

3.3. Mycotoxin Production in Non-decaffeinated and Decaffeinated Green Coffee Beans

Buchanan and Fletcher[188–190] showed that caffeine inhibited the formation of aflatoxins and ochratoxin A, and sterigmatocystin, citrinin and patulin production. These authors held the presence of caffeine directly responsible for the inhibition of mycotoxin production. They also suggested that the removal of caffeine might heighten the risk of stored decaffeinated coffee beans becoming mould infested. In order to test the caffeine hypothesis, Nartowicz *et al.*[191] in 1979 inoculated green and roasted coffee beans with *Aspergillus parasiticus*. They also inoculated decaffeinated green and roasted coffee beans with the same mould in order to test the aflatoxigenicity of caffeine. Ten grams of whole coffee beans were tested in flask cultures inoculated with 2·0 ml of a spore suspension containing 10^6 spores per millilitre. The flasks were incubated (no agitation) for 21 days at 28°C at different moisture levels. The caffeine-containing beans showed mould growth at 28% moisture content (green beans) and 48% moisture content (roasted beans). No aflatoxins were detectable. No moulds grew in the green caffeine-containing beans at moisture levels of 17 and 23%, and the roasted version did not show mould growth at moisture levels of 34 and 42%. These data suggest that caffeine-containing beans do not support the growth of *Aspergillus parasiticus*.

Aflatoxins, however, were detected in the laboratory-decaffeinated incubated coffee-bean samples. The samples of green beans at moisture levels of 37 and 40% showed aflatoxins ranging from 32·4 to 94·7 $\mu g\ g^{-1}$. On incubation roasted decaffeinated coffee developed only a trace of aflatoxins (0·3 $\mu g\ g^{-1}$). Decaffeinated commercial coffee beans under the same incubation conditions produced only 1·97 to 7·61 $\mu g\ g^{-1}$. The authors suggested that this difference might have been the result of two different methods of decaffeination. Factory-decaffeinated beans have a dark, semi-roasted appearance, whereas laboratory-decaffeinated beans look pale. Higher levels of free phenols in the factory-decaffeinated beans may inhibit the growth of moulds. Nartowicz *et al.* concluded that 'decaffeination increases the aflatoxigenic potential of coffee', and recommended that 'extra precaution should be taken to assure the absence of aflatoxin in decaffeinated coffee beans'.

Tsubouchi and co-workers,[192] studying *Aspergillus ochraceus*, also dealt with the subject of caffeine as a mould growth inhibitor. The data obtained in this 1985 study contribute to the evidence that moulds of the same species often behave in opposite ways. To one strain of *A. ochraceus* caffeine acts as a growth inhibitor, whereas to other strains it acts as a growth and mycotoxin production promoter. Eleven strains of *A. ochraceus* were isolated from green coffee beans or from polished rice. Nine strains were stimulated by caffeine to produce mycotoxin, whereas for two of the 11 strains caffeine acted as an inhibitor for ochratoxin A production. Several strains of *A. elegans* and one of *A. sclerotium* (both species are ochratoxin producers) were severely inhibited by caffeine in their ability to produce ochratoxin A. It should be emphasised that most of the *Aspergillus ochraceus* strains digested the caffeine without being inhibited in the formation of the nephrotoxic secondary metabolite ochratoxin A. This result is in contrast to that obtained by Buchanan and Fletcher[188] where caffeine inhibited the formation of ochratoxin A.

A 1983 publication by Stack *et al.*[193] describes the formation of ochratoxins A and B, and of xanthomegnin, viomellein and vioxanthin (see Fig. 11), by *Aspergillus ochraceus* isolated from green coffee beans. Mould spores isolated from 120 green coffee bean samples obtained from 23 different coffee-producing countries were inoculated on to a sterile rice/water mixture and incubated for 20 days at 22 ± 2°C. After 20 days the

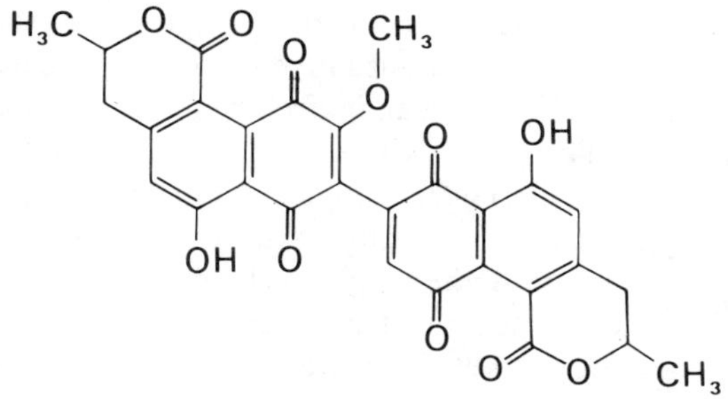

Xanthomegnin

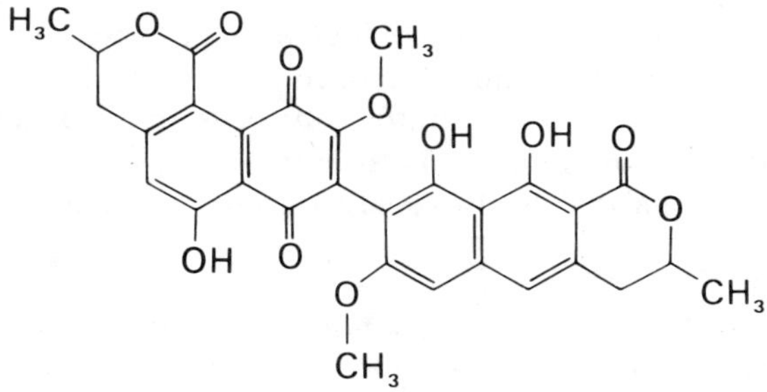

Viomellein

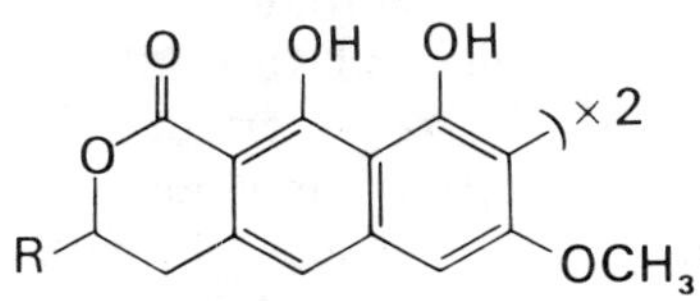

Viriditoxin R = CH_2COCH_3
Vioxanthin R = CH_3

Fig. 11. Structural chemical formulae for some mould toxins.

mycotoxins produced were separated from the rice cultures by extracting them with chloroform and subsequent filtration of the chloroform extract. The filtrates, without further purification or concentration, were injected directly into a high-pressure liquid chromatograph. The various mycotoxins mentioned above were separated as well resolved peaks. Quantitation was carried out with standards and peak height measurements. For additional assurance, 10-μl filtrates were also subjected to thin-layer chromatography. Identity was obtained by co-chromatographing the corresponding standards. This study shows that moulds isolated from green bean coffee samples from various regions of the world differ drastically in their ability to produce mycotoxins on rice as an incubation medium. Even different green bean coffee samples from one and the same country carried mould spores with widely differing abilities to produce mycotoxins. These differences ranged from zero production to a production as high as 2088 μg of ochratoxin A per gram of culture medium (Uganda). For ochratoxin B, the range was from zero to 3375 $\mu g\, g^{-1}$ (Costa Rica), and for xanthomegnin, from zero to 1562 $\mu g\, g^{-1}$ (Rwanda). Viomellein production levels ranged between zero and 2514 $\mu g\, g^{-1}$ (Rwanda), and the production range for vioxanthin was zero to 2054 $\mu g\, g^{-1}$ (Uganda).

One hundred and twenty *Aspergillus ochraceus* isolates from coffee samples stemming from 22 different coffee bean producing countries were tested for the production of ochratoxin A, ochratoxin B, xanthomegnin, viomellein and vioxanthin using a sterile rice/water (1:1) culture medium. The results, summarised in Table 11, are interesting for the following reasons:

(a) It is once again shown that not all of the *A. ochraceus* isolates produce toxins (or they may have lost their capability to produce toxin during transfer from the coffee beans to the rice cultures).
(b) More isolates actually produce (on rice cultures) more xanthomegnin, viomellein and vioxanthin than they produce ochratoxin A or B. This is important because it suggests that the presence of these three compounds may be more important for nephrotoxicity and endemic nephropathy than that of the well publicised ochratoxin A (provided similar production rates exist for corn, the medium mainly responsible for the production of ochratoxin A).

It should be clarified whether or not actual coffee bean growth media duplicate the data obtained from rice cultures. There are indications in the literature that rice is a much more effective medium for production of ochratoxin A than coffee beans. In 1985 Tsubouchi *et al.*[192] tested isolates of *A. ochraceus* strains side by side on rice and green coffee bean cultures. The rice cultures (with 11 *A. ochraceus* isolates) produced an average of 244 μg ochratoxin A per gram of rice culture, whereas the green bean cultures produced only 61 μg per gram. This result demonstrates that wet coffee beans provide a 75% lower incentive for production of ochratoxin A as compared with rice as a growth medium.

Table 11
Mycotoxin production on rice cultures by *Aspergillus ochraceus* isolated from green coffee beans

Country of origin	Number of isolates	Production of mycotoxins in $\mu g\, g^{-1}$ of culture															Samples tested
		Ochratoxin A[a]			Ochratoxin B			Xanthomegnin			Viomellein			Vioxanthin[b]			
		Non-active	High	Av.	Non-active	High	Av.	Non-active	High	Av.	Non-active	High	Av.	Non-active	High	Av.	
Argentina	1	1	0	0	1	0	0	1	0	0	1	0	0	Not analysed			
Brazil	39	15	783	179	16	188	25	4	584	172	4	358	195	2	556	224	14
Burundi	6	4	609	171	4	51	13	4	117	290	2	1 948	675	2	1 370	856	3
Colombia	14	8	552	113	8	412	49	1	1 460	279	1	2 194	446	1	363	203	6
Costa Rica	2	0	62	61	1	893	446	1	127	63	1	107	53	Not analysed			
Dom. Republ.	4	4	0	0	4	0	0	4	372	206	1	168	94	0	314	198	3
Ecuador	5	3	102	75	3	10	2	0	532	185	1	816	284	0	278	224	2
El Salvador	2	2	0	0	2	0	0	0	125	113	0	82	62	Not analysed			
Ethiopia	1	0	826	—	0	59	—	0	665	—	0	991	—	0	642	—	1
Guatemala	3	2	196	65	2	5	2	0	583	302	0	949	431	Not analysed			
Haiti	9	6	1 363	284	6	230	51	2	532	177	2	1 226	346	0	1 198	410	5
Honduras	6	4	232	65	4	38	6	1	411	190	1	839	303	1	364	183	4
India	2	2	0	0	2	0	0	2	0	0	2	0	0	Not analysed			
Indonesia	6	6	0	0	6	0	0	3	279	112	3	185	75	0	235	181	2
Ivory Coast	1	1	0	0	1	0	0	0	70	—	0	124	—	Not analysed			
Kenya	1	0	478	—	0	54	—	0	235	—	0	680	—	Not analysed			
Nicaragua	1	0	254	—	0	38	—	0	178	—	0	572	—	Not analysed			
Panama	1	1	0	0	1	0	0	0	157	—	0	244	—	Not analysed			
Paraguay	2	2	0	0	2	0	0	0	78	58	0	109	91	Not analysed			
Peru	9	6	422	79	6	22	2	1	1 524	369	1	2 247	536	1	64	32	2
Rwanda	1	0	1 421	—	0	1 238	—	0	1 562	—	0	2 514	—	Not analysed			
Uganda	1	0	2 088	—	0	204	—	0	1 117	—	0	2 150	—	0	2 054	—	1
Zaire	3	2	168	56	2	tr	tr	0	931	419	0	567	410	—	514	—	1

[a] LD_{50} ochratoxin methyl or ethyl ester about 150 μg per animal (duckling). Ochratoxin A about the same. Lactone important for activity.
[b] Vioxanthin mice 3–3·5 mg LD_{50}.

3.4. Sterigmatocystin in Coffee Beans

Sterigmatocystin in the biogenetic synthesis pathway is a precursor of aflatoxin (see Fig. 12). Since aflatoxins can be found on spoiled green coffee beans, it is not surprising that the occurrence of sterigmatocystin on green coffee beans has also been reported. In 1973 Purchase and Pretorius[194] identified sterigmatocystin in one sample of condensed, strongly mouldy, green coffee beans (1100 $\mu g\,kg^{-1}$).

A similar report was conveyed to Levi[22] in 1980 by Illicaffe SpA for a two-year-old green coffee bean sample that was heavily mould infested. A very high level of sterigmatocystin, estimated at 12 $mg\,kg^{-1}$, was found. In

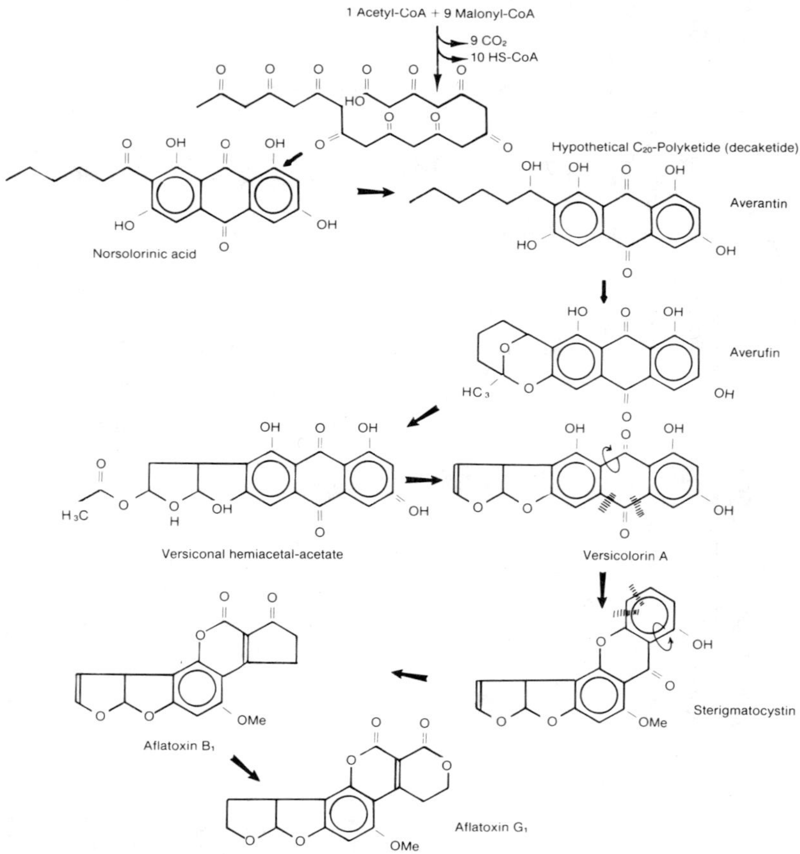

Fig. 12. Biosynthetic pathway of aflatoxins.

contrast to these two incidents, in 1975 Levi[195] had analysed 15 coffee samples for sterigmatocystin, ten of which were described as 5 to 20% mouldy or up to 100% mouldy. However, not even the very mouldy samples showed any sterigmatocystin (see Table 12).

Aspergillus versicolor, after isolation from green coffee beans and inoculation on to sterilised, moist, green coffee beans (moisture content 30%, 28°C), produced no analysable sterigmatocystin after 14 days' incubation. After about 28 days production of sterigmatocystin had reached a maximum (about 3000 $\mu g\ kg^{-1}$). On extending the incubation to 35 days no additional sterigmatocystin was produced.

The data available for sterigmatocystin in green coffee beans show that this mycotoxin, in similar fashion to the aflatoxins, ochratoxin and related compounds, is produced only when extremely favourable moisture, temperature and time conditions coincide.

Levi[22] found, when screening 60 *Aspergillus flavus* isolates from green coffee beans on synthetic rice medium, that only 13 of these isolates produced aflatoxins. When inoculating sterilised, green coffee beans with *A. ochraceus* isolates at a 30% moisture level and 28°C, it took 13 days to produce a maximal quantity of 450 μg of ochratoxin per kilogram of culture medium. This quantity of ochratoxin is about ten times higher than that found by Gallaz and Stalder,[196] who measured only 40 μg of ochratoxin when inoculating sterilised green coffee beans with a toxin-producing strain of *A. ochraceus*.

Nine different samples of *A. versicolor*, isolated by Levi[22] from green coffee beans, were all able to produce sterigmatocystin on a synthetic (rice) culture. When testing 14 different green coffee bean isolates of *A. flavus* side by side on a rice culture and on sterilised green coffee beans, enormous differences in aflatoxin production were found in 11 of the isolates. The most extreme isolate (from the Ivory Coast) produced 90 000 $\mu g\ kg^{-1}$ on sterilised rice, whereas the same isolate on sterilised green coffee beans produced only 20 $\mu g\ kg^{-1}$ (4500 times more on rice). The least dramatic difference, but still 33 times more for the rice culture, was 3000 μg aflatoxin B_1 versus 90 μg per kilogram produced on the coffee bean substrate. Only one isolate (from Angola) showed comparable aflatoxin B_1 production, namely 47 000 μg per kilogram of rice culture and 52 000 μg per kilogram of green coffee bean culture. A second isolate (also from Angola) produced no aflatoxin B_1 either on rice or on green coffee. A third isolate from Angola fell within the expected difference, namely it produced 11 000 μg aflatoxin B_1 per kilogram of rice culture, whereas only 10 μg was measured per kilogram of coffee bean culture.

Table 12
Mould toxins determined in green coffee beans[a]

Coffee samples analysed	*Aflatoxin* B_1	*Ochratoxin A*	*Sterigmatocystin*	*Reference*
640 commercial	0	NA	NA	1964–69 Procter & Gamble (Heusinkveld *et al.*[197])
58 very mouldy	0	NA	NA	1964–68 (Levi and Borker[17])
2 commercial	0	NA	NA	1968 (Scott[176])
18 commercial	0	NA	NA	1969 (Boutibonnes and Jacquet[177])
2 very mouldy	0	NA	(1) 1 100	1973 (Purchase and Pretorius[194])
267 hand cleaned from mouldy lots	NA	(1)–360 (2)–20+ (16)–20	NA	1974 General Foods (Levi *et al.*[186])
68 commercial	NA	(3) 20, 20, 80	NA	1974 (Levi *et al.*[186])
15 (5 no mould, 10 mouldy)	NA	NA	0	1975 (Levi *et al.*[197])
502 (2 years old, one sample mouldy)	0	0	(1)–12 000	Illicaffe (via Levi[22])
201 commercial	(2)[b] 3; 12	(2) 96; 24	0	FDA USA, Stoloff, L. and Francis, O., via Levi[22]
65 commercial	(28) < 3	NA	NA	Tsuboi *et al.*[198]
22 commercial	0	(4) 9·9–46	NA	Tsubouchi *et al.*[187]
1 commercial	0	NA	NA	Kiermeyer[180]

[a] All quantities in $\mu g\,kg^{-1}$ or $ng\,g^{-1}$.
[b] Numbers in parentheses indicate the number of samples infested with mycotoxin.
NA = Not analysed.

Table 12 represents a summary of the analytical data produced on the mycotoxins on green and roasted (one study) coffee beans. To summarise all the data obtained for mycotoxin production on green coffee beans, whether they pertain to sterigmatocystin, aflatoxins, ochratoxins, xanthomegnin, viomellein or vioxanthin, very drastic growth conditions are required. First, without sufficient moisture (ranging from higher than 13·4 to borderline 15%), no mould growth will occur. If sufficient moisture is present, the mould will have to grow for about 10 days, and in some instances even longer than 10 days, before the mould changes over from primary metabolite production (growth phase) to the secondary metabolite (idiolite) production (the production of mycotoxins) which happens to coincide with the production of mould spores.

Since coffee beans, after production of visible mould growth, emanate a mouldy odour and also undergo a drastic flavour change, it is highly unlikely that mouldy beans go undetected. Nevertheless, it is a wise precaution to keep the beans under very tight scrutiny at all times after plant entry. If mouldy beans were to reach the roasting stage, could they survive the roasting without decomposition? This question is dealt with in the next section.

3.5. Fate of Mycotoxins during Roasting

Levi *et al.*[186] followed the course of ochratoxin disappearance in simulated roasting experiments on green coffee beans that were artificially contaminated with ochratoxin A. The beans were heated to 200°C and the remaining level of ochratoxin A was determined after 5, 8, 15 and 20 min 'roasting' time. The results indicate that ochratoxin A is rather heat unstable, since after 5 min heating 78·6% of the ochratoxin had disappeared. Prolonged heating for up to 20 min removed an additional 8·6%, indicating that most of the ochratoxin disappears during the first 5 min of roasting and that prolonged heating (roasting) removes only small additional quantities of ochratoxin A. The actual quantity of ochratoxin A placed on to green coffee beans was 1635 $\mu g\,kg^{-1}$. Roasting for 5 min diminished mycotoxin to 245 $\mu g\,kg^{-1}$. Roasting for 20 min decreased the mycotoxin down to 210 $\mu g\,kg^{-1}$.

The data produced by Levi *et al.*[188,195] are essentially supported by those produced by Micco *et al.*[199] Micco and co-workers studied artificially and naturally contaminated samples and a naturally contaminated sample that was subjected, before roasting, to a solvent decaffeination process. The artificially contaminated samples consisted of Santos, Costa Rica and a 1:1 blend of Brazil and Zaire green coffee beans. The contamination levels were

46, 42 and 47 μg kg^{-1}, respectively. The three different samples were subjected to three roasting temperatures each: 200, 205 and 210°C. The Santos sample uniformly lost 87% of the mycotoxin at the three roasting temperatures. The Costa Rica sample, with increasing roasting temperature, lost 48, 52 and 54% of the mycotoxin. The Brazil/Zaire blend sample uniformly lost 80% for all three temperatures. The beans that were naturally contaminated with ochratoxin A (Zaire, 86 μg kg^{-1}; Conillon, 4·0 μg kg^{-1}), when subjected to the same three roasting temperatures, lost 90–99+% of the ochratoxin. These studies indicate that ochratoxin A shows a great sensitivity to heat. Provided the heat degradation products are biochemically inactive, the roasting process may constitute a convenient way of destroying small quantities of ochratoxin rather effectively.

3.6. Incidence of Moulds in Coffee Beans (Surface and Subsurface Populations)

In 1983 Mislivec *et al.*[200] reported the results of an extensive study to determine the presence of toxigenic and other moulds on the surface and inside coffee beans. A total of 944 coffee samples obtained from 31 different coffee-producing countries were examined by the Food and Drug Administration, Division of Microbiology, Washington, DC. This monumental study was undertaken to obtain a broader view of the mycotoxin picture which, up to that point, had been restricted to the natural occurrence of aflatoxins, ochratoxin A, sterigmatocystin and the moulds identified by Levi *et al.* between 1974 and 1980,[22,186] by Corte de Santos *et al.*[182] in 1971 and by Poisson *et al.*[184] in 1975. The study was carried out in two phases, namely: (1) to determine the mould population located mainly on the surface of the native green beans; and (2) to determine the mould population present on the inside of the coffee beans after disinfecting the beans with a 5% NaOCl solution for 1 min followed by three rinses with sterile water. All samples were plated aseptically on malt/salt agar (20 g malt extract, 75 g of NaCl, 40 mg of chlortetracycline HCl and 15 g of agar per litre). The antibiotic and salt were added to effectively inhibit bacterial growth and to retard the growth of the fast-growing Mucorales. All plates were incubated for 14 to 21 days at 22–25°C. The moulds were then counted by documented, routine procedures and identified using various specific identification methods.

The main mould flora, regardless of whether the beans were surface disinfected or not, was dominated by seven species of *Aspergillus*: *A. niger* showed an overall occurrence of 58·7% on the surface and 29·5% on the inside (58·7/29·5); *A. ochraceus* of 71·8/11·5; *A. glaucus*, 38·3/19·9; *A. flavus*, 10·6/3·2; *A. tamarii*, 28·0/11·2; *A. wentii*, 11·8/1·2; *A. versicolor*, 21·9/1·2; the *A. glaucus* group consisting of *A. repens*, *A. ruber*, *A. amstelodami* and *A. chevalieri* showed a distribution of 38·3% on the surface and 19·9% on the inside of the coffee beans.

Seven additional *Aspergillus* spp. were identified at the less than 1% level on beans that were not surface disinfected, namely: *A. candidus*, *A. fumigatus*, *A. nidulans*, *A. restrictus*, *A. sydowi*, *A. terreus* and *A. ustus*. The genus *Penicillium* was encountered regularly, but at a much lower percentage than *Aspergillus* spp. An unspecified *Penicillium* spp. amounted to 7·8% for surface-non-disinfected beans and to 1·5% for surface-disinfected beans, *P. frequentans*, 5·1/0·3%; *P. citrinum*, 4·4/1·8%; *P. brevicompactum*, 1·9/0·5%; *P. cyclopium*, 1·8/0·2%; and *P. expansum*, 1·6/0·4%. *P. citrinum* was the only species that averaged, over the 31 countries, higher than 1% for the surface-disinfected coffee beans. The authors conclude from this that the *Penicillium* species generally do not penetrate into the bean internally as much as the *Aspergillus* species. Three additional *Penicillium* species were found on the surface at levels below 1%: *P. chrysogenum*, *P. islandicum* and *P. viridicatum*.

Finally, some additional genera were also found and estimated at the less than 1% level. These are: *Alternaria alternata*, *Botrytis cinerea*, *Circinella* spp., *Cladosporium herbarum*, Dematiaceae, *Fusarium stilboides*, Moniliaceae, *Mucor* spp., *Rhizopus* spp., *Scopulariopsis brevicaulis*, *Stemphylium* spp., *Syncephalastrum* spp. and *Trichoderma viride*.

The lopsided distribution of the *Aspergillus* spp. versus *Penicillium* spp. is explained as being due to the tropical and subtropical climates of coffee-growing nations. Most of the *Penicillium* mould species do not grow well at elevated temperatures. The same is true, in part, for the genera *Alternaria* and also *Fusarium*.

The authors summarise as follows. The only toxin-producing genus that was identified and quantified in substantial quantities on the surface and the inside of the coffee bean was *Aspergillus*. Toxin-producing food-borne mould genera such as *Penicillium*, *Fusarium* and *Alternaria* were present at significantly lower levels inside coffee beans; therefore it is not very likely that any of the mycotoxins connected with these genera will play important roles.

The principal mould species found by Mislivec *et al.*[200] to invade coffee beans was *Aspergillus ochraceus*. This mould is capable of producing five important mycotoxins, ochratoxin A and B, xanthomegnin, viomellein and vioxanthin, as already mentioned.

3.7. Mould and Other Microbial Growth Potential on Coffee Beans from Production Site to Roaster

3.7.1. Production Site (Mainly Dry Processing)

The growth of mould organisms on coffee beans or on coffee cherries can begin in the field at the harvest stage. For instance, sometimes—especially in dry processing—it happens that picked coffee cherries have to wait in transport sacks, storage bins, etc., before they can be spread out on the drying floors. Such waiting periods will occur whenever rainy weather sets in. Beans, heaped or bunched up, like all vegetable matter (cocoa beans, cut grass, silage, green tobacco stems and tobacco leaves), will heat up as a result of the prolific growth of bacteria and moulds (*Aspergillus*, *Penicillium*, *Mucor*, *Rhizopus*). If such waiting periods are prolonged, the beans will heat up to 40–50°C or even higher. At this stage the thermophilic bacteria and moulds will prevail and reproduce. *Lactobacillus delbrueckii*, *L. leichmannii* and the mould *Aspergillus fumigatus* are known thermophilic microorganisms. Coffee cherries selectively contaminated by thermophilic microorganisms will most likely then continue on the drying floor to allow the proliferation of the thermophilic species. Since bacteria have a much more limited lifespan than moulds because of their survival requirements for higher water activities than moulds, it is to be assumed that after a relatively short, successful, drying period only the thermophilic moulds will survive. If rainfall remoistens the coffee beans, then bacteria may continue to grow again.

The beans on the drying floor are pushed into heaps during the later afternoon/early evening and then covered with tarpaulins. This is done for the purpose of moisture equalisation of coffee cherries with non-equal moisture content (as pointed out by Sivetz and Desrosier[201]). In those regions with cool nights, water condensation is likely to occur under the tarpaulin and remoisten the drying coffee cherries. This moisture, of course, will continue to stimulate further bacterial reproduction and much more so the continued growth of mould mycelia. It is also predictable that coffee plantations with low-lying drying floors (valley locations) will have more water condensation during the night than drying floors on the tops of hills or on ridges. The drying rates of coffee cherries are therefore expected to be slower in low-lying areas than those at elevated locations. These considerations agree with the 1987 recommendations of Vincent[202] for African plantations, namely that 'drying areas should be located in well-ventilated areas; clearings in equatorial forests should not be used as there is not enough ventilation and as stagnant fogs can form'.

Rapid delivery of freshly picked cherries and predrying (or total drying) in mechanical dryers should alleviate the problem of microbial growth to the extent of preventing the formation of off-flavours. After the cherries have been dried out to a moisture content ranging from 10 to 12% w/w, no microbial growth will occur as long as no exposure to moisture (rain or condensation in confined areas) or high humidity at elevated temperatures is permitted.

3.7.2. Production Site (Wet Processing)

In wet processing the freshly picked coffee cherries are sorted (into sinkers and floaters), sized (larger and smaller beans) and depulped either by hand or in a pulp-removal machine. Depulping removes most of the pulp (the exocarp and part of the mesocarp) except for a 1- to 2-mm thick layer of mucilage. This layer adheres to the coffee bean fairly tenaciously. It is removed by one of several methods:

(1) *Natural fermentation.* The depulped coffee cherries are flushed into concrete or wooden tanks and then the water is drained away. (For detailed information on size separation see references 201 and 202.) The native enzymes present in the moist mucilage break down the mucilage for easy removal. It is at this stage that microorganisms (bacteria and moulds) will start to grow. If this growth is permitted for extended periods, then the microbial metabolites will taint the flavour of the coffee beans. After natural fermentation, the mucilage no longer adheres to the beans and can be washed off.

Since mucilage fermentation normally takes overnight but seldom longer than 2 days, mould growth, to the extent of forming secondary metabolites, is unlikely. However, obviously, the growth of microorganisms during delayed mucilage removal can be sufficiently extensive as to impart undesirable flavours into the bean.

(2) *Fermentation with added enzymes.* For accelerated mucilage removal, the natural fermentation can be aided by addition of extrinsic enzymes. These enzymes are complex mixtures and are capable of breaking ester links as well as glucosidic bonds. Pectin-degrading enzymes are isolated from moulds (mostly *Aspergillus* spp.); and digestion, according to Sivetz and Desrosier,[201] can be considerably shortened, to 5–8 h. Such a short processing time is highly desirable since the growth of microorganisms is kept to a minimum. This is contrary to the natural fermentation process where microorganisms, depending on length of fermentation, have ample chance to impart their specific fermentation metabolites into the mucilage and coffee beans, and thus decrease the flavour quality.

(3) *Chemical mucilage removal.* Pulped coffee is treated with either dilute acids (HCl, H_2SO_4) or alkali (NaOH, Na_2CO_3, $Ca(OH)_2$). Mucilage depolymerisation is highly accelerated. According to Sivetz and Desrosier,[201] only 20 min to 1 h is required for successful mucilage removal. However, if the concentrations of the additives are not properly adjusted, then off-flavours as well as appearance damage will result.

(4) *Warm water treatment* for mucilage removal consists of mixing equal weights of water with depulped coffee and heating of the water to about 50°C. According to Sivetz and Desrosier,[201] mucilage removal can be achieved under these conditions in less than 5 min.

(5) *Mucilage removal by mechanical attrition.* Sivetz and Desrosier[201] describe a number of machines that either remove mucilage by attrition (Raoeng pulper, Hess washer, HAES washer used in the Kona coffee region) or depulp *and* remove mucilage in one process (Raoeng pulper). See also Vincent[202] for additional equipment and details of mucilage removal.

All these various mucilage removal processes are geared to the shortening of processing time and thus decrease the chances for microbial bean damage. Although the beans in the wet processing mode may come through the pulp and mucilage removal processes without microbial deterioration, the subsequent drying process of the resulting parchment coffee may still subject the parchment coffee to microbial damage. Such damage will occur when, during the sun drying process, the semi-dry beans are remoistened by sudden rains.

3.7.3. Production Site (Dry Processing)

Drying of wet-processed coffee beans is carried out on drying floors or in mechanical dryers. Natural drying on drying floors under favourable weather conditions will take about 5 to 7 days according to Sivetz and Desrosier[201] or 10 to 15 days according to Vincent.[202] Five to seven days are apparently too short a period for the formation of secondary metabolites (mycotoxins), but a drying time of 15 days is to be considered borderline for formation of secondary metabolites.

The general statement concerning processing time by Sivetz and Desrosier[201] is valid when applied to microbial growth and the quality of coffee beans: 'Coffee does not improve with increase of processing time—the best way to forestall [microbial] deterioration is to carry out all steps of processing [especially] including the drying as rapidly as possible.'

3.8. Storage and Transport of Coffee Beans (Storage at Production Site)

During unfavourable conditions of storage and transport in respect of temperature and humidity, appearance/odour changes in green beans and flavour deterioration (in brews from the roasted beans) can occur. From a microbiologist's point of view, these changes are indicators of mould growth, though not necessarily of the presence of mould toxins. In 1987 Clarke[203] summarised the published data on the relationships of critical relative humidity and temperature for mould growth. From these data it can be seen that green coffee beans, though dried to safe water contents (say below 13% w/w corresponding to an equilibrium relative humidity of approximately 67% at 28°C), will still absorb, at sufficiently high relative humidity and temperature, adequate moisture for mould growth.

Detailed studies on storage stability in respect of flavour stability of Kenya arabica coffee were reported by Stirling[204] in 1980, in relation to both hulled green coffee and green coffee still in parchment (dry). These show that, at 28°C, with 72% relative humidity (r.h.) for the former and 77% r.h. for the latter, deterioration in flavour becomes marked after about two months. This therefore suggests that mould growth is underway. This paper should be consulted for full details, and also those of Multon *et al.*[183] and Cortes dos Santos[182] (see also page 292).

The conclusions drawn by Stirling[204] are as follows. Most arabica bean-producing regions are located at high altitude tropical/subtropical locations. Such regions have relatively cold night temperatures and maximum high daytime temperatures. There are also prolonged rainy periods, during which consequently high humidities result. Such combinations of high humidity and high temperature conditions will cause the deterioration of coffee beans if stored for any length of time. Furthermore, the coffees destined for export will be stored in sea-level warehouses at ports with nearly constantly extreme humidities and high temperatures.

The coffee beans in Kenya and in many tropical warehouses of the world are stored in bags. Most of these warehouses have an open ridge vent running the entire length of the roof. Stirling[204] measured the effectiveness of this natural ventilation design and found that it 'does little or nothing to improve the internal storage environment'.

One important point must be stressed, though. Since the coffee bean, like all dried agricultural grains and seeds, is hygroscopic in nature, the water condensing on to the bean surface will rapidly be absorbed. Since water sorption is a faster process than desorption, and releases energy, unlike desorption which is an energy-requiring process, it is clear that the moisture content of coffee beans under such fluctuating temperature conditions at high humidity can be pumped up to levels of water activities where mould growth becomes possible, and simultaneously will also provide a deep-reaching moisture zone for extended mycelial penetration to the inside of the coffee bean.

The final conclusions drawn by Stirling were as follows. If the initial high quality of coffee beans is to be preserved during extended storage periods, then the beans must be kept cool. The temperature should not be higher than 20°C and the relative humidity should not be higher than 70%. Such stringent conditions call for air-conditioned warehouses. In non-humid tropical zones (higher altitudes), ventilated housing of green coffee beans packed in sacks stored in insulated buildings is a successful alternative solution for storing coffee for a period of one year without quality deterioration.

There is a strong economical driving force for improvement of green coffee bean quality. Stirling states that in Kenya a capital investment of about £3 million into improved coffee storage buildings and electronic coffee sorting equipment has yielded an extra £100 per ton on a crop of 70 000 per year.

3.9. Sea Transport of Green Coffee Beans

Transport of green coffee beans in sacks in holding bins of ocean freighters, depending on a combination of unfavourably wet weather conditions and on extended travel time, sometimes leads to mould spoilage of coffee beans. Such spoilage is highly undesirable, especially for the coffee roaster and also for the grower. In order to avoid or reduce such spoilage, the containerisation of green coffee beans for maritime transport has been described by Jouanjan in 1980.[205]

3.9.1. The Traffic Zones

The maritime transport of green coffee beans involves the loading of green coffee on to the ocean freighter in tropical climate conditions, i.e. high humidity and high temperature. Coffees shipped from tropical/subtropical areas are destined for cooler climates such as those of North America, Europe or Japan. Water condensation on to coffee beans is predictable, depending on the length of the trip and the extent of the climatic differences to which the coffee beans are subjected.

Interestingly, coffee beans exposed to the longest maritime travel time or shipment in closed containers (see Table 12, and also Tsuboi *et al.*,[198] Tsubouchi *et al.*[187] and Micco *et al.*[199]) show a higher incidence of secondary mould metabolites than coffees with shorter travel times and open/ventilated storage.

Before the bagged green coffee gets into the ship's hold it rests on stacked pallets on the pier of the harbour. If it should happen to rain and the sacks are not immediately covered, the beans obviously get wet. Such rains at the receiving harbour, during or before loading into the ship's hold or unloading from the ship's hold, may well explain the sporadic mouldiness of 'coffee that had got wet' (see Table 9, 267 hand-cleaned samples from mouldy lots[186]).

3.9.2. Use of Ventilated Container Liner Systems

Significant condensation (and thus mould growth) problems for shipping green coffee beans are now being alleviated by the introduction of ventilated cargo liner system.[206] The liner can be adapted to any 20- to 40-foot container, is suitable for both bulk and bagged shipments, protects the contents in the event of exterior damage to the container, and prevents insect infestation and also any water from entering the liner, etc.

A controlled experiment was carried out to test the superiority of the ventilated liner over a sealed container lined with kraft paper. Green coffee beans were shipped in the two different containers from Buenaventura, Colombia, on 5 February 1986. Both temperature and relative humidity fluctuated that day from 27 to 38°C and from 80 to 100%, respectively.

Loose bagged coffee (250 bags) was placed into the sealed container lined with kraft paper, and the ventilated container with the special liner was loaded with a further 250 sacks. The beans were removed from the sacks and conveyed into the liner. The ventilation system was installed and the container was sealed subsequently. Both containers were equipped with temperature and humidity recording/measuring devices. The ship left the harbour two days after loading and reached the Panama Canal on the eighth day after loading. On the voyage to Europe, the boat met a severe storm on day 21 and the temperature dropped to below freezing. The temperature after reaching the destination port on the 27th day had risen to 7°C. The containers were unloaded and opened, the instrument readings were compared and the quality of the coffee was inspected.

The coffee shipped in the bulk cargo liner showed no trace of condensation. Also the quality of the beans after 27 days of shipping showed a quality comparable to that of the coffee before shipping and gave the impression of an air-freighted coffee. The kraft-paper-lined container had massive condensation which had totally saturated and destroyed the paper liner. This happened after the freeze and warm-up period (warm-up to about 7°C). When flavour-tested, the coffee beans did not show as fresh a condition as the cargo-liner-shipped coffee. Visible mould growth as a result of the relatively short time period from water condensation of unpacking had not apparently had time to develop.

This experiment and the mould contamination resulting from sealed container shipments indicate that this form of shipment requires improvement. Obviously the use of a ventilated container, including a sealed liner, may constitute one solution to shipping coffee beans while avoiding water condensation problems.

3.9.3. Storage at Receiving Harbours

Coffee beans after unloading are stored in harbour warehouses until they are shipped by truck, etc., to the final destination site for roasting, grinding, packing or instant extraction.

More modern port unloading sites are equipped with covered unloading zones, modern warehouses with special coffee ventilation systems, and wooden infrastructures to avoid the water condensation that occurs on metal beams and to obviate water dripping on to coffee sacks.[207] Water dripping during storage on to one and the same spot will produce very localised mould growth. The mould mycelia grow only in the wet areas and may take the form of a stalactite due to the mould mycelia holding/clumping the wet beans together.

4. CONCLUSIONS

Mycotoxins and allergens found in association with green and roasted coffee beans have been discussed in this chapter, and historical and scientific developments have been briefly outlined. Related issues from other industries and the environment have been presented to provide a more meaningful background and context against which coffee-related subjects have been illuminated. We are able to state that the coffee industry does not contribute any additional or increased hazard to any of the areas discussed beyond what is naturally present in the human environment.

Contamination of green coffee beans with moulds, especially after maritime transport, is possible, especially when rain, high humidity and high temperatures prevail for prolonged periods of time. The coffee industry is in the fortunate position that mould contamination spoils coffee by producing strong off-flavours if mouldy coffee beans should ever be roasted and brewed. No development of secondary metabolites (mycotoxins) is required to occur to produce such pronounced off-flavours. This is in contrast to the many mould-contaminated food products where no deterring off-flavours are produced (e.g. mouldy rice, causing cardiac beri-beri; or mouldy corn, causing endemic Balkan nephropathy in humans or nephropathy in domesticated animals). The oldest known mycotoxicosis, ergotism, was already recognised by the Chinese 5000 years ago, halted warring armies about 1500 BC and affected those afflicted up to modern times in a very dramatic way. The mycotoxins produced by *Fusarium* moulds were capable of eradicating large percentages of the population of Russian villages (long before the Second World War) after villagers ingested bread produced from grain that had overwintered in the fields. Mycotoxin contamination of human food and animal feed is a fact of nature. Even in modern times, with all the improvements of harvesting and controlled silo storage or cold storage, moulds will continue to grow if given even only a remote chance to do so. As evidenced by finds of fossilised moulds, these organisms have been in existence long before the advent of mankind. And mankind has been able to develop and survive in spite of the presence of moulds, or indeed perhaps because of the presence of moulds if one considers the many storage-stable foods and beverages that are produced with the help of certain of these organisms.

It will, however, require an educational effort to ensure that the consumer becomes sufficiently knowledgeable, especially about the formation of moulds on food products in general and the physiological activities of moulds and mould metabolites (mycotoxins). The food and beverage industry delivers safe products to the consumer; it depends on the storage conditions of the food and beverage products in the consumer's household as to whether or not they will remain safe to consume, as is evidenced and demonstrated by the many different instances described in this chapter.

As the nearly 3000 green bean coffee samples analysed and referenced in this chapter indicate, mycotoxin contamination of green coffee beans from the grower countries reaching the consumer is very remote, especially when one considers the analytical data on coffees shipped to the US. There is some mould infestation and mycotoxin formation in the case of green beans shipped to Europe and to Japan. This is due to the fact that green coffee bean maritime shipment to these countries not only takes longer but also crosses climate zones that are conducive to water pick-up by the hygroscopic coffee beans. Furthermore, the roasting process (if ever any contaminated coffee beans should reach the coffee roaster) decreases the level of most mycotoxins quite effectively. Therefore the danger of accidental mycotoxin ingestion via the coffee beverage is extremely remote. However, there is no doubt that improved analytical procedures will discover additional secondary metabolites, and that more effective propagation of and growing conditions for laboratory cultures will reveal an even greater array of moulds and their corresponding secondary metabolites.

Green coffee bean allergens cause allergic reactions in atopic green bean coffee workers and may also be capable of sensitising non-atopic workers after long-term chronic exposure. Microscopic examination of the green bean coffee dust reveals that most of the coffee dust is of non-coffee origin. The following components have been identified (in decreasing order of occurrence): sack fibres, mould conidia, pollen and green coffee bean fragments. The description 'green coffee bean allergens' is therefore a misnomer. They should be more correctly described as 'allergens associated with green coffee beans'. The many studies reviewed in this chapter clearly demonstrate that the chlorogenic acids, among the main components of the green coffee bean and previously falsely accused of being allergens, do not provoke any allergic symptoms. The fact that chlorogenic acid (with its congeners) is found in many fruits, tubers, plant leaves and seeds is an indication that these compounds are beneficial and are required for the survival of eukaryotic cell systems. No scientific studies of any weight have been carried out to demonstrate the almost predictable *beneficial* effects of chlorogenic acids *for* human physiology.

On the positive side, literature studies reveal that the xanthine derivatives caffeine and theophylline (both contained in the coffee bean and in the coffee beverage) have a direct (*in vitro*) effect on maintaining a high level of c-AMP by blocking the activity of phosphodiesterase in both mast cells and eosinic (basophilic) cells, thus, in general, inhibiting the release of numerous anaphylactic mediators contained in the various basophilic cells. Such results, it is hoped, will encourage further studies to be carried out, to provide additional and more effective agents to block allergic reactions in the atopic individual and to keep the non-atopic individual from becoming sensitised to allergens after chronic exposure. The negative physiological effects of allergens attached to green coffee beans can be effectively eliminated by simply wearing dust masks and/or by installing efficient exhaust and filter systems, thus preventing the allergens from entering the nasal and pulmonary tracts.

ACKNOWLEDGEMENTS

The author is grateful to Debby Boyles, Shirley Heismann, Jan King, Stella Lewis, Mary Ann Siklosi, Arleen Thyen and Mike Vatter for their untiring assistance in the preparation of the manuscript of this chapter. Thanks are also due to Dr David Montezinos for his preparation of the coffee-dust photographs. Drs Jim McCarthy and Mario E. Tremblay carried out the analytical determinations on the green bean coffee dust samples.

REFERENCES

1. Golumbic, C., in *Mycotoxins in Foodstuffs*, Ed. G. N. Wogan, MIT Press, Cambridge, Mass., 1965, 49–67.
2. Hiscocks, E. S., in *Mycotoxins in Foodstuffs*, Ed. G. N. Wogan, MIT Press, Cambridge, Mass., 1965, 15–26.
3. Cahoon, G. A., *Am. Wine Soc. J.*, 1984, Fall Issue, 82–94.
4. Bischoff, R., *Bild Wiss.*, 1987, March, 46–57.
5. Bargel, G., in *Ergot and Ergotism*, Guerney and Jackson, London, 1931, 1–279.
6. Marth, E. H., in *Mycotoxins, a Review of Fungi and Foods, 7th Annual Symposium, New York Stock Experimental Station*, Cornell University, October 1972, 8–13.
7. Joffe, A. Z., in *Microbial Toxins, Vol. VII*, Eds S. Kadis, A. Ciegler and S. Ajil, Academic Press, New York, 1971, 139–189.
8. Miyaki, I., Naito, H. and Tsunodai, H., *Beikuko Riyo Kenkyujo Hokoku*, 1940, **1**, 1–29.
9. Tsunodai, H. and Haruki, Y., *Rep. Food Res. Inst.*, 1949, **2**, 157–164.
10. Graham, R., *Vet. Med.*, 1938, **XXXI**, 46–50.
11. Barnes, J. M., Carter, R. L., Peristianis, G. C., Austwick, P. K. C., Flynn, F. V. and Aldridge, W. N., *Lancet*, 1977, **1**, 671–675.
12. Shank, R. C., Bourgeois, C. H., Keschamras, N. and Chandavimol, P., *Food Cosmet. Toxicol.*, 1971, **9**, 501–507.
13. Shank, R. C., in *Mycotoxins in Human Health*, Ed. I. F. H. Purchase, Macmillan, London, 1971, 245–262.
14. Blount, W. P., *Turkeys*, 1961, **9**, 52, 55 and 61.
15. Lillehoj, E. B., Ciegler, A. and Detroy, R., in *Essays in Toxicology, Vol. 2*, Ed. F. R. Blood, Academic Press, New York, 1970, 48.
16. Wolf, H. and Jackson, E. W., *Science*, 1963, **42**, 676–678.
17. Levi, C. P. and Borker, E., *J. Assoc. Offic. Anal. Chem.*, 1968, **51**, 600–602.
18. Heim, R., *C. R. Acad. Agric. Fr.*, 1946, 1–5.
19. Natarajan, C. P. and Majumder, S. K., *Food Sci. India (Mysore)*, 1961, **X**, 315–321.
20. Majumder, S. K., Muthu, M., Srinivasan, K. S., Natarajan, C. P., Bhatia, D. S. and Subrahmanyan, V., *Food Sci. India (Mysore)*, 1961, **X**, 321–338.
21. Nicot, J., *Catalogue des Collections Vivantes, Herbiers et Documents, Mus. Nat. Hist. Nat.*, 1953, **VI**, 24–29.

22. Levi, C. P., *J. Assoc. Offic. Anal. Chem.*, 1980, **63**, 1282–1285.
23. Whittaker, R. H., *Science*, 1969, **163**, 150–160.
24. Rippon, J. W., *Medical Mycology*, 2nd edn, W. B. Saunders Co., Philadelphia, PA, 1982, 118.
25. Brock, T. D., *Biology of Microorganisms*, 3rd edn, Prentice-Hall, Englewood Cliffs, NJ, 1984.
26. Hesseltine, C. W., *Mycologia*, 1965, **57**, 149–197.
27. Satchell, J. E., in *Productivity of Forest Ecosystems*, Ed. P. Duvigneaud, UNESCO, Paris, 1971, 619–630.
28. Lacey, J., in *Biology of Conidial Fungi, Vol. 1*, Eds G. T. Cole and B. Kendrick, Academic Press, New York, 1981, 373–410.
29. Pitt, J. I., in *Biology of Conidial Fungi, Vol. 2*, Eds G. T. Cole and B. Kendrick, Academic Press, New York, 1981, 111–147.
30. Perlman, D., in *The Fungi, Vol. 1*, Eds G. C. Ainsworth and A. S. Sussman, Academic Press, New York, 1965, 479–489.
31. Strobel, R. G., Quinn, H. and Lange, W., *Can. J. Microbiol.*, 1967, **13**, 121–136.
32. Ormerod, J. G., *Arch. Mikrobiol.*, 1967, **56**, 31–39.
33. Pitt, J. I., in *Water Relations of Foods*, Ed. R. B. Duckworth, Academic Press, New York, 1975, 273–307.
34. Fries, N., in *The Fungi, Vol 1*, Eds G. C. Ainsworth and A. S. Sussman, Academic Press, New York, 1965, 491–532.
35. Scott, W. J., *Adv. Food Res.*, 1957, **7**, 83–127.
36. Pitt, J. I. and Hocking, A. D., *J. Gen. Microbiol.*, 1977, **101**, 35–40.
37. Hocking, A. D. and Pitt, J. I., *Trans. Br. Mycol. Soc.*, 1979, **73**, 141–145.
38. Troller, J. A. and Christian, J. H. B. (Eds), in *Water Activity and Food*, Academic Press, New York, 1978, 215–216.
39. Snow, D., *Ann. Appl. Biol.*, 1949, **XXXVI**, 1–13.
40. Schneider, R., *Phytopathol. Z.*, 1954, **21**, 63–78.
41. Moss, M. O., *Assoc. Promot. Ind. Agric.*, *Paris*, 1984, 205–225.
42. Martin, J. F. and Demain, A. L., *Microbiol. Rev.*, 1980, **44**, 230–251.
43. Bu'Lock, J. D., in *The Filamentous Fungi, Industrial Mycology, Vol. 1*, Eds J. E. Smith and D. R. Berry, Edward Arnold, London, 1975, 35–58.
44. Bu'Lock, J. D., Hamilton, D., Hulme, M. A., Powell, A. J., Smalley, H. M., Shepherd, D. and Smith, G. N., *Can. J. Microbiol.*, 1965, **11**, 765–778.
45. Bu'Lock, J. D., *Essays in Biosynthesis and Microbial Development*, Wiley, New York, 1967.
46. Aharonowitz, Y. and Demain, A. L., *Biotechnol. Bioengng*, 1980, **22** (Suppl. 1), 5–9.
47. Zähner, H., Anke, M. and Anke, T., in *Secondary Metabolism and Differentiation of Fungi*, Eds J. W. Bennett and A. Ciegler, Marcel Dekker, New York, 1983, 35–54.
48. Bennett, J. W., in *Advances in Biotechnology III: Fermentat. Prod. C*, Eds C. Vecina and K. Singh, Pergamon, Toronto, 1981, 409–415.
49. Turner, W. B., in *Fungal Metabolites*, Academic Press, London, 1971.
50. Reiss, J. (Ed.), *Mycotoxine in Lebensmitteln*, Gustav Fischer Verlag, Stuttgart, 1981, 119–142.
51. Smith, J. E. and Moss, M. O., in *Mycotoxins: Formation, Analysis, and Significance*, John Wiley, Chichester, 1983, 31.

52. Drew, S. W. and Wallis, D. A., in *Secondary Metabolism and Differentiation of Fungi*, Eds J. W. Bennett and A. Ciegler, Marcel Dekker, New York, 1983, 35–54.
53. Campbell, I. M., in *Secondary Metabolism and Differentiation of Fungi*, Eds J. W. Bennett and A. Ciegler, Marcel Dekker, New York, 1983, 55–72.
54. Nover, L. and Luckner, M., *Biochem. Physiol. Pflanz.*, 1974, **166**, 293–305.
55. Luckner, M., Nover, L. and Böhm, H. (Eds), *Secondary Metabolism and Cell Differentiation*, Springer-Verlag, Berlin, 1977, 1–130.
56. Martin, J. F. and Demain, A. L., in *The Filamentous Fungi, Developmental Biology, Vol. 3*, Eds J. E. Smith and D. R. Berry, Edward Arnold, London, 1978, 436–450.
57. Walker, J. B., *Meth. Enzymol.*, 1975, **43**, 451–458.
58. Satoh, A., Ogawa, H. and Satomura, Y., *Agric. Biol. Chem.*, 1976, **40**,191–196.
59. Aharonowitz, Y. A., in *Genetics of Industrial Microorganisms*, Eds O. K. Sabek and A. I. Laskin, American Society of Microbiology, Washington, DC, 1979, 210–217.
60. Aharonowitz, Y. A. and Demain, A. L., *Can. J. Microbiol.*, 1979, **25**, 61–67.
61. Matteo, C. C., Cooney, C. L. and Demain, A. L., *J. Gen. Microbiol.*, 1976, **18**, 1045–1048.
62. Pier, A. C., Richard, J. L. and Thurston, J. R., *J. Am. Vet. Med. Assoc.*, 1980, **176**, 719–724.
63. Bryden, W. L., *Food Technol. Australia*, 1982, **34**, 216–223.
64. Smith, J. E. and Moss, M. O., in *Mycotoxins: Formation, Analysis, and Significance*, John Wiley, Chichester, 1985, 57.
65. Sarkisow, A. Ch., *Mycotoxicoses*, Seljchosisdat, Moscow, 1954.
66. Wong, J. J. and Hsieh, D. P. H., *Proc. Natl Acad. Sci. USA*, 1976, **27**, 2241.
67. Wong, J. J., Sing, R. and Hsieh, D. P. H., *Mutation Res.*, 1977, **44**, 447–450.
68. Reiss, J. (Ed.), *Mycotoxine in Lebensmitteln*, Gustav Fischer Verlag, Stuttgart, 1981, 200.
69. Garner, R. C. and Wright, C. M., *Chem. Biol. Interact.*, 1975, **11**, 123–131.
70. Garner, R. C., *Chem. Biol. Interact.*, 1973, **36**, 261–264.
71. Essigman, J. M., Croy, R. G., Nadzan, A. M., Bushby, W. F., Reinhold, V. N. and Wogan, G. N., *Proc. Natl Acad. Sci. USA*, 1977, **74**, 1870–1874.
72. Lin, J. K., Miller, J. A. and Miller, E. C., *Proc. Natl Acad. Sci. USA*, 1977, **74**, 1870–1874.
73. Miller, J. A., *Cancer Res.*, 1970, **30**, 559–576.
74. Shank, R. C., in *Mycotoxic Fungi, Mycotoxins, Mycotoxicoses, Vol. 3*, Eds T. D. Wyllie and L. G. Morehouse, Marcel Dekker, New York, 1978, 1 and 9.
75. Patterson, D. S. P., *Biochem. Soc. Trans.*, 1973, **1**, 917–922.
76. Campbell, T. C. and Hayes, J. R., *Toxicol. Appl. Pharmacol.*, 1976, **35**, 199–222.
77. Croy, R. G., Essigman, J. M., Reinhold, V. N. and Wogan, G. N., *Proc. Natl Acad. Sci. USA*, 1978, **75**, 1745–1749.
78. Patterson, D. S. P., *Ann. Nutr. Aliment.*, 1977, **31**, 901–910.
79. Peers, F. G. and Linsell, C. A., *Ann. Nutr. Aliment.*, 1977, **31**, 1005–1017.
80. van Rensburg, S. J., in *Mycotoxins in Human and Animal Health*, Eds J. V. Rodricks, C. W. Hesseltine and M. A. Mehlmann, Pathotox Publishers, Park Forest South, Ill., 1977, 699–711.

81. van Rensburg, S. J., in *Mycotoxins and Phytotoxins*, Eds P. S. Steyn and R. Vieggar, IUPAC Symposium on Mycotoxins and Phycotoxins, Pretoria, July 1985, Elsevier, Amsterdam, 1986, 483–494.
82. Bourgeois, C., Olson, L., Comer, D., Evans, H., Keschamras, N., Colton, R., Grossman, R. and Smith, T., *Am. J. Clin. Pathol.*, 1971, **56**, 558–571.
83. Gröger, D., in *Microbial Toxins, Vol. 8*, Eds S. Kadis, A. Ciegler and S. J. Ajl, Academic Press, New York, 1972, 321–373.
84. Mirocha, C. J., in *The Applied Mycology of Fusarium*, Eds M. O. Moss and J. E. Smith, Cambridge University Press, Cambridge, 1984, 141–155.
85. Booth, C., in *The Applied Mycology of Fusarium*, Eds M. O. Moss and J. E. Smith, Cambridge University Press, Cambridge, 1984, 2–13.
86. Ueno, Y., Ishii, K., Sakai, K., Kanaeda, S., Tusunoda, H., Tanaka, T. and Enomoto, M., *Japan. J. Exp. Med.*, 1972, **42**, 187–203.
87. Dounin, M., *Phytopathology*, 1926, **16**, 305–308.
88. Mirocha, C. J., in *Assoc. Promot. Ind. Agric. (Paris)*, CPC1A, B001, 1984, 51–130.
89. Mirocha, C. J., Pawlosky, R. A., Chatterjee, K., Watson, S. and Hayes, W., *J. Assoc. Offic. Anal. Chem.*, 1984, **66**(6), 1485–1499.
90. Nesterov, A. I., *Arthritis and Rheumatism*, 1964, **7**, 29–40.
91. Kurmanov, I. A., translated by Novacky, A., in *Mycotoxic Fungi, Mycotoxins, Mycotoxicoses, Vol. 2*, Eds T. D. Wyllie and L. G. Morehouse, Marcel Dekker, New York, 1978, 85–110.
92. Scott, P. M., van Walbeek, W., Harwig, J. and Fennel, D. I., *Can. J. Plant Sci.*, 1970, **L**, 583–585.
93. van Walbeek, W., Scott, P. M., Harwig, J. and Lawrence, J. W., *Can. J. Microbiol.*, 1969, **XV**, 1281–1285.
94. Krogh, P. and Elling, F., *Lancet*, 1976, **2**, 40.
95. Maebayashi, Y., Miyaki, K. and Yamazaki, M., *Chem. Pharm. Bull. (Tokyo)*, 1972, **20**, 2172–2175.
96. Reiss, J. (Ed.), *Mycotoxine in Lebensmitteln*, Gustav Fischer Verlag, Stuttgart, 1981, 126.
97. Cole, G. T. and Samson, R. A., in *Mould Allergy*, Eds Y. Ar-Doorey and J. F. Domson, Lea & Febiger, Philadelphia, PA, 1984, 66–103.
98. Totora, G. J. and Anagnostakos, N. P., in *Principles of Anatomy and Physiology*, 2nd edn, Canfield, San Francisco, CA, 1978.
99. Austwick, P. K. C., in *The Fungus Spore*, Ed. M. F. Madelin, Butterworths, London, 1966, 321–338.
100. Mitchell, R. I., *Am. Rev. Resp. Dis.*, 1960, **82**, 627.
101. Figley, K. D. and Rawling, F. F. A., *J. Allergy*, 1950, **21**, 545–553.
102. Vernon, D. R. H. and Allan, F., *Clin. Allergy*, 1980, **10**, 217–227.
103. Bellanti, J. A., in *Immunology*, 2nd edn, W. B. Saunders Co., Philadelphia, PA, 1978, 471–621.
104. Roitt, I. M., *Essential Immunology*, 3rd edn, Blackwell Scientific Publications, Oxford, 1977, 1–324.
105. Benacerraf, G. and Uname, E. R., *Textbook of Immunology*, Williams and Wilkins, Baltimore, MD, 1979.
106. Coombs, R. R. A. and Gell, P. G. H. (Eds), *Clinical Aspects of Immunology*, Davis, Philadelphia, PA, 1963, 17.

107. Austen, K. F., *Hosp. Practice*, 1982, **17**(11), 98.
108. O'Flaherty, J. T., *Lab. Invest.*, 1982, **47**, 314.
109. Leung, D. Y. M., Frankel, R., Wood, N. and Geha, R. S., *J. Clin. Invest.*, 1986, **77**, 952–957.
110. Cotran, R. S., *J. Cell Biol.*, 1982, **93**, 357.
111. Sandberg, A. L., in *Cellular Functions in Immunity and Inflammation*, Eds J. Oppenheim and D. L. Rosenstreich, Elsevier/North-Holland, Amsterdam, 1981, 373–395.
112. Ward, P. A., in *Cellular Functions in Immunity and Inflammation*, Eds J. Oppenheim and D. L. Rosenstreich, Elsevier/North-Holland, Amsterdam, 1981, 153–177.
113. Bruun, E., *Acta Allergol.*, 1957, **11**, 150–154.
114. Freedman, S. O., Krupey, J. and Schon, A. H., *Nature (London)*, 1961, **192**, 241–243.
115. Freedman, S. O., Siddiqi, A. L., Krupey, J. and Schon, A. H., *Trans. Assoc. Am. Phys.*, 1962, **75**, 99.
116. Freedman, S. O., Schulman, R. and Krupey, J., *J. Allergy*, 1964, **35**, 108–116.
117. Freedman, S. O., Schulman, R. and Krupey, J., *J. Allergy*, 1964, **35**, 97–107.
118. Layton, L. L., Greene, F. C. and Panzani, R., *J. Allergy*, 1965, **36**, 84–91.
119. Layton, L. L., Panzani, R., Greene, F. C. and Corse, J. W., *Int. Arch. Allergy*, 1965, **28**, 116–127.
120. Lowell, F. C., *J. Allergy*, 1966, **38**, 276.
121. Freedman, S. O., Krupey, J. and Schon, A. H., *J. Allergy*, 1966, **38**, 277–279.
122. Layton, L. L. (reply letter), *J. Allergy*, 1966, **38**, 279.
123. Layton, L. L., Panzani, R. and Cortese, T. A., *Int. Arch. Allergy*, 1968, **33**, 417.
124. Turula, M., Aho, J., Taipale, S. and Förström, L., in *XVth Int. Congr. Occup. Health, Vol. II*, H. Egerman, Vienna, 1966, 845–848.
125. VanToorn, D. W., *Thorax*, 1970, **25**, 399–405.
126. Somazzi, S. and Wüthrich, B., *Méd. Hyg.*, 1975, **33**, 677–683.
127. Bernton, H. S., *J. Am. Med. Assoc.*, 1973, **223**, 1146–1147.
128. Karr, R. M., Lehrer, S. B., Butcher, B. T. and Salvaggio, J. E., *J. Allergy Clin. Immunol.*, 1978, **62**, 143–148.
129. Lehrer, S. B., Karr, R. M. and Salvaggio, J. E., *Clin. Allergy*, 1978, **8**, 217–226.
130. Osterman, K., Johansson, S. G. and Zetterström, O., *Allergy*, 1978, **33**, 350.
131. Zuskin, E., Vàlić, F. and Skurić, Z., *Br. J. Ind. Med.*, 1979, **36**, 117–122.
132. Zuskin, E. and Bouhuys, A., *J. Allergy Clin. Immunol.*, 1976, **57**, 473–479.
133. Zuskin, E., Vàlić, F. and Bouhuys, A., *Lung*, 1976, **154**, 17–24.
134. Cox, J. S., *Br. J. Dis. Chest*, 1971, **65**, 189–204.
135. Zuskin, E., Vàlić, F. and Kanceljak, B., *Thorax*, 1981, **36**, 9–13.
136. Peters, J. M., Mead, J. and VanGanse, W. F., *Am. Rev. Resp. Dis.*, 1969, **99**, 617–622.
137. National Institute for Occupational Safety and Health, *Occupational Exposure to Cotton Dust: Criteria for a Recommended Standard*, HEW Publication (NIOSH), Washington, DC, 1974, 75–118.
138. Osterman, K., Zetterström, O. and Johansson, S. G., *Allergy*, 1982, **37**, 313–322.
139. Osterman, K., Johansson, S. G. and Zetterström, O., *Allergy*, 1985, **40**, 336–343.

140. Gerrard, J. W., Cockroft, D. W., Mink, J. T., Cotton, D. J., Poonawala, R. and Dosman, J. A., *Am. Rev. Respir. Dis.*, 1980, **122**, 577–581.
141. Becker, C. G., Van Hamont, N. and Wagner, M., *Blood*, 1981, **58**, 861–867.
142. Becker, C. G., Dubin, T. and Wiedemann, H. P., *Proc. Natl Acad. Sci. USA*, 1976, **73**, 1712–1716.
143. Becker, C. G. and Dubin, T., *J. Exp. Med.*, 1977, **146**, 457–467.
144. Malley, A., Baecher, L. and Begley, D., in *Int. WHO–IABS Symposium on Standardization and Control on Allergens Administered to Man*, Devel. Biol. Stand., WHO, Geneva, Basel, Karger, 1975.
145. Pinckard, R. N., Halonen, M. and Meng, A. L., *J. Allergy Clin. Immunol.*, 1972, **49**, 301–310.
146. Levin, J. and Baug, F. B., *Thromb. Diath. Haemorrh.*, 1968, **19**, 186–197.
147. Levi, R., Zavecz, J. H., Holland, B. and Becker, C. G., *Fed. Proc.*, 1978, **37**, 590.
148. Becker, C. G., Dubin, T. and Glenn, F., *J. Exp. Med.*, 1980, **151**, 81–90.
149. Zuskin, E., Duncan, P. G. and Douglas, J. S., *Br. J. Ind. Med.*, 1983, **40**, 193–198.
150. Kalsner, S., *Life Sci.*, 1977, **20**, 1689–1696.
151. Zuskin, E., Kanceljak, B., Skurić, Z. and Butković, D., *Br. J. Ind. Med.*, 1985, **42**, 415–420.
152. Van den Bosch, J. M., Van Toorn, D. W. and Wagenaar, S. S., *Thorax*, 1983, **38**, 720.
153. Weill, H., Diem, J., Hughes, J. and Jones, R., *Chest*, 1984, Suppl. 6, 125.
154. Smith, D., Brott, K. and Koki, G., *Br. J. Ind. Med.*, 1985, **42**, 495–498.
155. Kovats, F. and Bugyi, B., *Occupational Mycotic Diseases of the Lung*, Akad. Kiado, Budapest, 1968, 1–212.
156. Schuyler, M. and Salvaggio, J. E., *Semin. Respir. Med.*, 1984, **5**, 246–254.
157. Pepys, J., *Monographs in Allergy, Vol. 4*, Karger, Basel, 1969.
158. Wang, J. L., Patterson, R., Mintzer, R., Roberts, M. and Rosenberg, M., *J. Pediatr.*, 1979, **94**, 376–381.
159. Wang, J. L. F., Patterson, R., Roberts, M. and Ghory, A. C., *Am. Rev. Respir. Dis.*, 1979, **120**, 87–92.
160. Godard, P., Chaintreuil, J., Damon, M., Coupe, M., Flandre, O., Crastes DePaulet, A. and Michel, F. B., *J. Allergy Clin. Immunol.*, 1982, **70**, 88–93.
161. Golbert, T. M. and Patterson, R., *Ann. Int. Med.*, 1970, **72**, 395–403.
162. Stankus, R. P. and Salvaggio, J. E., *Clin. Chest Med.*, 1983, **4**, 55–62.
163. Rippon, J. W., *Medical Mycology*, 2nd edn, W. B. Saunders Co., Philadelphia, 1982, 507–508.
164. Oddes, F. C., *Candida and Candidosis*, University Park Press, Baltimore, MD, 1979.
165. Winner, H. L. and Hurley, R., *Candida albicans*, Little, Brown and Co., Boston, Mass., 1964.
166. Goodpasture, E. W., *J. Med. Sci.*, 1919, **158**, 163.
167. Pepys, J., *Ann. NY Acad. Sci.*, 1974, **221**, 27–31.
168. van den Zijden, A. S. M., Blanche Koelensmid, W. A. A., Bolding, J., Barrett, C. B., Ord, W. O. and Philip, J., *Nature (London)*, 1962, **195**, 1060–1062.
169. Nesbitt, B. F., O'Kelly, J., Sargeant, K. and Sheridan, A., *Nature (London)*, 1962, **195**, 1062–1063.
170. Hartley, R. D., Nesbitt, B. F. and O'Kelly, J., *Nature (London)*, 1963, **198**, 1056–1058.

171. Asao, T., Buchi, G., Abdel-Kader, M. M., Chang, S. B., Wick, E. L. and Wogan, G. N., *J. Am. Chem. Soc.*, 1963, **85**, 1706–1707.
172. Chang, S. B., Abdel-Kader, M. M., Wick, E. L. and Wogan, G. N., *Science*, 1963, **142**, 1191–1192.
173. Asao, T., Buchi, G., Abdel-Kader, M. M., Chang, S. B., Wick, E. L. and Wogan, G. N., *J. Am. Chem. Soc.*, 1965, **87**, 882–886.
174. Levi, C. P., *9th Int. Coll. ASIC*, 1980, 125–133.
175. Levi, C. P., *J. Assoc. Offic. Anal. Chem.*, 1969, **LII**, 1300–1303.
176. Scott, P. M., *J. Assoc. Offic. Anal. Chem.*, 1968, **5**, 609.
177. Boutibonnes, P. and Jacquet, J., *C.R. Scéances Soc. Biol. (Paris)*, 1969, **163**, 1119–1124.
178. Jacquet, J., Boutibonnes, P. and Teherani, A., *Ind. Alim. Agric.*, 1971, **5**, 689–697.
179. Beebe, R. M., *J. Assoc. Offic. Anal. Chem.*, 1978, **61**, 1347–1352.
180. Kiermeyer, F., *Z. Lebensmitt. Unters. Forsch.*, 1985, **180**, 389–393.
181. Majumder, S. K., Narasimham, K. S. and Parpia, H. A. B., in *Mycotoxins of Foodstuffs*, Ed. G. N. Wogan, MIT Press, Cambridge, Mass., 1965, 27–47.
182. Corte de Santos, A., Hahn, D., Cahagnier, B., Drapron, R., Guilbot, A., Levebvre, J., Multon, J. L., Poisson, J. and Trentesaux, E., *Café, Cacao, Thé*, 1971, **XV**(4), 329–339.
183. Multon, J. L., Poisson, J., Cahagnier, B., Hahn, D., Barel, M. and Corte de Santos, A., *6th Int. Coll. ASIC*, 1973, 268–277.
184. Poisson, J., Cahagnier, B., Multon, J. L., Hahn, D. and Corte de Santos, A., *7th Int. Coll. ASIC*, 1975, 311–321.
185. Rao, N. G., Balachandran, A., Natarajan, C. P. and Sanfaran, A. N., *J. Food Sci. Technol. (Mysore)*, 1971, **8**, 174–176.
186. Levi, C. P., Trenk, H. L. and Mohr, H. K., *J. Assoc. Offic. Anal. Chem.*, 1974, **57**, 866–870.
187. Tsubouchi, H., Yamamoto, K., Hisada, K. and Sakabe, Y., *Proc. Jap. Assoc. Mycotoxicol.*, 1984, **19**, 14–21.
188. Buchanan, R. L. and Fletcher, A. M., *J. Food Sci.*, 1978, **43**, 654–655.
189. Buchanan, R. L. and Fletcher, A. M., *J. Food Sci.*, 1981, **47**, 319–321.
190. Buchanan, R. L. and Fletcher, A. M., *J. Food Sci.*, 1983, **48**, 1226–1228.
191. Nartowicz, V. B., Buchanan, R. L. and Segall, S., *J. Food Sci.*, 1978, **44**, 446–448.
192. Tsubouchi, H., Terada, H., Yamamoto, K., Hisada, K. and Sakabe, Y., *Mycopathologia*, 1985, **90**, 181–186.
193. Stack, M. E., Mislivec, P. B., Denizel, T., Gibson, R. and Pohland, A. E., *J. Food Prot.*, 1983, **46**, 965–968.
194. Purchase, I. F. and Pretorius, M. A., *J. Assoc. Offic. Anal. Chem.*, 1973, **56**, 225–226.
195. Levi, C. P., *7th Int. Coll. ASIC*, 1975, 287–294.
196. Gallaz, I. and Stalder, R., *Chem. Microbiol. Technol. Lebensm.*, 1976, **4**, 147–149.
197. Heusinkveld, M. R., Shera, C. C. and Baur, F. C., *J. Assoc. Offic. Anal. Chem.*, 1965, **48**, 448–449.
198. Tsuboi, S., Kawamura, K. and Iwamura, N., *Ann. Int. Cent. Med. Res.*, 1983, **3**, 109–116.

199. Micco, C., Grossi, M., Miraglia, M. and Brera, C., *12th Int. Coll. ASIC*, 1987 (in press).
200. Mislivec, P. B., Bruce, V. R. and Gibson, R., *J. Food Protect.*, 1983, **46**, 969–973.
201. Sivetz, M. and Desrosier, N. W., in *Coffee Technology*, AVI Publishing Company, Westport, CT, 1979, 128.
202. Vincent, J. C., in *Coffee, Vol. 2: Technology*, Eds R. J. Clarke and R. Macrae, Elsevier Applied Science Publishers, London, 1987, 3.
203. Clarke, R. J., in *Coffee, Vol. 1: Chemistry*, Eds R. J. Clarke and R. Macrae, Elsevier Applied Science Publishers, London, 1987, 42–82.
204. Stirling, H., *9th Int. Coll. ASIC*, 1980, 189–200.
205. Jouanjan, F., *9th Int. Coll. ASIC*, 1980, 177–188.
206. Felise, G., *Tea and Coffee Trade J.*, 1986, **160**, July, 22–27.
207. Bell, J., *Tea and Coffee Trade J.*, 1987, **159**, August, 10–12.
208. Persson, C. G. A., *J. Allergy Clin. Immunol.*, 1986, **78**, 780–787.
209. Gergen, P. J. and Turkeltaub, P. C., *National Health Survey, Series 11, No. 235*, US Dept of Health and Human Services, National Center for Health Statistics, Hyattsville, MD, July 1986, 1–60.
210. Frierson, M. R., Klopman, G. and Rosenkranz, H. S., *Experimental Mutagenesis*, 1986, **8**, 283–327.

Chapter 10

Polycyclic Aromatic Hydrocarbon Contaminants in Coffee

R. G. K. STROBEL
The Procter & Gamble Company, Cincinnati, Ohio, USA

1. INTRODUCTION

1.1. General

The coffee bean roasting process requires the heating of green coffee beans up to 200°C or higher depending on the degree of roasting desired. Depending on the nature of the roasting process, roasting may take from about a minute (in a spouting bed roaster) to about 10 min or more (in a conventional roaster). During this roasting process, not only are the very stimulating aroma and flavour compounds generated, but also very small quantities of polycyclic aromatic hydrocarbons are produced. This chapter deals with these polycyclic aromatic hydrocarbons produced during coffee roasting, some of which are reported to have carcinogenic properties. The history of the discovery of polycyclic aromatic hydrocarbons and also the ancient and more recent history of cancer will be touched on briefly. Analytical data on polynuclear aromatic hydrocarbons in various coffee products, a number of food products and the human environment are presented.

The physiological transformation from non-reactive compounds to polynuclear aromatic hydrocarbons to physiologically reactive compounds and their *in vivo* enzymatic detoxification will be described.

1.2. Early History of Cancer

The oldest document concerning cancer is, according to Graffi and Bielka,[1] the Indian script 'Ramajana' (~2000 BC). Hippocrates (460–375 BC) seems to have introduced the name 'carcinoma'. In his *Definitiones medicae*, Galenus (AD 131–203) explains that the word *carcinoma* (Greek) or *cancer* (Latin) was chosen because some cancers, especially cancer of the breast, sometimes have a striking resemblance to the animal called *cancer* (the crab). The extirpation of cancer tissue with a knife or glowing iron was used in India by 2000 BC. Hippocrates thought that cancer and all other diseases are caused by a disturbance of the normal ratios of blood, mucins, and the black and yellow gall bladder content. Galenus[1] stated the opinion that cancer is caused by the black gall bladder content, or 'melancholia' (melancholic disposition).

Cancer, formerly also called 'the wrath of gods', was first described as being caused by extrinsic agents by Theophrastus B. von Hohenheim[2] in his *Die drei Bücher des Opus Paramirum* (1531). In writing about the *morbis mineralium*, said to be caused by sulphur-bearing minerals, he wrote: '... wie das sie mancherlei Form machen als in Krebsen, Fisteln ... Darnach weiter aus Vitriolo kommen dise Krankheiten ... Item aus Vitriolo albo, was do Cancrena, etc.... Aus disem [sulphure] fressen und nagen da entspringen die Ulcerationes, Cancer und Cancrena, etc.' In the three volumes of *Von der Bergsucht und anderen Bergkrankheiten*, written in 1533–34, von Hohenheim[3] describes in very vivid detail from observations made in the mines of Tyrol the difference in severity of symptoms according to whether the *Bergkrankheit* was brought about by arsenic sulphide (realgar) and arsenic (As), or by antimony (Sb), and whether present in miners actually working in the mines inhaling the mine dust or in smelters inhaling the *Hüttenrauch* (smelting fumes). His descriptions of the *Bergkrankheit* are congruent with those by Härting and Hesse[4] provided 345 years later.

Nasal cancer was for the first time associated with the habit of snuffing tobacco by Dr Hill in 1761.[5] In 1761 John Hill found several subjects who developed 'all the frightful symptoms of an open [nasal] cancer after long and immoderate (vast quantities) use of snuff'.[5] In 1775 Sir Percival Pott[6] associated the high incidence of scrotal and skin cancer in English chimney sweeps with their constant exposure to soot. Twenty years thereafter Sömmering[7] reported the obvious association between pipe smoking and cancer of the lip.

In 1875 Volkmann[8] pointed out the frequent occurrence of skin cancer in German coal workers. Seventeen years later Butlin[9] demonstrated that Scottish shale oil workers were at risk. It was in 1879 that Härting and

Hesse[4] had identified lung cancer in the miners of Schneeberg. This is the identical disease already described in precise detail in 1533–34 by von Hohenheim as *Bergkrankheit*. Additional occupational cancers are listed in Table 1.

These observations indicated the presence of a carcinogenic factor in the various materials. Brosch,[10] in 1900, was apparently one of the first investigators to be involved in experimental carcinogenesis. He used repeated local physical injury and applied organic solvents and also tobacco juices to guinea pigs and observed epithelial proliferation. In 1915 the Japanese investigators Yamagiwa and Ichikawa[11] described how they produced experimental cancer by repeatedly painting the ears of rabbits with tar. Bloch and Dreifuss[12] reported in 1921 that the carcinogen in tar does not contain nitrogen or sulphur; however, it forms picrates and is probably an aromatic hydrocarbon.

Table 1
Occupational cancers[13]

Year	*First reported by:*	*Agent or process*	*Site of activity*
1531	von Hohenheim[2]	Mine dust, *Hüttenrauch* (smelting fumes)	Lung (*Bergkrankheit*), skin
1533–34	von Hohenheim[3]		
1761	Hill[5]	Tobacco snuff	Nasal cavity
1775	Pott[6]	Chimney soot	Scrotum
1795	Sömmering[7]	Tobacco pipe smoking	Lip
1822	Paris	Arsenic	Skin, scrotum
1875	Volkmann[8]	Crude wax from coal	Skin
1876	Bell	Shale oil	Skin
1879	Härting and Hesse[4]	Ionising radiation	Lung
1892	Butlin[9]	Shale oil	Skin
1894	Unna	Ultraviolet radiation	Skin
1895	Rehn[14]	Aromatic amines	Bladder
1898	Mackenzie	Creosote	Skin
1906	Frieben	X-rays	Skin
1910	Wilson	Shale oil; mineral oil lubricants	Skin
1911	Pfeil[15]	Chromate production	Lung
1917	Legmann	Crude anthracene	Skin
1922	Southam, Wilson	Spindle oil	Scrotum
1926	Prunes	Salt petre	Skin
1928	Dolore, Borgomano	Benzene	Blood-cell-forming tissues (leukaemia)
1929	Martland	Radium	Bone
1932	Grenfell	Nickel refining	Lung and sinuses
1935	Lynch and Smith	Asbestos production	Lung
1952	Weil *et al.*	Isopropyl manufacture	Larynx

1.3. Discovery of 3,4-Benzpyrene

In 1922, in England, Kennaway[16] started an extensive search for the carcinogenic factor in tar. He tested (on mouse skin, without success) a large number of hydrocarbons known at this time to occur in tar.

In collaboration with Goulden he found that gas-works pitch, when heated to 500°C, yielded a highly carcinogenic distillate. Furthermore, carcinogenic materials were produced when non-carcinogenic petroleum, isoprene, acetylene, cholesterol, yeast, human skin, hair and muscle were heated in the presence of hydrogen according to a procedure of Berthelot[17] (to produce benzene, vinylbenzene, fluorescing hydrocarbons, etc., from acetylene). In addition to the carcinogenic Berthelot preparations, the production of which required a temperature of ~450°C, Kennaway was looking for a reaction that would produce hydrocarbons at body temperature. This was realised when Kennaway read a paper by Schroeter[18] describing the production of many cyclic compounds, including benzene and octahydroanthracene (**I**), by treating tetralin (**II**) with $AlCl_3$ at 30–40°C.

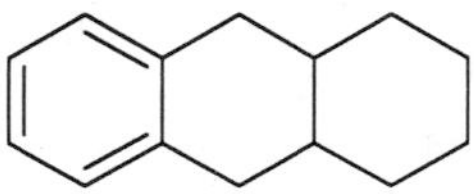

I

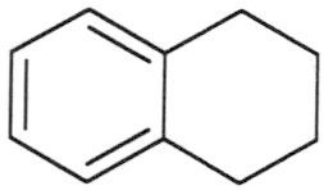

II

The result produced by heating and then testing the effects of such Schroeter mixtures was extremely encouraging. The mixtures were obviously less complex than tar and strongly carcinogenic, and in addition were vividly fluorescent, not only under UV but also in daylight. This fluorescence led the physicist M. V. Mayneord (who had collaborated with Kennaway since 1926) to examine the spectroscopic properties. In collaboration with I. Hieger he tested all of the Berthelot tars and Schroeter mixtures by means of spectroscopy. All the carcinogenic materials had one spectroscopic property in common, namely they showed two distinct bands in the blue and violet regions of the spectrum. Next they tried to find the compounds causing these 'cancer bands', but none of the hydrocarbons on hand showed such bands. In 1928 Hieger obtained some polycyclic hydrocarbon samples from J. W. Cook, who was working on the synthesis of hydrocarbons. Fortunately one of these samples, 1,2-benzanthracene (**III**) (now called benzo[*a*]anthracene), exhibited a spectrum almost identical to that of the carcinogenic mixtures.

The next move was to look for hydrocarbons of the 1,2-benzanthracene

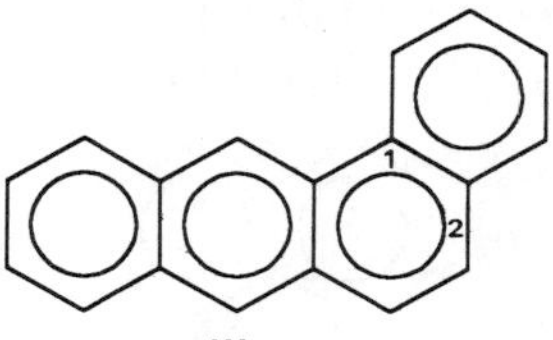

III

type. Luckily Kennaway found a paper by Clar[19] on the synthesis of 1,2:5,6-dibenzanthracene (**IV**) (now called dibenz[*a,h*]anthracene). However, the spectrum of this compound was still not quite the expected one, but upon cancer testing both benzanthracenes proved active. Thus the first cancer-causing synthetic chemical compounds had been synthesised. Then

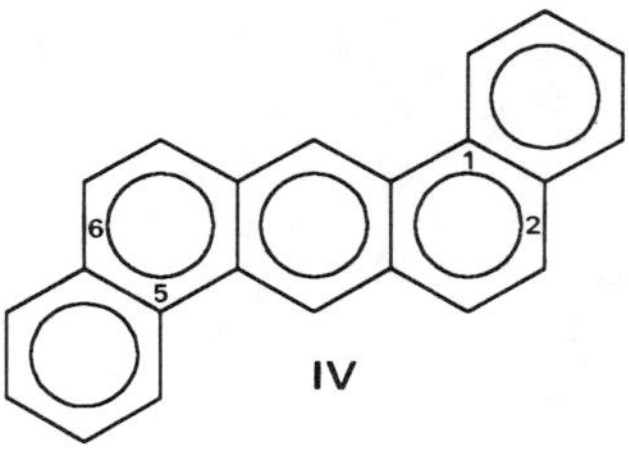

IV

in 1929 J. W. Cook joined Kennaway's group, and numerous anthracene and chrysene derivatives were synthesised. Many of the compounds were active carcinogens but none showed the desired spectrum. Early in 1930 Hieger was able to sharpen the 'cancer bands' of gas-works tar by alcohol extraction. Two tons of gas-works tar were extracted, and, by the autumn of 1931, 7 g of a yellow crystalline powder, which exhibited a melting point of 116°C, had been obtained. The powder was highly carcinogenic and showed the required 'powerful fluorescence' in the two spectral regions.

Another batch of this material was fractionated by C. L. Hewett, who had joined the group in 1931. He isolated 2,3-benzcarbazole (**V**), chrysene (**VI**) and phenanthrene (**VII**).

The remaining material was taken over by Cook, who was able to isolate

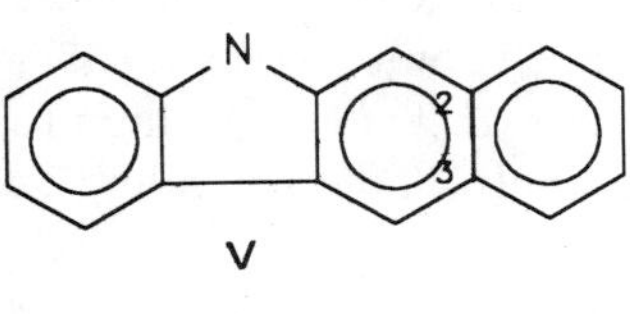

V

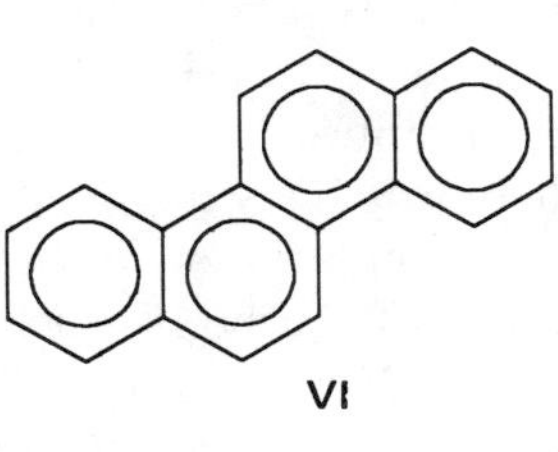

VI

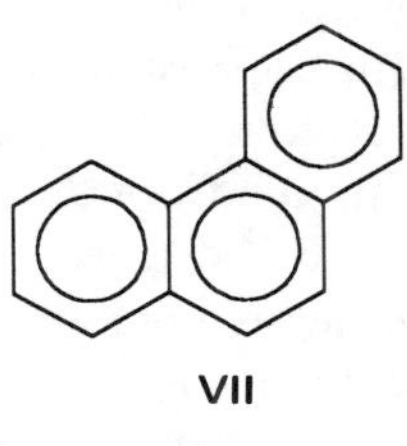

VII

two isomeric materials melting at 176 and 187°C. Both materials appeared isomeric with the pentacyclic hydrocarbon, perylene (**VIII**).

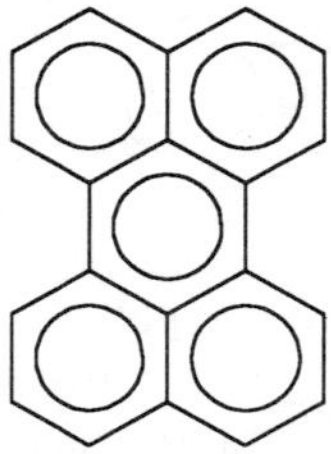

VIII

The major component (m.p. 176°C) exhibited exactly the desired characteristic spectrum and now the syntheses of the then unknown pentacyclic compounds benzo[*e*]pyrene (**IX**) and benzo[*a*]pyrene (**X**) were carried out. The synthetic material melted at 177°C.

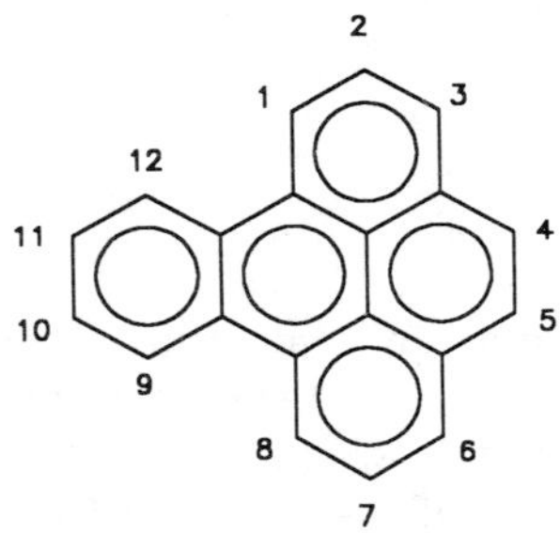

IX

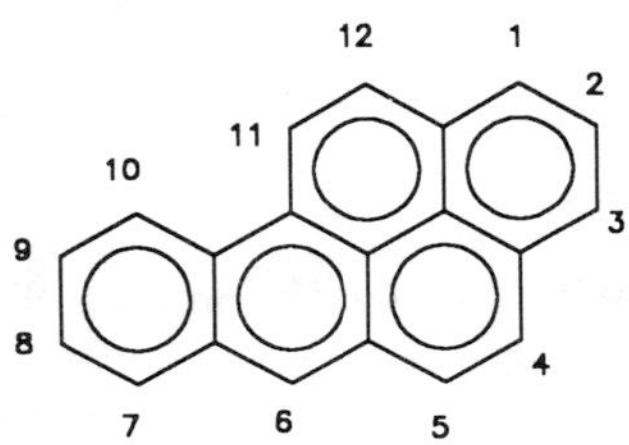

X

When the latter compound exhibited the same spectral properties seen in the pitch isolates, and in addition showed a strong carcinogenic effect, the search for the long-sought carcinogen was crowned in 1932 with the discovery of the benzo[*a*]pyrene.[20]

1.4. Physical, Viral and Chemical Carcinogens

All cancers, however diverse they may be, share the common feature of uncontrolled cell growth. Benign tumours remain confined within an envelope of fibrous tissue at the site of the physical, viral or chemical reaction starting site. In contrast, malignant tumours are unrestrained and migrate from the site of origin (as single cells or cell complexes) via either the bloodstream or the lymph system, or both, in a 'metastasis' fashion to other regions or organs of the body. Such uncontrollable spreading of carcinogenic growth usually ends in death by cancer.

The cancers caused by *physical carcinogens* span radioactive gases (radon), neutron and gamma-ray emission (as affected atom bomb victims), long-wave and short-wave Röntgen rays (X-rays), wood dust (especially that of cedars), asbestos fibres, silica crystals of millet flour[21,22] and pipe smoking (from the pressure of the hot pipe stem on lip and tongue).

The emphasis in cancer research shifted in the 1960s towards the direction of *viral* aetiology. After the discovery of the direct involvement of nucleic acids in the neoplastic transformations, however, attention paid to viral cancers was negligible. However, interest was to erupt again after it was found that Epstein–Barr virus causes Birkett's lymphoma in the peoples of the African malaria belt, and nasopharyngeal carcinoma in south-east China. There is also some evidence that Herpes Simplex Virus Type II may lead to cervical cancer. Carcinoma of the penis often occurs when large condylomata appear (wart-like inflammatory growths often present in the secondary stage of syphilis). The outbreak of Kaposi's sarcoma in male homosexuals is brought about as a result of the acute suppression of the immune system through the use of amyl nitrite. The human T-cell Leukaemia Virus (HTLV) is held responsible for certain leukaemias and lymphomas, and for Acquired Immune Deficiency Syndrome (AIDS). For references see the reviews by Cartwright[22] and by Blackburn and Kellard.[23]

The frequency of cancers caused by chemical agents by far outweighs both physical and viral causes. About 20% of all chemicals acting as carcinogens towards humans are alkylating agents. Some of these compounds, without chemical or enzymatic modification, will alkylate nucleophilic sites (DNA, RNA, proteins) directly and are therefore termed *primary carcinogens*. Some of these primary carcinogens are mono-functional, difunctional or even trifunctional, i.e. they have either one, two or even three alkylating sites per molecule, respectively. This enables these alkylating molecules to interact in monofunctional fashion; alternatively the latter two are capable of undergoing crosslinking reactions. Examples of the monofunctional alkylating reagents are compounds **XI** to **XVI**.

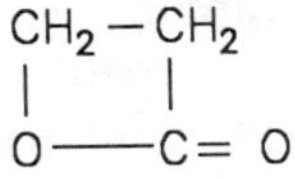

β-Propiolactone (**XI**)

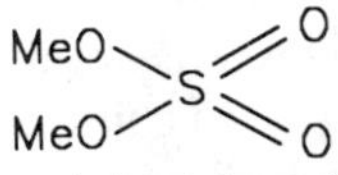

Dimethyl sulphate (**XII**)

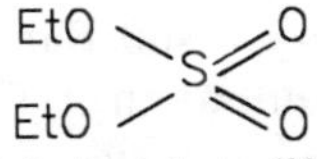

Diethylsulphate (**XIII**)

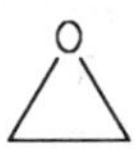

Oxyrane (**XIV**)

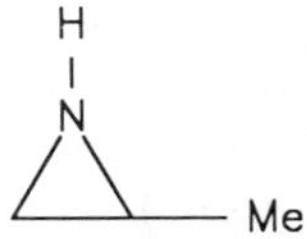

2-Methylaziridine (**XV**)

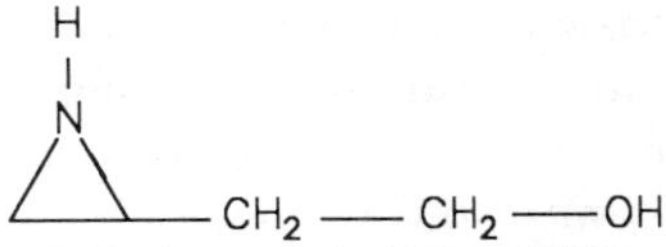

2-Hydroxyethylaziridine (**XVI**)

Examples of the bifunctional alkylating agents are compounds **XVII** to **XXI**.

CL—CH_2—CH_2—S—CH_2—CH_2—CL

Di-(2-chloroethyl)-methylamine (**XVII**)
(mustard gas)

CL—CH_2—CH_2—N(Me)—CH_2—CH_2—CL

Di-(2-chloroethyl)-methylamine (**XVIII**)
(nitrogen mustard)

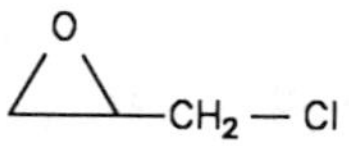

Epichlorohydrin (**XIX**)

Cl—CH_2—CH_2—O—CH_2—CH_2—Cl

Bischloromethylether (**XX**)

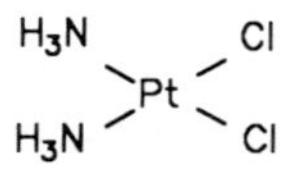

cis-Dichloroplatinum diamine (**XXI**)

Thiotepa (**XXII**) is an example of a trifunctional alkylating agent.

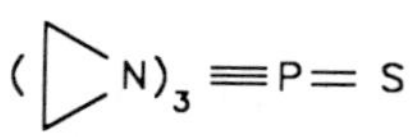

XXII

Dimethyl nitrosamine (**XXIII**) is described as a secondary carcinogen.

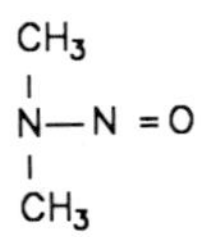

XXIII

Dimethylnitrosamine was first described as a potent carcinogen by Magee and Barnes[24] in 1956. This was soon confirmed by Schmähl and Preußmann.[25] The amount of research being carried out on nitrosamines is ever-increasing, according to Preußmann and Stewart.[26] About three publications a day are produced covering the carcinogenicity, embryotoxicity, mutagenicity, teratogenicity, toxicity, occurrence, analysis and biochemistry of these compounds, and their *in vivo* and *in vitro* formation for precursors. (See also the review by Preußmann and Eisenbrand[27] on N-nitroso carcinogens in the environment.)

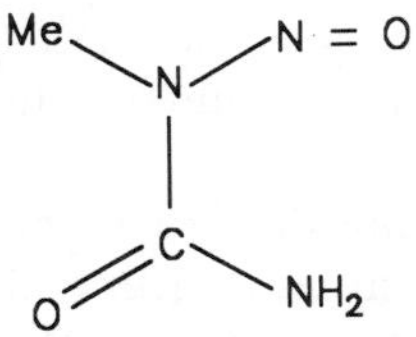

Methylnitroso urea

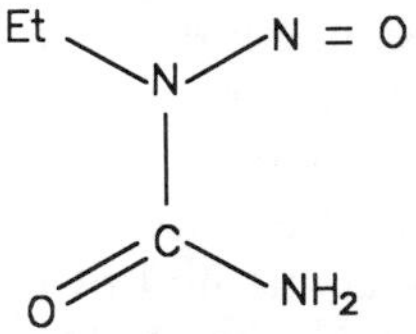

Ethylnitroso urea

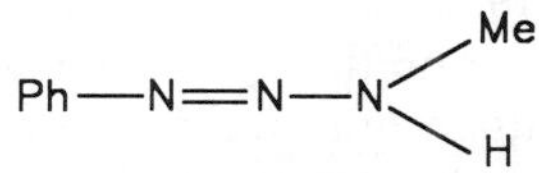

Methylphenyltriazole

Methylnitroso urea, ethylnitroso urea and methylphenyltriazole are additional examples of secondary carcinogens. As a result of acidic or basic hydrolysis, methylphenyltriazole is transformed into methyldiazo-hydroxide:

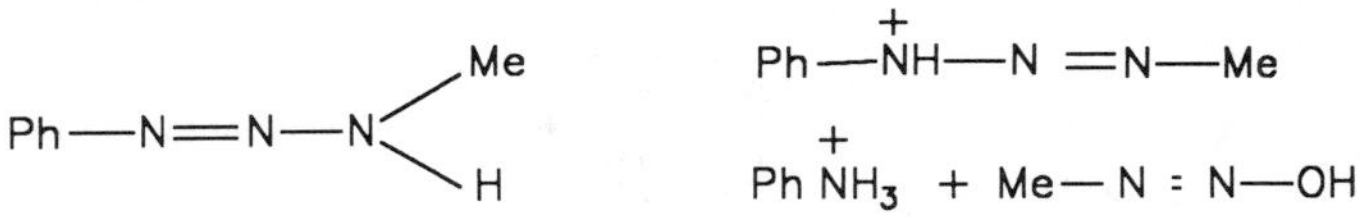

Methyldiazohydroxide is the active carcinogen resulting from this activation reaction. For additional examples of secondary non-polycyclic carcinogens see the review by Blackburn and Kellard.[23]

One typical example of the enzymatic transformation of a secondary carcinogen into a primary carcinogen is the activation of dimethyl-nitrosamine:

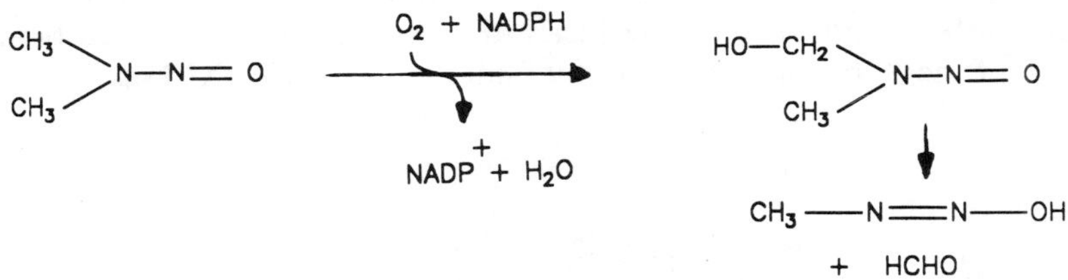

The reaction is triggered by α-hydroxylation by cytochrome P-450. The resulting *N*-hydroxo-methyl-*N*-methylnitrosamine spontaneously loses formaldehyde. The separation of HCHO from the original dimethyl-nitrosamine produces the active carcinogen, namely, in this case, methyldiazohydroxide.

1.5. Carcinogenic Activity and Structure of Polynuclear Aromatic Carcinogens

Ten years after the finding of the carcinogenic activity of 1,2-benz-anthracene (benzo[*a*]anthracene in the modern nomenclature) by the Kennaway group, Schmidt,[28] in 1939, attempted to correlate the carcinogenic activity with the electronic configuration of the carcinogens. Schmidt's theory was extended by Coulson[29] in 1953, by Pullman and Pullman[30] in 1955, and by Boyland[31] and others.[32]

The theory is a conglomerate of wave mechanical considerations, the quantum mechanical resonance method of describing large unsaturated molecules, the molecular orbital method, the numerical application of the latter two and, finally, the metabolic reactivity of carcinogenic hydrocarbons.

The theory describes the 1,2-benzanthracene (benzo[*a*]anthracene) as a molecule with various regions, namely:

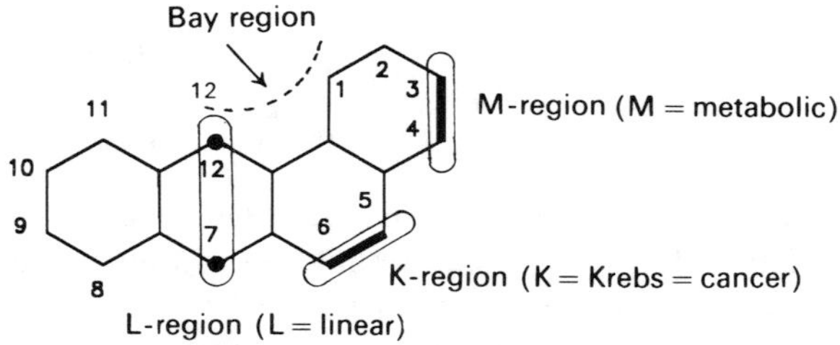

The K-region (occurring in all angular hydrocarbons) is the very important one. Good correlation was found in a series of carcinogenic hydrocarbons and some of their derivatives between the calculated bond order of the K-region and the *rate* of addition of OsO_4 (**XXIV**).

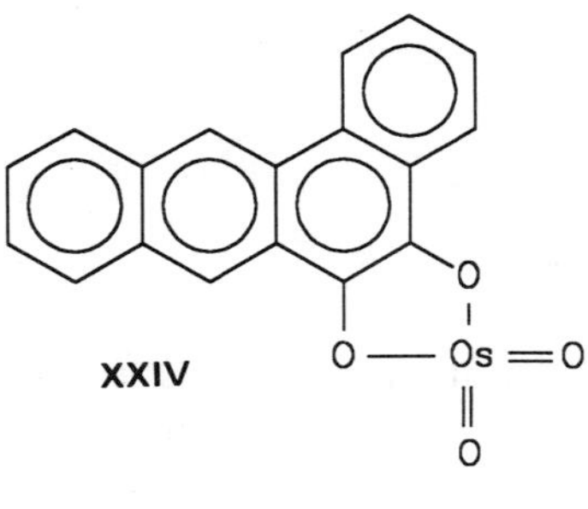

XXIV

The finding that OsO_4 adds to the 5,6-phenanthrenoid bond (the K-region) was taken as a model for the attachment of cell constituents to the hydrocarbon (**XXV**). Furthermore, the fact that metabolic hydroxylations take foremost place in the M-region again favours cell attachment in the K-region. Since such a system admits a spreading out of the π-electrons of the hydrocarbon into the bonds that unite it to the cell, the whole electronic cloud of the hydrocarbon would thus assume the configuration of an *ortho*-quinone (**XXV**).

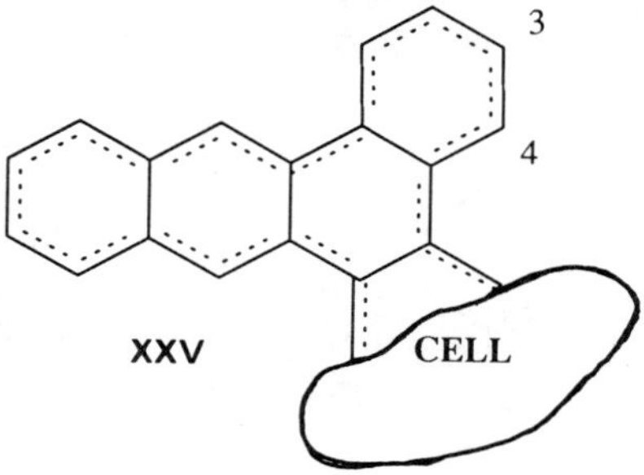

XXV

The calculation of the bond order for the benzo[*a*]anthracene attached to the cell in this way reveals at the 3-carbon the greatest concentration of electrical charge (binuclear carbons excepted), which thus becomes the centre for electrophilic attacks as outlined in formula sequence (**XXVI**) by Boyland[31] in 1950.

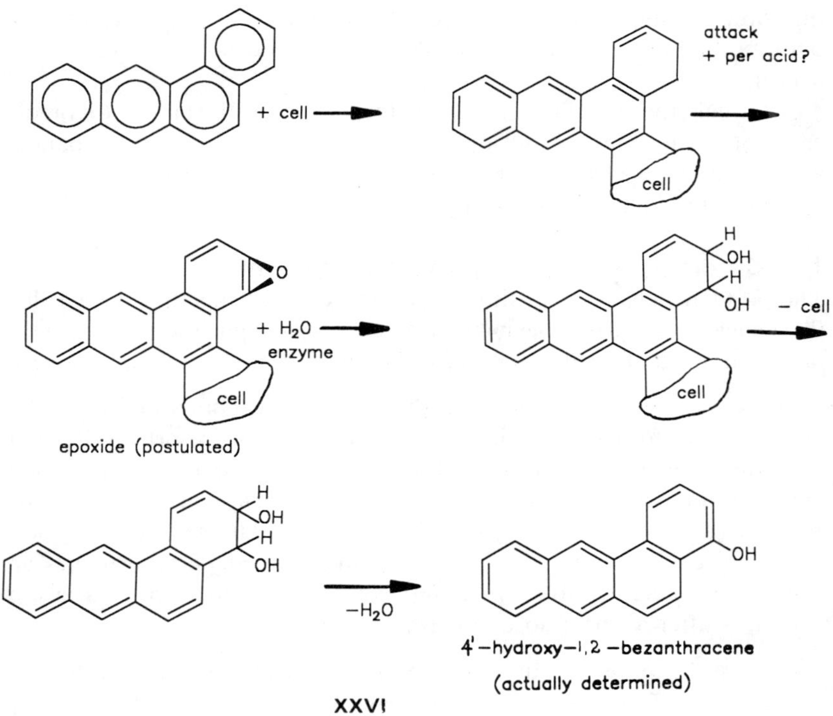

XXVI

Furthermore, the Pullmans[30] were able to predict the carcinogenic activity of benzo[*a*]acridine (**XXVII**) and many other benzacridine derivatives, and of dibenzo[*a*,*h*]pyrene (**XXVIII**).

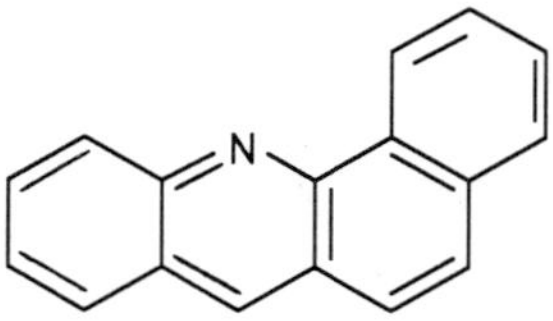

XXVII

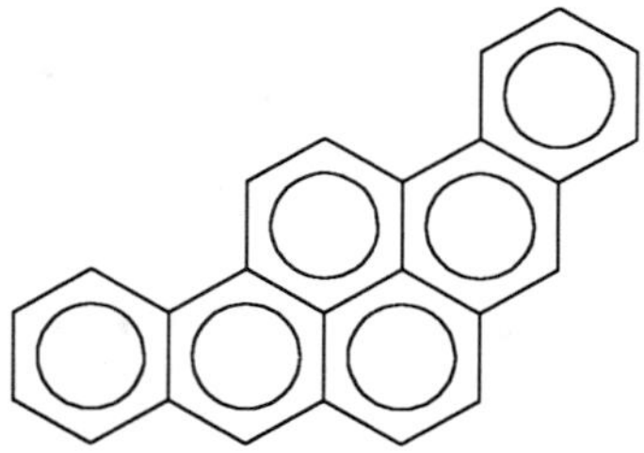

XXVIII

The latter compound serves as a good example to illustrate how other factors besides chemical reactivity can affect carcinogenic response, namely: it did not induce cancer in mice when tested by skin painting. However, when the treated area was covered with a plastic spray, or when saturated *o*-dichlorobenzene solutions were applied, or after subcutaneous injections, the material proved extremely carcinogenic.[33] Several months of painting human skin with dibenzo[*a*,*h*]pyrene did not, however, induce cancer.

Although the electronic theory helped in many instances to predict carcinogenic responses, Coulson finally stated: 'We are obliged to abandon any hope of a clear description of the relation between electronic configuration and carcinogenicity.'

1.6. Carcinogenic Activity and Metabolism

The peculiar behaviour of dibenzo[*a*,*h*]pyrene just mentioned regarding the dependence of carcinogenicity from the site of application is only one of many unexpected results that are summarised by Schoenthal,[34] namely:

> Shape, size, and steric factors of the carcinogenic compounds are all of importance for carcinogenic activity, but in the last resort this depends on the recipient tissue. It depends on the species, the particular strain of the test animal, its age, sex, its nutritional and hormonal state and on the phase of the mitotic activity of a particular cell. It has been recognized[35–37] that very young animals, especially newborn ones, are particularly susceptible to carcinogens and can develop a variety of tumors after a single dose of carcinogen.

It must be pointed out that the main reason for the inconsistent carcinogenic effects of various carcinogens in various recipient tissues may be the specific metabolism to which the administered compound is subjected. Thus in one case a carcinogen may be produced from a non-carcinogen (activation), or a carcinogen may also get inactivated after metabolic transformations by a specific tissue.

1.6.1. Activation

For instance, 2-acetylaminofluorene (**XXIX**) is not a carcinogen in guinea pigs; however, it produces cancer in mice.

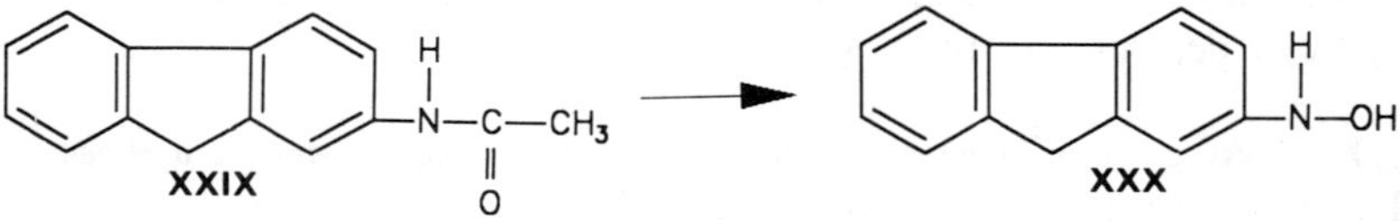

XXIX **XXX**

It was found that the compound in the mice was transformed into the *N*-hydroxy derivative (**XXX**), which also produced cancer upon subcutaneous injection.[38] Guinea pigs which are resistant to 2-acetylaminofluorene appear not to *N*-hydroxylate this compound.[39] Metabolic hydroxylation and simultaneous carcinogenicity was also found in the case of other amines, namely 4-acetylaminodiphenyl[40] and 2-naphthylamine.[41] This appears to be a general feature of the metabolism of aromatic amines in animals in which they induce cancer.[34]

Several of the aromatic amines, such as 2-naphthylamine (**XXXI**), benzidine (**XXXII**) and 4-aminodiphenyl (**XXXIII**), that are used for

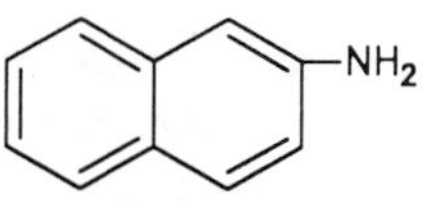

XXXI

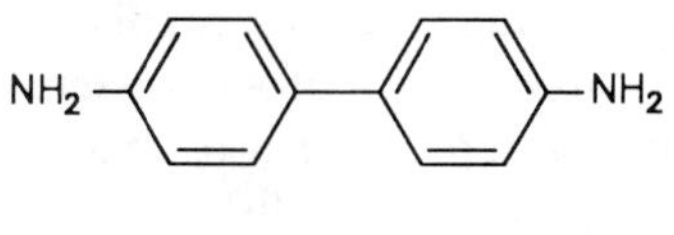

XXXII

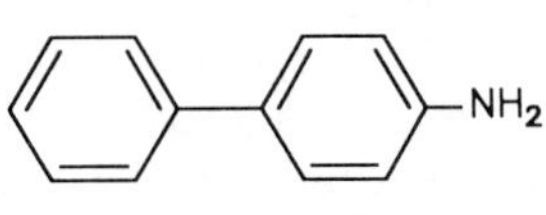

XXXIII

industrial purposes (as antioxidants and hardeners, and in dye manufacture) are also carcinogens. They are responsible for a high incidence of bladder tumours among workers exposed to these compounds.[42] The development of tumours in organs distant from the site of application (e.g. by inhalation or percutaneous absorption) again suggests that the aforementioned amines are not active as such, but that carcinogens are formed from them in the course of metabolic transformation.

1.6.2. Inactivation

Benzo[*a*]anthracene, one of the classical carcinogens for mice, does not affect rabbits. The rabbit's metabolism transforms this compound into the 2′-hydroxy derivative (and 4′-hydroxy derivative) and causes it to be excreted as the inactive sulphate or glucuronic acid conjugate.[43]

Additional examples of the deactivation of carcinogenic compounds via hydroxy derivatives are shown in Table 2.

1.7. Two-stage Theory and Quantitation of Carcinogenesis

The work of Rous and co-workers[44–48] on rabbit skin, and that of Berenblum[49] and Kline and Rush[50,51] on mouse skin, has led to theories that carcinogenesis occurs in two steps, namely the normal cell is converted (mutated) by the carcinogen to an intermediate called the 'sensitised' or 'latent' tumour cell. The latent tumour cell then develops under the continued action of the carcinogen, a co-carcinogen or promoters into the visible tumour. (The term 'co-carcinogen' was introduced by Shear[52] in 1938 to describe the enhancement of the carcinogenic effect of 3,4-benzpyrene by a basic fraction obtained from creosote.) The terminology of Rous describes the cellular changes as follows: the conversion from the normal to the latent tumour cell is called 'initiation' and the change from the latent tumour cell to the visible tumour is called 'promotion'.

Berenblum,[49] in his version of this two-stage theory, defines the initiation as an irreversible change of a few 'stem cells' of the tissue to latent tumour cells, which do not proceed at once to form a tumour only because they are surrounded by normal cells. Promoting agents have two capabilities, namely increasing mitotic frequency and delaying cell maturation (whereas non-promoting irritants have only the former which ceases after removal of the agent). All the stem cells are induced to divide rapidly, and the maturation of their daughter cells is delayed so that more than the usual 50% of them divide again before differentiating. Thus the relative and absolute numbers of stem cells increase, as Glücksmann[53] showed by actual cell counts, and among these the latent tumour cells form compact clusters which reach a critical size and then proliferate without further stimulus.

In an attempt to obtain a quantitative measure for the relation between the initiation and promotion of cancer, Berenblum and Shubik[54] suggested that the initiating agent converted a defined number of normal cells into

Table 2
Comparison of carcinogenic activity before and after hydroxylation[34] (old nomenclature kept for reasons of convenience)

Compound	*Carcinogenic activity*	*OH substitution*	*Carcinogenic activity of OH derivative*
1,2-Benzanthracene	+	4′	−
(benzo[*a*]anthracene)	−	3	−
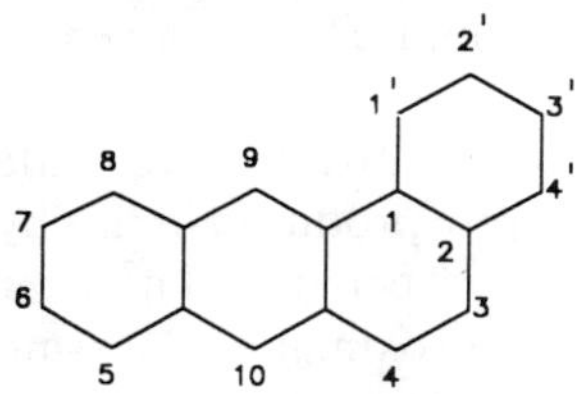		8	+
		10	−
1,2,5,6-Dibenzanthracene	+ + +	9	−
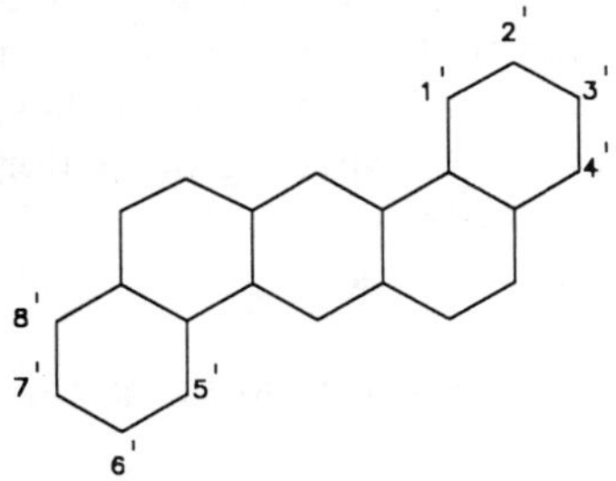		3′,7′-di-	−
		4′,8′-di-	−
		2′,6′-di-	−
3,4-Benzpyrene	+ + +	2′	−
(benzo[*a*]pyrene)		3′	−
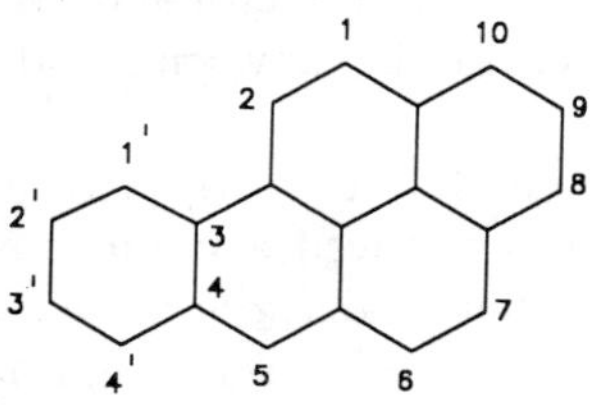		4′	−
		2	−
		8	−
4-Dimethylamino azobenzene	+ + +	2′	−
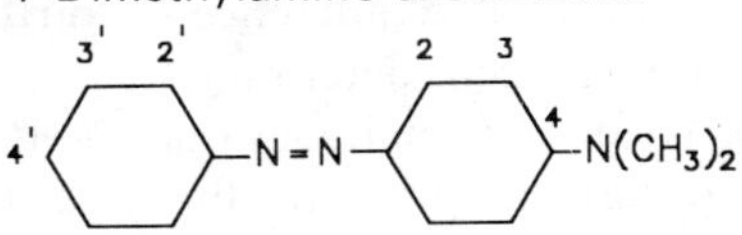		3′	−
		4′	−
		2	−
		3	−

latent tumour cells, the number depending on its potency. The promoting agent, on the other hand, was unable to produce initiation but changed the latent tumour cells into a tumour, and delaying application of the promoting agent would delay the appearance of the tumours but have no effect on the percentage of animals producing tumours. These predictions hold true for experiments where croton oil was used as the promoting agent and 1,2:5,6-dibenzanthracene, 3,4-benzpyrene and 9,10-dimethyl-1,2-benzanthracene were used as initiating agents. Although this theory has its merits for certain given experimental conditions, pure initiators and promoters, according to Clayson,[55] have yet to be found. This statement is no longer true.

Attempts to assess the activity of carcinogens quantitatively were made by Druckrey and co-workers.[56–59] Liver tumours were produced in rats by applying 3 to 30 mg of 4-dimethylaminoazobenzene per day. The liver tumours appeared after a total application of approximately 1000 mg irrespective of whether the compound was applied in larger amounts over a short period of time or in smaller amounts over a longer period of time. Upon reduction of the daily doses to 1 mg the tumours developed after application of only 700 mg, and so on. Mathematical treatment of these results yielded the following expression, namely the carcinogenic action (W) is a function of the sum of the single doses (d):

$$W = f(\sum d)$$

Beyond this summation the time factor (t) is also of importance, thus refining the mere summation equation into

$$C \, . \, t^n = \text{constant}$$

whereby C is the concentration of the carcinogen, t is a time factor and n is a number >1 which was found for most experiments to be between 2 and 3 but for some as high as 60.

A more sophisticated model for carcinogenesis was proposed by Moolgavkar and Knudson in 1981.[60] They also reviewed a number of attempts at models by other authors. In essence, their model adheres to the two-stage proposal consisting of initiation (first stage) and mutation (second stage). The basic theory explains that the 'initiation' alone does not suffice to produce a malignant tumour stem cell. The difference of 'birth rate' (α_2) and 'death rate' (β_2) of intermediate stem cells determines the rate of proliferation of intermediate cells (IC). The model envisages that the IC have to mutate (go into the second stage) at the same site as the first one on the homologous chromosome, and also a homologous chromosomal exchange is proposed to be required for the conversion of an IC to a malignant cell (MC). The model for carcinogenesis is shown in Fig. 1.

Moolgavkar and Knudson suggest that 'many of the biologic implications of the model follow immediately without the intervention of any mathematics. However, if the model is to be tested quantitatively against human epidemiology and animal experimental data, it must be translated into mathematical terms.' Under the condition that μ_2 is a small value, then the carcinogenic incidence function $I(t)$ for tissue cells is reflected in the approximate mathematical model expression that combines deterministic as well as stochastic elements as shown in the following formula:

$$I(t) \approx \mu_1 \mu_2 \int_0^t X(s) \exp\,[(\alpha_2 - \beta_2)(t-s)]\, \mathrm{d}s$$

The units μ_1 and μ_2 are multiplicative factors that measure the transition rates of normal cells (NC) to intermediate cells (IC) and the transition rates of IC to malignant cells (MC), respectively, per time unit (cell year). $X(s)$ represents the number of SC in a tissue at time s. Moolgavkar and Knudson state that 'the number of IC that have arisen from SC by the time t is a random variable that has a Poisson distribution with expectation $\int_0 \mu_1 X(s)\,\mathrm{d}s$, where μ_1 is a constant that measures the rate of transition of NC (or SC) to IC. The constants μ_1 and μ_2 are the transition rates per cell per year.' The factor α_2 represents the 'birth' rate of IC per cell per year and β_2 represents the 'death' rate of IC per cell per year. A clone of IC has a positive growth rate when $\alpha_2 > \beta_2$. However, when $\alpha_2 < \beta_2$, then the clone

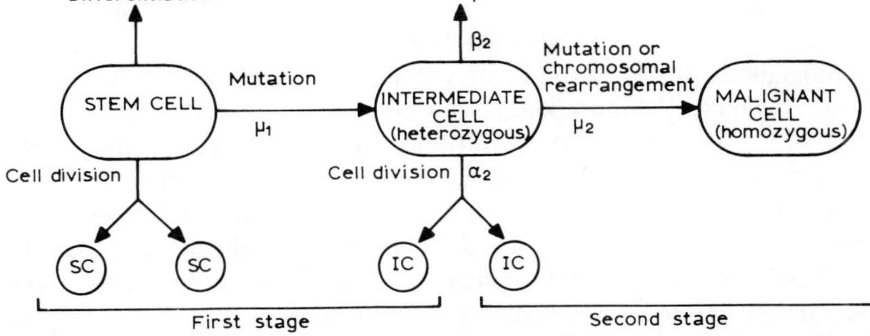

Fig. 1. Two-stage biochemical model for carcinogenesis (according to Moolgavkar and Knudson[60]).

of IC decreases in size. The integral represents the numerical number of IC available for the second stage of mutation (or chromosomal rearrangement) to yield malignancy. The mathematical expression can be adapted to fit the many different forms of cancer production and induction times quite well.

The model clearly supports a two-stage event (initiation–promotion) for carcinogenesis. It also implies that an agent causing the initiation event (hit) can also take the function of a promoting agent.

1.8. Enzymatic Transformation of Polynuclear Aromatic Hydrocarbons into Carcinogens

None of the PAHs are primary carcinogens. Before they can interact with biopolymers such as DNA, proteins, etc., they have to be transformed into nucleophilic compounds. As always, the search for long-sought key reactions is often painful but nevertheless fascinating, and the final discovery of the important missing pieces of a scientific puzzle is reserved for those few investigators who are untiring, thorough and lucky. Boyland[31] came close to finding the missing pieces in 1950, but he lacked some important experimental tools to put the right pieces together.

In 1965 Boyland and Sims,[61] using rat liver slices and homogenates, found that PAHs are converted, *in vivo* and *in vitro*, into glutathione conjugates, *trans*-dihydrodiols and phenolic compounds. Since none of these transformation derivatives showed any significant carcinogenicity, they concluded that PAH metabolism is primarily a detoxification/excretion mechanism. This time they came even closer, but failed again to make the key observations.

Miller,[62] using fluorescence techniques, showed in 1951 that benzo[*a*]-pyrene is strongly bound to mouse skin proteins. Heidelberger and Moldenhauer[63] in 1956 found, using ^{14}C-labelled PAH, that the intensity of *in vitro* protein binding was directly equivalent to the degree of carcinogenicity. Brookes and Lawley[64] in 1964 found that the PAH is also bound to DNA of mouse skin. These findings permitted the important conclusion to be drawn that the initiation of carcinogenesis is the result of covalent binding of the carcinogen to cellular macromolecules. One significant intermediate discovery was made independently by Grover and Sims in 1968[65] and Gelboin[65] a year later, when they demonstrated *in vitro* that (labelled) PAHs are metabolically induced, in the presence of rat liver homogenates, to form covalent bonds with DNA.

Borgen *et al.*[67] in 1973 studied the binding capacities of benzo[*a*]pyrene and various *trans*-dihydrodiol metabolites of benzo[*a*]pyrene to DNA. The relative binding force for the 9,10-dihydrodiol was 3, for the 4,5-

dihydrodiol it was 5, for benzo[*a*]pyrene it was 11, but for the 7,8-dihydrodiol it was 168! Borgen *et al.* concluded that the 7,8-dihydrodiol was an important intermediate step for the activation of benzo[*a*]pyrene to bind covalently to DNA; however, a further step is required to yield the active alkylating agent. Sims *et al.*[68] in 1974 suggested and demonstrated that, upon chemical oxidation of the *trans*-7,8-dihydrodiol with *m*-chloroperoxybenzoic acid, the 7,8-dihydrodiol-9,10-epoxide of benzo[*a*]-pyrene is produced. It is this epoxide that reacts with DNA in a covalent fashion. Sims *et al.* then also showed that the 7,8-dihydrodiol-9,10-epoxide reaction product with DNA was identical to the *metabolic* reaction product obtained from benzo[*a*]pyrene and DNA in rodent embryo cells (see Fig. 2). It took 24 years from the first proposal that an epoxide is involved in the carcinogenicity of hydrocarbons until its demonstration experimentally.

1.8.1. Cytochrome P-450

As is shown in the hydrocarbon reaction sequence (see Fig. 2), cytochrome P-450 possesses oxidative catalytic activity against thousands of xenobiotic organic compounds, including the polycyclic aromatic hydrocarbons, and also against endogenous physiological compounds such as steroids, prostaglandins, biogenic amines and fatty acids, etc.

The word 'cytochrome' comes from the Greek words meaning 'coloured substance in the cell'. Cytochrome P-450 is a reddish pigment that has its major absorption maximum at 450 nm, when it is in the reduced form and combined with carbon monoxide. Like haemoglobin it contains the haem ring structure.

A general consensus has existed that there are about 20 forms of cytochrome P-450.[69] However, Nebert[70] proposed a genetic variation in the formation of cytochrome P-450, causing it to adapt structurally to the many thousands of cytochrome inducers. This proposal parallels the genesis of immunoglobulins of the immune system that are precisely adapted to the mechanism by which B-lymphocytes, on command, with a small number of genes, can produce many millions of different types of antibodies (as discovered by Tonegawa between 1976 and 1979).[71] The absorption maximum of the reduced carbon monoxide complex of different cytochrome P-450s ranges from 447 to 451 nm. Their subunit molecular weights vary from 44 000 to 60 000. Also their substrate specificities as well as immunological properties differ from one another. Although all of the cytochrome P-450 will catalyse the oxidation of benzo[*a*]pyrene, the metabolism of benzo[*a*]pyrene, i.e. the rate of its conversion (turnover number) by various P-450 types, ranges from 0·03 to 4·1. High-pressure liquid chromatography of benzo[*a*]pyrene metabolites showed that

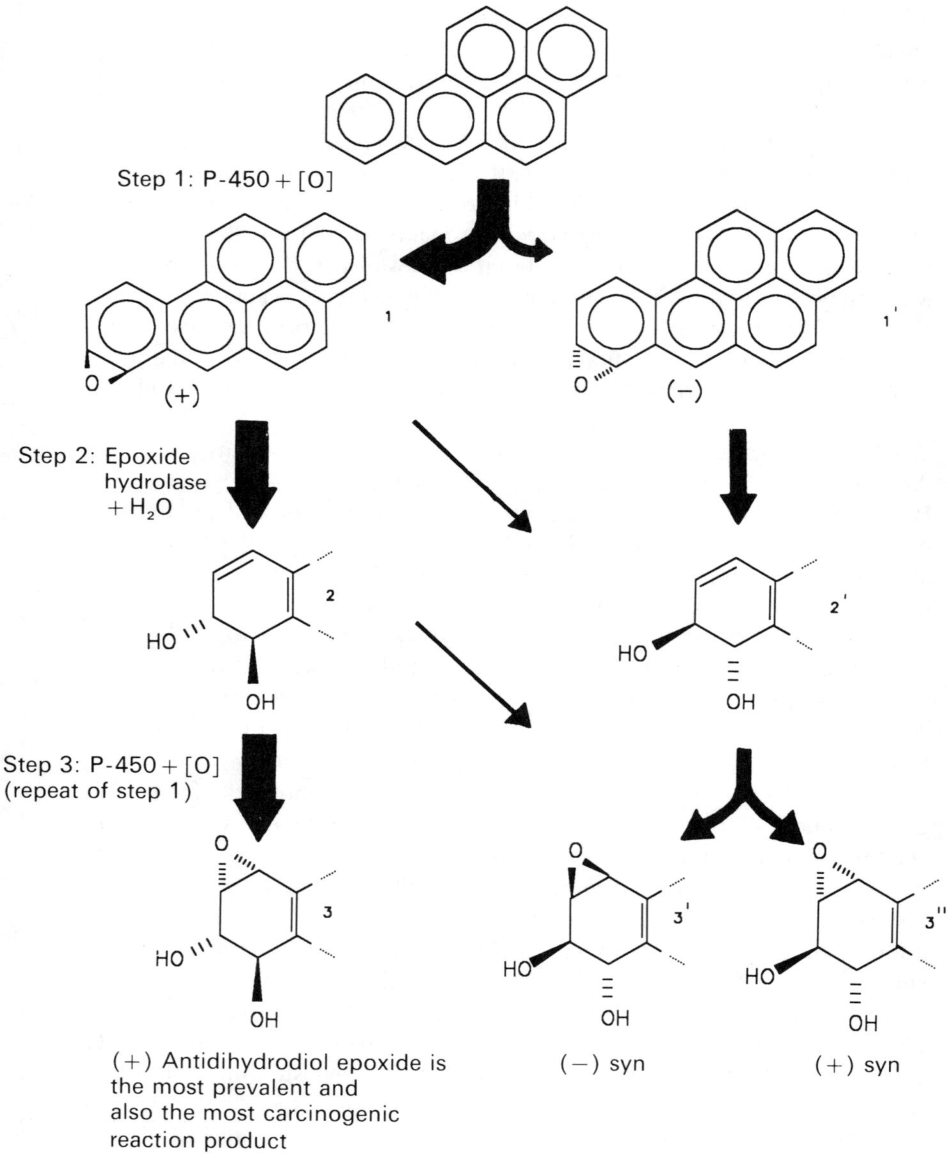

Fig. 2. Metabolic transformation of benzo[*a*]pyrene.

different types of cytochrome P-450 also attack the polycyclic hydrocarbon at different sites. Finally, the functionality of cytochrome P-450 depends on the presence of NADPH-cytochrome P-450 reductase (flavoproteins) and phosphatidylcholine. The combined complex is called the mixed function oxidase system. This enzyme system is primarily associated with microsomal preparations, but its presence has also been shown in nuclei. Cytochrome P-450-mediated mono-oxygenase activities are found in virtually all bacteria, plants and animals. The reactions for the P-450-mediated mono-oxygenase system range from oxidation/reduction to hydrolysis reactions. They include the insertion of one atom of atmospheric oxygen into suitable substrates. Other reactions are: hydroxylation of aliphatic and aromatic carbons; dealkylation of *N*-, *O*- and *S*-compounds; *N*-oxidation, *N*-hydroxylation and *S*-oxidation; desulphuration, dehalogenation and deamination; oxidation of certain purines and amino compounds; and azo-, nitro-group, arene oxide and *N*-hydroxyl reductions. According to Pelkonen and Nebert,[72] many of these reactions can result in the formation of reactive intermediates, e.g. dealkylation of some of these compounds can lead to the alkylation of nucleic acids and proteins which thereby enter the carcinogenesis reactions.

1.8.2. Epoxide Hydrolase

Epoxide hydrolase, like the mixed function oxidase system, is also a membrane-bound microsomal enzyme. This enzyme, however, has no prosthetic group and has an approximate monomolecular weight of ~53 000. It exists in aggregates of molecular weights of about 600 000. The enzyme has a very broad substrate specificity. The purpose of this enzyme is to transform the epoxide to a *trans*-dihydrodiol (see Fig. 2).

The dihydrodiol may then undergo conjugation with gluconic acid, glutathione or sulpho groups. These reactions are catalysed by microsomal UDP-glucuronyl transferases or cytosolic enzymes such as sulphotransferases or glutathione-S-transferases (see Fig. 3). The conjugation reactions result in a more water-soluble hydrocarbon metabolite which is now transportable for subsequent excretion.

The exocyclic amino groups of some nucleic acids are the dominant sites for reacting with the bay region dihydrodiol epoxide of benzo[*a*]pyrene. The main product of benzo[*a*]pyrene binding with DNA is apparently the result of a *trans*-addition of the exocyclic amino group of deoxyguanosine residues to the C-10 position of the (+)-anti-dihydrodiol epoxide in most systems.

Benzo[*a*]pyrene binding to mouse skin,[73,74] hamster embryo cells[75] and

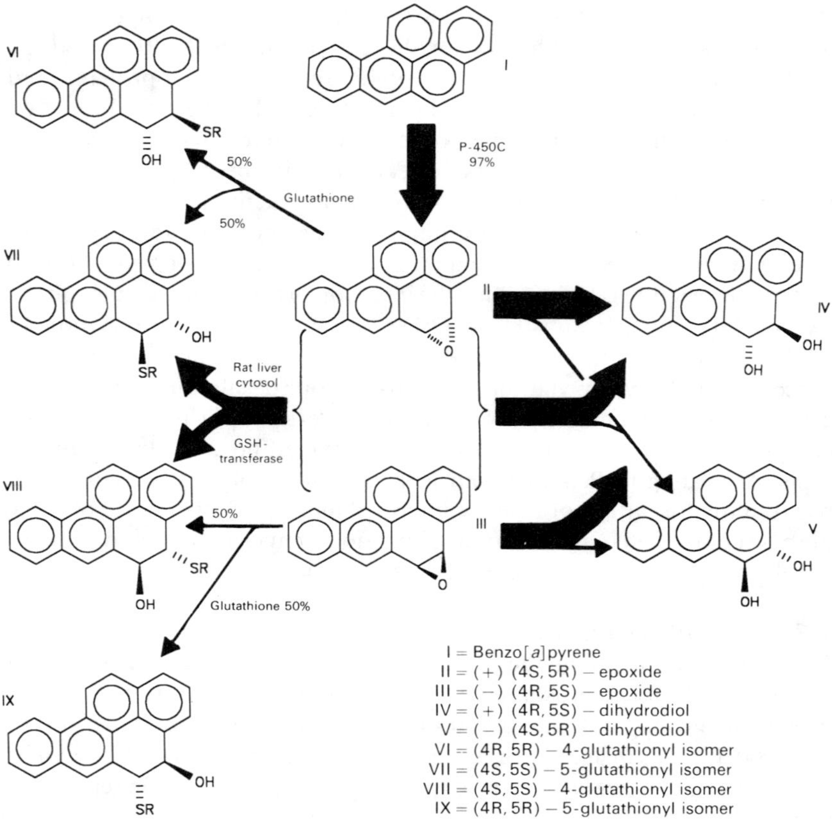

Fig. 3. Interaction of cytochrome P-450C (P-488) with benzo[*a*]pyrene. (Redrawn from Dipple *et al.*[77])

human bronchial tissue[76] is the result of *trans*- and *cis*-addition of the exocyclic amino group of guanosine residues to the (+)-anti-enantiomer, respectively, and also of *trans*-addition to the (+)-syn-dihydrodiol epoxide (see Fig. 4). This strongly suggests that the epoxide of benzo[*a*]pyrene is to be considered, in these cases, the ultimate carcinogen. However, in not all instances is the epoxide the ultimate carcinogen. In mice the 6,7-dihydrodiol of benzo[*a*]pyrene is considered the ultimate carcinogen. Dipple *et al.*[77] in addition suggest that the ultimate carcinogen may be some reactive intermediate generated in the course of the metabolic conversion of the dihydrodiol to the dihydrodiol epoxide. They see the formation of a benzylic carbonium ion as one such potential intermediate for the carcinogenic or tumour-initiating effect.

1.9. Detoxification and Nucleic Acid Repair and Inhibition

As was indicated earlier, cells are exposed constantly to physical stress (mechanical, UV, Röntgen rays, neutron bombardment, γ-rays) and to primary and secondary chemical carcinogens, viruses, etc.—i.e. to man-made and natural attacking sources of DNA damage. The reason for the astonishing cell survival rate is attributable to a very functional and well developed DNA repair process system that acts against nucleic acid base modifications by alkylation:

(1) 3-Methyladenine residues are repaired by enzymatic hydrolysis of the glycosidic bond.
(2) The depurinated sugar is then excised by hydrolytic cleavage of the neighbouring phosphate esters.
(3) The gap thus resulting is repaired by inserting a new deoxy-nucleotide which complements exactly the base in the opposite strand of the double helix.

According to Karran and Lindahl,[78] this same mechanism is also applied for other methylation products such as 3-methylguanines, 2-O-methyl-cytosines and 2-O-methylthymine residues. All these compounds have the methyl group located in the minor groove of the DNA B helix. This same mechanism also works for some of the radiation-damaged bases.

In the case of metabolic substitution of DNA with bulky reactants (e.g. polycyclic aromatic hydrocarbons, pyrimidine photodimers, aromatic amines), the repair system is different. In these cases the entire substituted nucleotide is excised in one piece. This excision process utilises many enzymes requiring the involvement of several human genes for their production.

One other repair process works by directly eliminating the offending attached methyl or ethyl substituents. This repair process works for deletion of O-methyl or O-ethyl groups from guanine and thymine bases. Yeast cells have such a repair system. The enzyme carrying out the demethylation has two active sulphydryl groups. One SH-group is located near the terminal carboxy-end residue in the enzyme protein chain and the other one is located near the amino-end residue. The SH-group near the C-terminus will demethylate O^6-methylguanine or O^4-methylthymine. The SH-group located near the N-terminus is capable of demethylating DNA-phosphate triesters provided they have the *S*-configuration at the

STRUCTURE A

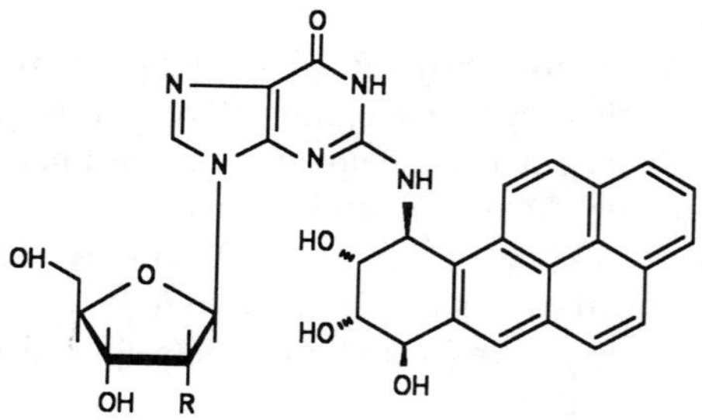

STRUCTURE C

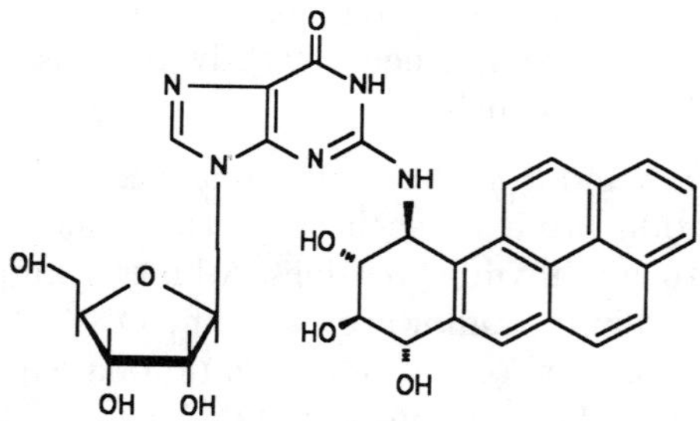

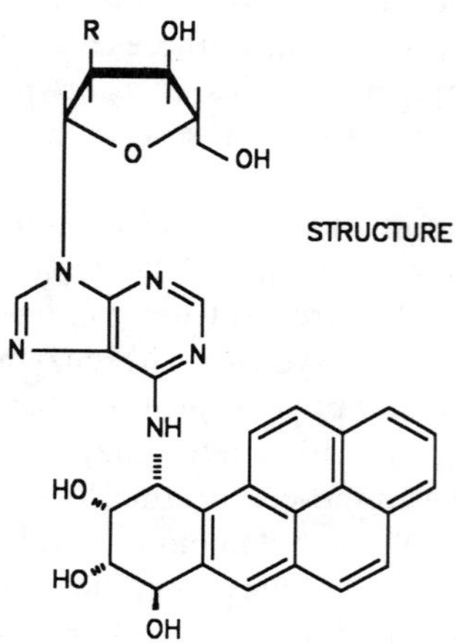

STRUCTURE E

STRUCTURE B

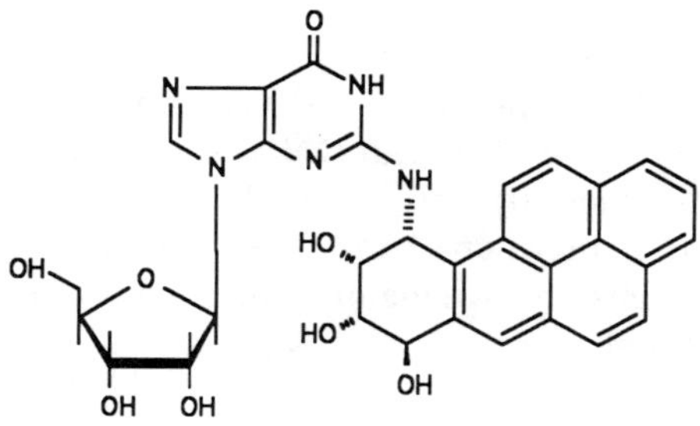

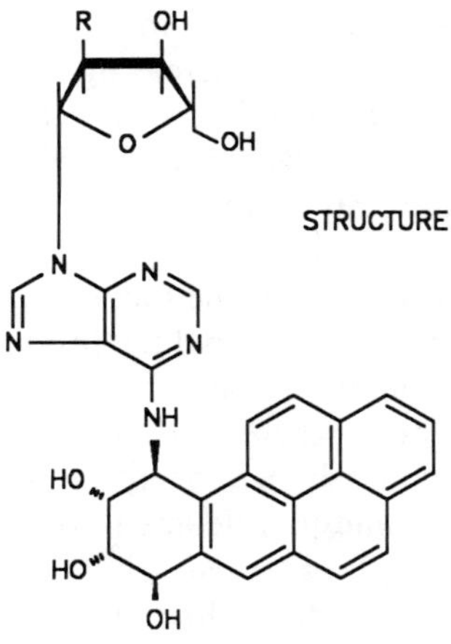

STRUCTURE D

Fig. 4. *Cis-* and *trans*-addition products of benzo[*a*]pyrene with N^2 of guanosine, deoxyguanosine, N^6 of adenosine and N^6 of 2′-deoxyguanosine (after Dipple *et al.*[77]). A, Product of *trans*-addition of N^2 of guanosine (R = OH) and 2′-deoxyguanosine (R = H) to the (+)-anti-dihydrodiol epoxide of benzo[*a*]pyrene. B, Product of *cis*-addition of N^2 of guanosine to the (+)-anti-dihydrodiol epoxide of benzo[*a*]pyrene. C, Product of *trans*-addition of N^2 of guanosine to the (+)-syn-dihydrodiol epoxide of benzo[*a*]pyrene. D, Product of *trans*-addition of N^6 of adenosine (R = OH) and 2′-deoxyadenosine (R = H) to the (+)-anti-dihydrodiol epoxide of benzo[*a*]pyrene. E, Product of *cis*-addition of N^6 of adenosine (R = OH) and 2′-deoxyadenosine (R = H) to the (+)-anti-dihydrodiol epoxide of benzo[*a*]pyrene.

phosphorus. The corresponding human demethylation enzyme only attacks O^6-methylguanine. Every normal human cell contains an estimated 10 000 to 25 000 such demethylating enzymes, thus providing a good possibility of DNA repair.

Blackburn and Kellard[79] point out that, in comparison with rodent cells, human cells are much more efficient at DNA repair and therefore are much more resistant to attack by mutagens. This also indicates that any carcinogenic risks to man, when extrapolated from experiments of rodent exposure to carcinogens (man-made or natural/environmental), are likely to result in overestimates.

1.10. Carcinogenic Polynuclear Hydrocarbons Present in the Human Environment

1.10.1. General

A major reason for the continuing interest in analysis for carcinogens in the human environment is the steadily increasing rate of lung cancer. The carcinogenic polynuclear hydrocarbons are thought to be the main cause of this, since these compounds have been found in all particulate air samples as a result of industrial, automotive and residential effluents and tobacco smoke.

The carcinogenic hydrocarbons in the human environment seem to be formed mainly at high temperatures. Coal tar, produced by heating of coal to 300–450°C, consists mainly of non-carcinogenic paraffins, cycloparaffins, olefins and phenols, whereas pyrolysis at higher temperatures yields carcinogenic tars with a high concentration of benzo[*a*]pyrene.[80] Furthermore, non-carcinogenic petroleums have been shown to become carcinogenic after heating.[80] Scottish shale oil was found to be carcinogenic—and to contain benzo[*a*]pyrene—only after heating.[81] The occurrence of benzo[*a*]pyrene in soot, in exhaust gases of automobiles, in carbon blacks and in tobacco tar, etc., can likewise be attributed to high-temperature reactions involving incomplete combustion.

It also has been found that benzo[*a*]pyrene and many other polycyclic compounds are formed by the pyrolysis of acetylene,[82] butadiene, butylbenzene and tetralin,[83] etc., and also from aliphatic hydrocarbons extracted from tobacco.[84]

1.11. Formation of 3,4-Benzpyrene

According to Badger *et al.*,[85,86] the formation of 3,4-benzpyrene follows the scheme shown in Fig. 5, namely: (1) from two-carbon fragments by

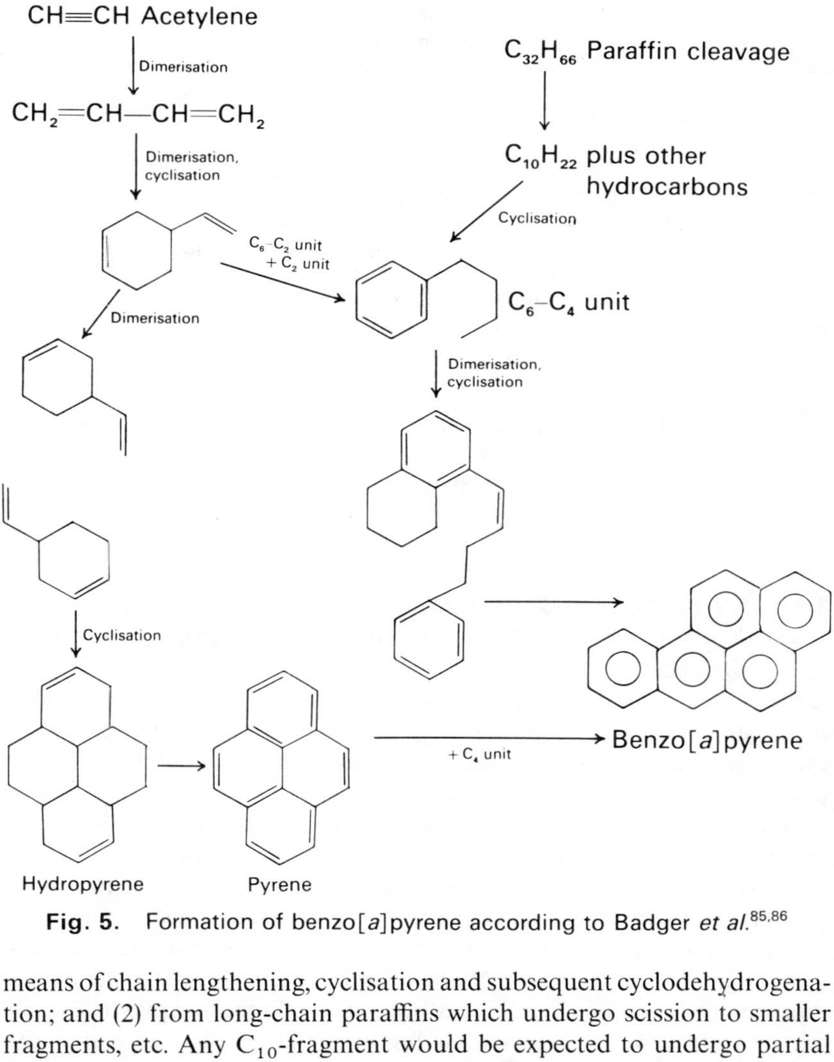

Fig. 5. Formation of benzo[*a*]pyrene according to Badger *et al.*[85,86]

means of chain lengthening, cyclisation and subsequent cyclodehydrogenation; and (2) from long-chain paraffins which undergo scission to smaller fragments, etc. Any C_{10}-fragment would be expected to undergo partial cyclisation to a C_6–C_4 unit (such as butylbenzene or tetralin) to yield benzo[*a*]pyrene (among other products). In order to test the validity of the mechanism, acetylene, butadiene, styrene, vinylcyclohexene, ethylbenzene, phenylbutadiene, butylbenzene, isoprene and tetralin were pyrolysed at 700°C and the resulting tars were analysed. Benzo[*a*]pyrene was identified in all these tars.[85,86]

Table 3
Carcinogens in tobacco smoke
(original nomenclature maintained)

Arsenous oxide
Chrysene
1,2-Benzanthracene (benz[*a*]anthracene)
6,7-Cyclopentene-1,2-benzanthracene
5,6-Cyclopentene-1,2-benzanthracene
1,2:5,6-Dibenzanthracene (dibenz[*a,h*]anthracene)
3-Methylpyrene
3,4-Benzpyrene (benzo[*a*]pyrene)
1,2:3,4-Dibenzpyrene
3,4:9,10-Dibenzpyrene
1,12-Benzperylene
10,11-Benzfluoroanthrene
3,4-Benzfluoroanthrene
1,2:5,6-Dibenzacridine
1,2:7,8-Dibenzacridine
3,4:5,6-Dibenzcarbazole

1.12. Carcinogenic Constituents of Coal Tar, Industrial, Automotive and Residential Effluents and Tobacco Smoke

It must be stressed that 3,4-benzpyrene is not the only carcinogenic hydrocarbon that is produced during the pyrolysis reaction. Coal tar, for example, contains three additional carcinogens, namely 3,4:8,9-dibenzpyrene, 3,4:9,10-dibenzpyrene and 3,4-benzfluoroanthrene,[87–91] all of which evidently contribute to the carcinogenic activity. At least 16 different carcinogens have been found in tobacco smoke.[92–94] (See Table 3.)

However, the ubiquity and abundance of benzo[*a*]pyrene with its high carcinogenic activity (in animals) and its relative ease of detectability (it is 30 times more fluorescent than 1,2-benzanthracene and anthracene; ~200 times more intense than 1,2:5,6-dibenzanthracene; and 2000 to 6000 times more fluorescent than chrysene, 3,4-benzphenanthrene and pyrene[95]) in all materials that were pyrolysed to higher or lesser degrees makes this compound the most convenient and logical tracer compound to analyse whenever the carcinogenic property of a certain material has to be evaluated.

Thus the carcinogenic activity of atmospheric dust in cities has been demonstrated on many occasions, and benzo[*a*]pyrene has always been identified as a constituent.[96–102] Waller[103] analysed the air drawn from

Table 4
3,4-Benzpyrene content of urban air (1959)[104]

City	*Country*	*μg Benzpyrene per 1000 m³ air*	*μg Benzpyrene per gram soot*
Los Angeles	USA	30–32·5	—
Reykjavik	Iceland	2·3	152–255
Bergen	Norway	5–19·0	155–233
Oslo	Norway	0·9–15·2	143–300
Copenhagen	Denmark	14·5	295
Sheffield	England	20–33·0	136–190
London	England	26–14·7	—

eight different cities in England and found that the amount of 3,4-benzpyrene due to coal burning rose sharply during winter.[103] The average 3,4-benzpyrene value of the air in these coal-burning towns was 150 to 300 μg per 1000 m^3 of air. Based on 20 m^3 of inhaled air per day, the benzo[*a*]pyrene inhaled in the winter months could amount to 90 to 180 μg per month. In petroleum-burning cities such as Los Angeles, the average concentration of 3,4-benzpyrene is less[10] (see Tables 4 and 5).

Like tobacco smoke, the air pollution particles contain, besides 3,4-benzpyrene, other carcinogens, namely alkyl derivatives of 3,4-benzpyrene, 1,2-benzanthracene and benz(*c*)acridine, and the heterocyclic compounds dibenz(*a,h*)acridine and dibenz(*a,g*)acridine. Sawicki and co-workers[105]

Table 5
Concentration of 3,4-benzpyrene in selected urban and non-urban areas within the US (1960)[105]

State	*μg 3,4-Benzpyrene per 1000 m³ air*	
	Urban	*Non-urban*
Alabama	24	0·076
Indiana	39	1·8
Maryland	14	0·70
Missouri	54	0·025
North Carolina	3·9	0·25
Oregon	0·8	0·01
Pennsylvania	61	1·9
South Carolina	24	1·1

Table 6
Carcinogens in residential and industrial effluents ($mg\ 100\ m^{-3}$)[106]

Compound	Stack gas of residential coal furnace	Coal-tar pitch polluted air sample	Industrial effluent incineration unit	
			Before	After
Benzo[*a*]pyrene	500	0·4	2 700	0·60
Benz[*a*]anthracene	1 300	0·7	270	0·10
Benz(*c*)acridine	15	0·12	60	—
Dibenz(*a,h*)acridine	17	0·01	0·7	—
Dibenz(*a,g*)acridine	2	0·001	1·8	—

determined the amounts of these carcinogens in three different sources of air pollution, namely from a stack of a residential coal furnace and from an 'industrial process unit' before and after an incineration-type air cleaning system. The results are listed in Table 6.

The results listed in Table 6 reconfirm that 3,4-benzpyrene (considering its strong carcinogenic effectiveness and its high fluorescence) is also present in industrial and residential effluents, and is the most important component to be analytically concerned about.

Furthermore, 3,4-benzpyrene was also determined in pyrolysis products such as special soots, charred matter from biscuits, condensed tobacco smoke and carbon black. The results are tabulated in Table 7.

The amounts of 3,4-benzpyrene found in automotive exhaust fumes are shown in Table 8 and require no further comment.

Table 7
Benzo[*a*]pyrene content in pyrolysis products

Source	3,4-Benzpyrene
Domestic soot[107]	300 000 parts per 10^9
Coal soot[108]	12 000–56 000 parts per 10^9
Wood soot[108]	1 700–36 000 parts per 10^9
Coffee soot[109]	200–440 parts per 10^9
Charred matter from biscuits[109]	10–72 parts per 10^9
Carbon black (additive for plastics)[110]	1–1 600 parts per 10^9
Condensed cigarette smoke[96]	0·9–10·4 $\mu g/100\ g$ cigarette tobacco
Condensed cigar smoke[111,112]	3·4–70·0 $\mu g/100\ g$ cigar tobacco
Condensed pipe smoke[112]	8·5 $\mu g/100\ g$ pipe tobacco

Table 8
3,4-Benzpyrene in automotive exhausts

Gasoline engine[113]	10–120 μg min^{-1}, depending on engine speed
Gasoline engine[114]	5 μg min^{-1} (car 8 000–12 000 miles (*c.* 12 900–19 300 km) old) 27 μg min^{-1} (car 29 000–33 000 miles (*c.* 46 600–53 000 km) old)
Diesel engine[115]	876–1 687 μg min^{-1} under maximum load at different rpm and inefficient operation. 'However, virtually undetectable under efficient operating condition regardless of load'

2. BENZO[*a*]PYRENE CONTENT OF COFFEE PRODUCTS AND BY-PRODUCTS

The discovery of benzo[*a*]pyrene and the facile conditions for formation at low and high temperatures (as shown by Kennaway and his colleagues in 1932) prompted Lickint,[116] in 1938, to postulate that benzo[*a*]pyrene was also likely to be present in roasted coffee.

Roffo,[117] in 1938, described a coffee tar he produced by heating coffee up to 350°C. As expected, this tar contained 3,4-benzpyrene. Brushing of rabbit skin with this tar produced epithelioma. It should be pointed out that coffee beans during roasting are not subjected to temperatures higher than 200–220°C. Therefore Roffo's experiment is of interest from a historical and academic perspective view, but has no practical validity.

Kuratsune and Hueper[109] found, in the soot of coffee roasting plants, a range of 200–400 μg kg^{-1} of 3,4-benzpyrene. Chassevant and Heros[117] were able to detect quantities of 3,4-benzpyrene as high as 150 μg kg^{-1} in the chaff separated from coffee beans during the roasting process. In coffee roasting soot they found a level of 200–440 μg kg^{-1}.

In 1960 Kuratsune and Hueper[119] found 4 μg kg^{-1} 3,4-benzpyrene only in coffee beans of a very dark, burnt roast. They found none in beans of a normal roast.

As the analytical procedures were made more sensitive, in 1966 Grimmer and Hildebrandt[120] and Fritz[121] found the 3,4-benzpyrene also in coffee beans of normal roast.

Also, in 1966 (published in 1973), Strobel,[122] with a procedure to detect picogram levels of 3,4-benzpyrene, found that 3,4-benzpyrene was present in measurable quantities in chromatographic paper, all analytical glassware and even in spectral grade solvents. Analytical procedures were developed to meticulously avoid any 3,4-benzpyrene contamination of the substrates to be analysed by the reagents or glass instruments used for analysis. (All solvents were triply distilled; all chromatographic paper or acetylated cellulose fibres were washed with hot solvent; all glassware was heated to 600°C in a glassblower's tempering oven.) 3,4-Benzpyrene at very low levels was detectable in instant coffee powder (0·02 to 0·06 $\mu g\,kg^{-1}$), coffee oil (0·1 to 0·5 $\mu g\,kg^{-1}$) and in the (made from roast and ground) coffee beverage (0·000 15 $\mu g\,kg^{-1}$).

D'Arrigo,[123] in 1971, reported the results of his analysis of green and roasted beans. His data for green beans ranged from 0·2 to 0·3 $\mu g\,kg^{-1}$, and for roasted beans from 1·1 to 1·9 $\mu g\,kg^{-1}$. Bracco,[124] in 1973, analysed coffee oil and green and roasted coffee beans, and found small quantities in both samples. In roasted coffee beans he found approximately 0·1 $\mu g\,kg^{-1}$, which was about ten times the quantity determined in green coffee beans. He found 0·01 $\mu g\,kg^{-1}$ of benzo[*a*]pyrene in coffee oil, which is less than Strobel's data indicate. Soos and Fozi,[125] in 1974, analysed green and roasted coffee beans and also coffee brew. Their data for benzo[*a*]pyrene are approximately one order of magnitude higher than those obtained by Bracco.[124]

Fritz[126] analysed coffee brew in 1975. His data fall between those of Bracco[124] and of Soos and Fozi.[125] In 1979 the analyses by Lintas and DeMattaeis[127] yielded 0·8 $\mu g\,kg^{-1}$ of benzo[*a*]pyrene for roasted coffee and 0·01 $\mu g\,kg^{-1}$ for coffee brew. Bories and Gasc,[128] in 1980, using a column chromatographic method, found $>0{\cdot}5\,\mu g\,kg^{-1}$.

Levi[129] presented data in 1980 for roasted beans, roasted to various degrees, and instant coffee. For roasted beans the range is from 0·5 to 0·3 $\mu g\,kg^{-1}$, and for instant coffee from 0·1 to 0·7 $\mu g\,kg^{-1}$.

The data of Guyot *et al.*[130] for regular roasted and dark roasted coffee beans are nearly congruent with the data obtained by Levi.[129] For roasted coffee of regular roast the range is from 0·2 to 0·5 $\mu g\,kg^{-1}$, and for the resulting brew 0·006 to 0·009 $\mu g\,kg^{-1}$. The darkly roasted beans showed a range from 1·8 to 3·2 $\mu g\,kg^{-1}$ and the resulting brew from the dark roast was correspondingly higher, namely 0·043 to 0·101 $\mu g\,kg^{-1}$.

In 1982 van der Stegen and van Overbruggen,[131] to dispel concerns that coffee might contain undue quantities of benzo[*a*]pyrene, analysed green beans, roasted beans and instant powder. Their results, obtained on green beans, are identical with those of D'Arrigo;[123] for roasted coffee with those of Fritz[126] and Guyot *et al.*;[130] and for instant powder a close match is achieved with Levi's data.[129] Undue quantities of benzo[*a*]pyrene were not found.

Ruschenburg and Jahr in 1986[132] reported on the analysis of benzo[*a*]pyrene for 20 commercial roasted coffee samples from various countries. The results vary between less than 0·1 to 0·35 $\mu g\,kg^{-1}$. Two espresso coffee samples from Italy show 0·32 and 0·55 $\mu g\,kg^{-1}$, respectively. Two instant coffee samples (one decaffeinated) both show less than 0·1 $\mu g\,kg^{-1}$. Two samples of a liquid coffee concentrate also show values of less than 0·1 $\mu g\,kg^{-1}$.

Hischenhuber and Stijve,[133] in 1987, analysed roasted coffee and coffee brews for benzo[*a*]pyrene. The method of isolation and analysis of benzo[*a*]pyrene was similar to those used by DeKruijf *et al.*[134] Fourteen samples were analysed, and attention was focused on the benzo[*a*]pyrene content in the coffee brew (prepared by the filter drip method) to determine how much of the polycyclic hydrocarbon is ingested by the consumer when drinking coffee. The authors also mentioned the solution effect of 15% caffeine solutions on benzo[*a*]pyrene. This effect had been exploited previously to remove benzo[*a*]pyrene from food extracts.[135–138] The original studies by Weil-Malherbe[139] in 1946 for utilisation of purines to solubilise benzo[*a*]pyrene should be mentioned for completeness.

Apparently, when caffeine is present at the comparatively low concentrations typical for a coffee brew, no solubilisation effect for benzo[*a*]pyrene occurs. This is evidenced by the fact that a dark roast of robusta beans with 2·4% caffeine against a medium roast of arabica beans with 1·3% caffeine did not carry any more benzo[*a*]pyrene into the coffee brew.[114] Both brews contained 2 ng of benzo[*a*]pyrene per litre of coffee brew. Furthermore, three decaffeinated coffees showed a range of 0·2 to 0·3 $\mu g\,kg^{-1}$ of benzo[*a*]pyrene in ground coffee, and eight commercial blends of regular coffee showed a median of 0·36 μg benzo[*a*]pyrene per kilogram of ground coffee. It is of interest to note that two roasts of 100% robusta beans (one light roast, one dark roast) both contained only 0·2 $\mu g\,kg^{-1}$, whereas two roasts of arabica beans (one medium, one dark) contained 0·4 and 0·5 $\mu g\,kg^{-1}$, respectively.

DeKruijf *et al.*[140] utilised a very appealing HPLC method (developed in 1980 at TNO-CIVO, Holland) for the separation and determination of benzo[*a*]pyrene, coupled to a Perkin–Elmer Model 650-10LC dual-beam monochromator fluorescence detector. Special care was taken to avoid a false-positive finding of benzo[*a*]pyrene. The detection limit for benzo[*a*]-pyrene for roasted coffee and spent coffee grounds was given as 0·1 $\mu g\,kg^{-1}$ and for coffee brew as 0·000 01 $\mu g\,kg^{-1}$. The quantity of benzo[*a*]pyrene they found for the coffee brew (namely, for very dark roasted coffee (over-roasted), 2 μg of benzo[*a*]pyrene per kilogram of coffee) agrees well with that found by Strobel[122] in 1966. The corresponding coffee brew (filter method) measured only 0·0008 μg kg^{-1}. The spent grounds resulting from the extraction showed an increase to 2·56 μg kg^{-1}. This is roughly as expected since one has to assume that hardly any benzo[*a*]pyrene was transported into the brew, and brew solids were removed from the spent grounds.

Brewing the same coffee with the traditional method (no filter used), the benzo[*a*]pyrene determined in the brew is 20% higher than that in the filter-brewed coffee: 0·001 vs 0·0008 μg kg^{-1}. The analysis on the spent

Table 9

3,4-Benzpyrene levels in coffee oil, soybean oil, corn oil, and other oils and fats

Substrate	N(X)[a]	*Used for analyses (g)*	*Unsaponifiables (%)*	*3,4-BP (ng g^{-1} of oil, etc.)*
Data of Strobel[122]				
Coffee oil	(1)	20	12·0	0·5
Coffee oil	(1)	20	11·3	0·55
Coffee oil	(1)	20	10·0	0·5
Coffee oil (distilled)	(1)	20	0·24	0·1
Soybean oil	(1)	20	0·30	0·1
Soybean oil	(1)	20	1·3	0·1
Corn oil	(1)	10	0·41	0·12
Data of Hopia et al.[141]				
Cooking margarines	(4)	40	—	0·56–2·2
Table margarines	(8)	40	—	0·05–1·9
Cold-pressed sunflower oil	(1)	40	—	0·16
Corn oil	(1)	40	—	0·02
Cold-pressed corn oil	(1)	40	—	1·2
Rapeseed/turnip seed oil	(1)	40	—	0·42
Soybean oil	(2)	40	—	0·74–1·3
Coconut fat	(1)	40	—	0·02
Crude coconut oil	(1)	40	—	24·0
Deodorised coconut oil	(1)	40	—	0·2
Data of Welling and Kaandorp[138]				
Salad oil	(5)	25	—	0·6–6·8
Sunflower oil	(4)	25	—	0·7–1·9
Olive oil	(1)	25	—	0·8
Safflower oil	(1)	25	—	0·7
Corn oil	(3)	25	—	0·5–26·1

[a] Number of samples taken.

grounds reflects that increase of benzo[*a*]pyrene in the brew by showing a slight decrease over the filter brew spent grounds: 2·40 vs 2·56 $\mu g\,kg^{-1}$.

The data for the coffee brew obtained by DeKruijf *et al.*[140] from a regular roast of coffee (50 g of roast and ground per 1 litre of water) are of the same order of magnitude as those obtained by Strobel[122] for a coffee brew from coffee of regular roast (0·0003 vs 0·000 15 $\mu g\,kg^{-1}$). A long-existing wide variation of benzo[*a*]pyrene coffee data was thus eliminated by these findings.

Some analytical results on the coffee oils are compared in Table 9 with commercial soybean oil, corn oil, and other oils and fats. All of the oils show very similar low levels.

Table 10 summarises the analytical data obtained on the benzo[*a*]pyrene content of green coffee beans, roasted coffee, silverskin, coffee soot, instant coffee, instant coffee beverage, beverage from regular brew and, finally, coffee oil. The transfer of benzo[*a*]pyrene from the roasted bean to the coffee brew occurs at a very low level (~1% transfer). The presence of caffeine in the coffee brew does not contribute to additional benzo-[*a*]pyrene transfer to the coffee brew. It should be pointed out that, for those results where highly meticulous clean-up procedures were applied to avoid false positives, the quantities of benzo[*a*]pyrene found in the coffee beverage are in good agreement.[122,134,140]

Table 11 summarises the various detection and quantitation procedures for benzo[*a*]pyrene that were applied during 24 years of benzo[*a*]pyrene analysis. Procedures utilising the pronounced fluorescence property of benzo[*a*]pyrene are far more sensitive than those utilising UV detection or strictly gas chromatographic procedures.

Table 12 gives a list of various foods and the 3,4-benzpyrene contents measured. From this table we can see that the level of 3,4-benzpyrene is relatively low in coffee products compared with many other foodstuffs. We can speculate that the value is so low because coffee is grown in parts of the world that are far away from smoke, factories, major highways and other forms of air pollution that put polycyclic hydrocarbons into the environment.

In 1983 Dennis *et al.*,[164] for the UK, calculated the ingested quantities of benzo[*a*]pyrene based on an estimated total food and beverage intake. The total calculated dietary load amounted to 0·25 μg per person per day or 91·25 μg per person on a yearly basis. The authors estimated that 50% of the total intake of benzo[*a*]pyrene is ingested with oils and fats, 30% via cereals, 10% via vegetables, 5% via fruit and sugar, 4% via meat and fish, 2% via milk and negligible quantities via beverages. The authors make

Table 10
Benzo[*a*]pyrene in coffee materials
($\mu g\,kg^{-1}$)

Authors	Coffee bean: Green	Coffee bean: Roasted	Coffee bean: Soot	Coffee bean: Chaff	Coffee bean: Brew	Instant: Powder	Instant: Brew	Coffee oil
1958 Kuratsune and Hueper[109]		3–4*						
1960 Kuratsune and Hueper[119]		4* (None found in samples of regular roast)						
1963 Chassevant and Heros[118]		1·2*	200–440	150				
1963 Pertoldi-Marletta[142]		1·0						
1966 Grimmer and Hildebrandt[120]		<0·4						
1966 Strobel[122]					0·000 15****	0·02–0·06	0·000 3**	0·1–0·5
1967 Calzolari and Pertoldi-Marletta[143]		2·9–6·9*						
1968 Fritz[144]		0·3–6·1*						
1969 Fritz[145]					0·2–0·3***			
1971 D'Arrigo[123]	0·6–2·3	1·1–1·9						
1973 Bracco[124]	<0·01	<0·1						0·01
1974 Soos and Fozi[125]	0·4	1·1–3·1			0·3–0·4			
1975 Fritz[126]		0·3–0·5	245 (tar)	120				
1979 Lintas and DeMattaeis[127]		0·8			0·01			
1980 Levi[129]		0·5–3*				0·1–0·7		
1980 Bories and Gasc[128]		<0·5						
1982 Guyot *et al.*[130]		0·2–0·5			0·006–0·009			
		1·8*–3·2*			0·043*–0·101*			
1982 van der Stegen and van Overbruggen[131]	0·2–0·3	0·1–0·6		22·7		0·1–0·5		
1986 Ruschenburg and Jahr[132]		0·1–0·55			<0·01			
1987 DeKruijf *et al.*[134]		<0·1–0·2 (2·0*)			0·000 8*			
1987 Hischenhuber and Stijve[133]		0·1–0·7 (decaffeinated 0·2–0·3)			0·002–0·007			
1987 DeKruijf *et al.*[134]		<0·2–1·2 (average 0·3)			0·000 3 (50 g roast and ground/1 litre water)			

*Very dark roast; **calculated value; ***no remarks = regular roast; ****65 g roast and ground/1 litre water.

Table 11
Comparative sensitivities of various procedures to determine 3,4-benzpyrene

Procedure	*Lower limit* (μg)	*Reference*
1. UV absorption, quantitative	10·0	1964 Sawicki *et al.*[146]
2. Gas chromatography, quantitative	1·0	1963 Lijinsky *et al.*[147]
3. Fluorometric measurement in solution, room temperature, quantitative	0·1	1964 Sawicki *et al.*[146]
4. Direct fluorometry on cellulose, qualitative	0·000 5[a]	1964 Sawicki *et al.*[148]
5. Fluorometric measurement at cryogenic temperature 78 K, quantitative	0·001 3[a]	1963 Eichhoff *et al.*[149]
6. Fluorometric measurement in solution with special adaptation	0·005[a]	1966 Fritz[150]
7. Direct fluorometry on acetylated cellulose (paper), quantitative	0·001	1966 Strobel[122]
8. Direct fluorometry on acetylated cellulose (powder on Teflon), qualitative	0·0001[a]	1966 Strobel[122]
9. SiO_2 column chromatography; GC/mass spec. (selected ion monitoring)	0·000 01	1986 Hopia *et al.*[141]
10. HPLC + fluorometry	0·000 01[a]	1987 DeKruijf *et al.*[134]

[a] Qualitative detection limit.

reference to other previous dietary studies, and it appears that these are in good agreement with the data presented by Dennis *et al.*

Depending on the location, the benzo[*a*]pyrene content in the air varies greatly. Considering that the average person inhales 20 m^3 air per day, then it follows that a person living in Los Angeles inhales 219 μg of benzo[*a*]pyrene per year. A dweller in urban Indiana will inhale 285 μg per year, whereas a person living in rural Indiana will inhale only 13 μg per year. These data suggest that the sum of dietary intake and inhalation of benzo[*a*]pyrene is at least 100 μg per year or maybe as much as ~400 μg per year or more (see also the estimates by van der Stegen and van Overbruggen.[131]

Using the averaged benzo[*a*]pyrene values determined in the coffee beverage by Strobel,[122] Hischenhuber and Stijve[133] and DeKruijf *et al.*,[134,140] then it follows that about 0·0007 μg of benzo[*a*]pyrene is imbibed on the assumption that 500 ml of coffee is imbibed per day. On a yearly basis this amounts to ~0·25 μg of benzo[*a*]pyrene. Comparing the quantity of benzo[*a*]pyrene imbibed per year by drinking coffee with those quantities contained in the air we breathe (~100 to >400 μg) or the food we eat (>90 μg per year), then the quantity of benzo[*a*]pyrene that we imbibe by drinking coffee (~0·25 μg per year) is minute indeed.

Table 12
3,4-Benzpyrene levels in various foods derived from animals and plants

Substrate	*3,4-BP* ($ng\,g^{-1}$)	*Authors*
Charcoal-grilled sirloin steak	11·1	1967 Lijinsky and Ross[151]
Charcoal-grilled T-bone steak	50·4	1967 Lijinsky and Ross[151]
Charcoal-grilled chicken	3·7	1967 Lijinsky and Ross[151]
Charcoal-grilled pork chops	7·9	1967 Lijinsky and Ross[151]
Charred matter from biscuits	10–72	1958 Kuratsune and Hueper[109]
Smoked trout	2·1	1958 Bailey and Dungal[152]
Smoked mutton	1·3	1958 Bailey and Dungal[152]
Smoked cod	0·5	1958 Bailey and Dungal[152]
Smoked ham	3·2	1966 Howard *et al.*[153]
Dried herring	<15·0	1976 Panalacks[154]
Smoked herring	1·0	1976 Panalacks[154]
Smoked meat products	0·2–2·9	1971 Filipovitch and Toth[155]
Industrial frying fat	4·0	1969 Berner and Biernoth[156]
Wheat	0·68	1965 Grimmer and Hildebrandt[157]
Flour	0·80	1965 Grimmer and Hildebrandt[157]
Bread	0·97	1965 Grimmer and Hildebrandt[157]
Wheat	1·44	1965 Grimmer and Hildebrandt[157]
Flour	1·60	1965 Grimmer and Hildebrandt[157]
Bread	1·63	1965 Grimmer and Hildebrandt[157]
Wheat grown remote from industry	0·079–0·17	1964 Bolling[158]
Wheat grown near industry	1·87	1964 Bolling[158]
Lettuce (dry weight) grown remote from and near industry	23–150·00	1971 Fritz and Engst[159]
Roasted malt	15·00[a]	1966 Fritz[160]
Roasted malt	1·0–0·9[b]	1968 Fritz[144]
Roasted malt	0·13–0·24	1969 Meier and Stender[161]
Drinking water	0·004 4	1983 Fritz[162]
Beer	·08	1983 Fritz[162]
Lettuce (fresh weight) grown under glass	Undetectable	1986 Wickström *et al.*[163]
Lettuce (fresh weight) grown in various areas	0·05–1·0	1986 Wickström *et al.*[163]

[a] Coal-fired roaster, no longer in use.
[b] Gas-fired roaster, used presently.

3. CONCLUSIONS

Exposure of humans to polycyclic hydrocarbons is unavoidable because of natural and man-made combustion processes (in volcanoes, forest fires, camp/cave fires, coal-fired generator stations, household combustion and motor-vehicle exhaust). This is also true for the ones that are transformed into carcinogens (e.g. benzo[*a*]pyrene) after physiological activation via cytochrome P-450 in lung or liver mitochondrial systems. Such materials are contained in the air, in the soil, and in our food and beverages. As proven by analysis, the quantities of polycyclic hydrocarbons that we ingest by drinking the coffee beverage are minute ($<0{\cdot}1\%$) when compared with the quantities contained in the air we breathe, in the food we eat or in some other beverages we drink. As stated in this chapter, the healthy human body is more capable and better equipped to detoxify carcinogenic compounds and repair damaged nucleic acids than are murine rodents.

However, there is no doubt that even the most effective physiological defence system against cancer will succumb under certain conditions. This happens when there is exposure to massive doses of primary or secondary chemical carcinogens, as was formerly experienced in occupational cancers ranging from *Bergkrankheit* to those resulting from the manufacture of isopropyl alcohol. Also physical carcinogens, like wilful UV radiation by direct and prolonged sunlight, involuntary exposure to neutron bombardment or exposure to gamma-rays, will increase with certainty the relative risk of carcinogenesis. And, of course, it must be pointed out that some tissue cell systems are genetically more prone to undergo the initiation and promotion process than the genetically more resistant cell system. Factors of nutritional, physical and mental dispositions of stress are also not conducive to an effective defence.

The opinion that the coffee beverage contains more benzo[*a*]pyrene than any other beverage does not correspond to the facts. This is evidenced by multiple analyses carried out by several groups belonging to governmental and industrial organisations in three different countries. This opinion was formed on the basis that caffeine, purposely added at high levels to, for example, food products ($>15\%$), functions as a convenient agent for removal of polycyclic hydrocarbons from such food products for analysis. It has been demonstrated, however, with coffees prepared from robusta beans containing about 2·4% caffeine and with arabica beans containing about 1·3% caffeine, that, regardless of caffeine content in the different beverages, equally low quantities of benzo[*a*]pyrene were transferred into the brew.

The scientists and engineers of the coffee industry, academia and governmental institutions have been actively engaged in research programmes spanning wide areas, and have produced important insights into all areas studied. There is a further need, though, for continued study to penetrate even more deeply into the secrets nature holds to improve the quality of the end-product and to ensure, on a continued basis, that the customer is provided at all times with safe products for consumption.

ACKNOWLEDGEMENTS

The author is grateful to Debby Boyles, Shirley Heismann, Jan King, Stella Lewis, Mary Ann Siklosi, Arleen Thyen and Mike Vatter for their untiring assistance in preparing the manuscript of this chapter. Mr Frank Krause is gratefully acknowledged for his assistance in the determination of benzo[*a*]pyrene.

REFERENCES

1. Graffi, A. and Bielka, H., *Probleme der Experimentellen Krebsforschung Leipzig*, 1959, Akademische Verlagsgesellschaft Geest & Portig KG.
2. von Hohenheim, T. B., in *Opus Paramirum*, 1531, 89, 107, 236, 237, 241.
3. von Hohenheim, T. B., in *Von der Bergsucht und anderen Bergkrankheiten*. Written 1533–34. First printed in 1567, reprinted 1925. Otto Wilhelm Barth Verlag GmbH, Munich, 463–544.
4. Härting, F. H. and Hesse, W., Der Lungenkrebs, die Bergkrankheit in den Schneeberger Gruben. *Vierteljahrsschr. f. gerichtl. Med. u. öffentl. Gesundheitswesen* (NF), 1879, **30**, 296–309; **31**, 102–129; **31**, 313–337.
5. Redmond, D. E., *New Engl. J. Med.*, 1970, **282**, 18–23.
6. Pott, P., in *Chirurgical Observations Relative to Cancer of the Scrotum*, Eds L. Hanes *et al.*, Collins, London, 1775, 53–68 (cited by Eckardt, R. E. in *Industrial Carcinogens*, Grune & Stratton, New York, 1959).
7. Sömmering, S. Th., in *De Morbis Vasorum Absorbentium Corporis Humani*, Varrentrappi et Wenneri. Traiectinae ad Moenum (Frankfurt/Main, Germany), 1795, 109 (cited by Wynder, E. L. and Hoffman, D., *Adv. Cancer Res.*, 1964, **8**, 250).
8. Volkmann, R., Über das häufige Auftreten von Hautkrebsen bei Braunkohlenarbeitern. *Beitr. Chirur.*, 1875, 370 (cited by Eckardt, R. E., in *Industrial Carcinogens*, Grune & Stratton, New York, 1959).
9. Butlin, H. J., *Brit. Med. J.*, 1892, **1**, 1341; **2**, 1 and 66.
10. Brosch, A., *Arch. Pathol. Anat. Physiol.*, 1900, **162**, Band II, 32–84 (cited by Wynder, E. L. and Hoffman, D., *Adv. Cancer Res.*, 1964, **8**, 250).
11. Yamagiwa, K. and Ichikawa, K., *Mitt. med. Fak. Tokio*, 1915, **15**, 295–344; *J. Cancer Research*, 1918, **3**, 1.
12. Bloch, B. and Dreifuss, W., *Schweitz. Med. Wochenschr. Woch.*, 1921, **51**, 1033.
13. Eckardt, R. E., in *Industrial Carcinogens*, Grune & Stratton, New York, 1959.
14. Rehn, L., *Arch. Klin. Chir.*, 1895, **1**, 588 (cited by Oettel, H., *Angew. Chem.*, 1958, **70**, 532).
15. Pfeil, E., *Dtch. Med. Wschr.*, 1911, **61**, 1197.
16. Kennaway, E., *Br. Med. J.*, 1955, 749–752 (review article).
17. Berthelot, M. C. R., *Acad. Sci. (Paris)*, 1866, **63**, 479 (cited by Kennaway, E., *Br. Med. J.*, 1955, 749 (review article)).
18. Schroeter, G., *Brennst. Chemie*, 1920, **1**, 39 (cited by Kennaway, E., *Br. Med. J.*, 1955, 749 (review article); *Ber. dtsch. Chem. Ges.*, 1924, **57**, 1990–2003).
19. Clar, E., *Ber. dtsch. Chem. Ges.*, 1929, **62**, 350.
20. Cook, J. W., Hewett, C. L. and Hieger, I., *J. Chem. Soc.*, 1933, 395–405.

21. O'Neill, C. H., Hodges, G. M., Riddle, P. N., Jordan, P. W., Newman, R. H. and Flood, R. J., *Int. J. Cancer*, 1980, **26**, 617–628.
22. Cartwright, R. A., in *Chemical Carcinogens, Vol. 1*, 2nd edn, Ed. C. E. Searle, ASC Monograph 182, American Chemical Society, Washington, DC, 1984, 1–39.
23. Blackburn, G. M. and Kellard, B., *Chemistry and Industry*, 20 Oct. 1986, 687–695.
24. Magee, P. N. and Barnes, J. M., *Br. J. Cancer*, 1956, **10**, 114–122.
25. Schmähl, D. and Preußmann, R., *Naturwissenschaften*, 1959, **46**, 175.
26. Preußmann, R. and Stewart, B. W., in *Chemical Carcinogens, Vol. 2*, 2nd end, Ed. C. E. Searle, ASC Monograph 182, American Chemical Society, Washington, DC, 1984, 643–828.
27. Preußmann, R. and Eisenbrand, G., in *Chemical Carcinogens, Vol. 2*, 2nd edn, Ed. C. E. Searle, ASC Monograph 182, American Chemical Society, Washington, DC, 1984, 830–868.
28. Schmidt, O., *Z. Phys. Chem.*, 1939, **42**, 83.
29. Coulson, C. A., *Adv. Cancer Res.*, 1953, **1**, 1.
30. Pullman, A. and Pullman, B., *Cancerisation par les Substances Chimiques et Structure Moleculaire*, Masson, Paris, 1955.
31. Boyland, E., *Biochem. Soc. Symp. (Cambridge, England)*, 1950, **5**, 40–54.
32. Pullman, A., *Biopolymers*, Symp. No. 1, 1964, 65–67 (review article).
33. Lacasagne, A., Zajdela, F., Buu-Hoï, N. P. and Chalvet, H., *Acad. Sci., Paris*, 1957, **244**, 273.
34. Schoenthal, R., in *Polycyclic Hydrocarbons, Vol. I*, Academic Press, London and New York, 1964, 143.
35. Klein, M., *Cancer Res.*, 1959, **19**, 1109.
36. Pietra, G., Spencer, K. and Shubik, P., *Nature (London)*, 1959, **183**, 1689.
37. Kelly, M. G. and O'Gara, R. W., *J. Natl. Cancer Inst.*, 1961, **26**, 651.
38. Miller, E. C., Miller, J. A. and Hartmann, H. A., *Cancer Res.*, 1961, **21**, 815–824.
39. Miller, J. A., Cramer, J. W. and Miller, E. C., *Cancer Res.*, 1960, **20**, 950–962.
40. Miller, J. A., Wyatt, C. S., Miller, E. C. and Hartmann, H. A., *Cancer Res.*, 1961, **21**, 1465.
41. Boyland, E., Hanson, D. and Neri, R., *Ann. Rep. Brit. Empire Cancer Campaign*, 1960, **38**, 52.
42. Walpole, A. L. and Williams, M. H. C., *Brit. Med. Bull.*, 1958, **14**, 141.
43. Harper, K. E., *Br. J. Cancer*, 1959, **XIII**, 746.
44. Rous, P. and Beard, J. W., *J. Exp. Med.*, 1935, **62**, 523.
45. Rous, P. and Kidd, J. G., *J. Exp. Med.*, 1939, **69**, 399.
46. Rous, P. and Kidd, J. G., *J. Exp. Med.*, 1941, **73**, 365.
47. Friedewald, W. F. and Rous, P., *J. Exp. Med.*, 1944, **80**, 101.
48. Friedewald, W. F. and Rous, P., *J. Exp. Med.*, 1950, **91**, 459.
49. Berenblum, I., *Cancer Res.*, 1941, **I**, 44; *Arch. Pathol.*, 1944, **38**, 233.
50. Kline, B. E. and Rush, H. P., *Cancer Res.*, 1944, **4**, 762.
51. Rush, H. P. and Kline, B. E., *Arch. Pathol.*, 1946, **42**, 445.
52. Shear, M. J., *Am. J. Cancer*, 1938, **33**, 499.
53. Glücksmann, A., *Cancer Res.*, 1945, **5**, 385.
54. Berenblum, I. and Shubik, P., *Brit. J. Cancer*, 1947, **1**, 383.
55. Clayson, D. B., in *Chemical Carcinogenesis*, Little, Brown & Co., Boston, Mass., 1962, 297.

56. Druckrey, H., *Klin. Wschr.*, 1943, **22**, 532.
57. Druckrey, H. and Küpfmüller, K., *Z. Naturforsch.*, 1948, **36**, 254.
58. Druckrey, H., Schmähl, D., Dischler, W. and Schildbach, A., *Naturwiss.*, 1962, **49**, 217.
59. Druckrey, H., Kruse, H., Preußmann, R., Ivanovic, S. and Landschütz, C., *Z. Krebsforsch.*, 1970, **74**, 241.
60. Moolgavkar, S. H. and Knudson, A. G. Jr, *J. Natl. Cancer Inst.*, 1981, **66**, 1037–1052.
61. Boyland, E. and Sims, P., *Biochem. J.*, 1965, **97**, 7–16.
62. Miller, E. C., *Cancer Res.*, 1951, **11**, 100–108.
63. Heidelberger, C. and Moldenhauer, G., *Cancer Res.*, 1956, **16**, 442–449.
64. Brookes, P. and Lawley, P. D., *Nature (London)*, 1964, **202**, 781–784.
65. Grover, P. L. and Sims, P., *Biochem. J.*, 1968, **110**, 159–160.
66. Gelboin, H. V., *Cancer Res.*, 1969, **29**, 1272–1276.
67. Borgen, A., Darvey, H., Castaglioni, N., Crocker, T. T., Rasmussen, R. E. and Wang, I. Y., *J. Med. Chem.*, 1973, **16**, 502–506.
68. Sims, P., Grover, P. L., Swaisland, A., Pal, K. and Hewer, A., *Nature (London)*, 1974, **252**, 326–328.
69. Kato, R., *Pharmacol. Ther.*, 1979, **6**, 41–98.
70. Nebert, D. W., *Mol. Cell. Biochem.*, 1979, **27**, 27–46.
71. Tonegawa, S., *Nature (London)*, 1983, **302**, 575.
72. Pelkonen, O. and Nebert, D. W., *Pharmacol. Rev.*, 1982, **34**, 189–222.
73. Moore, P. D., Koreeda, M., Wislocki, P. G., Levin, W., Conney, A. H., Yagi, H. and Jerina, D. M., in *Drug Metabolism Concepts*, Ed. D. M. Jerina, ACS Symp. Series No. 44, American Chemical Society, Washington, DC, 1977, 127–154.
74. Koreeda, M., Moore, P. D., Wislocki, P. G., Levin, W., Conney, A. H., Yagi, H. and Jerina, D. M., *Science*, 1978, **199**, 778–781.
75. Ivanovic, V., Geacintov, N. E., Yamasaki, H. and Weinstein, I. B., *Biochemistry*, 1978, **17**, 1597–1603.
76. Jeffrey, A. M., Weinstein, I. B., Jenette, K. W., Grzeskowiak, K., Nakanishi, K., Harvey, R. G., Autrup, H. and Harris, C., *Nature (London)*, 1977, **269**, 348–350.
77. Dipple, A., Moschel, R. C. and Bigger, C. A., in *Chemical Carcinogens, Vol. 1*, 2nd end, Ed. C. E. Searle, ACS Monograph 182, American Chemical Society, Washington, DC, 1984, 41–163.
78. Karran, P. and Lindahl, T., *Cancer Surveys*, 1985, **4**, 583.
79. Blackburn, G. M. and Kellard, B., in *Chemistry and Industry*, 17 Nov. 1986, 770–779.
80. Kennaway, E. L., *Br. Med. J.*, 1925, 564.
81. Berenblum, I. and Schoenthal, R., *Br. J. Exp. Pathol.*, 1943, **24**, 232.
82. Badger, G. M., Lewis, G. E. and Napier, M., *J. Chem. Soc.*, 1960, 2825.
83. Badger, G. M. and Spotswood, T. M., *J. Chem. Soc.*, 1960, **266**, 4431.
84. Lam, J., *Acta Path. Microbiol. Scand.*, 1956, **39**, 207.
85. Badger, G. M., Kimber, R. L. W. and Spotswood, T. M., *Nature (London)*, 1960, **187**, 663–665.
86. Badger, G. M., Donelly, J. K. and Spotswood, T. M., *Aust. J. Chem.*, 1966, **19**, 1023–1043.
87. Kruber, O. and Oberkobusch, R., *Chem. Ber.*, 1952, **85**, 433.

88. Schoenthal, R., *Nature (London)*, 1957, **180**, 606.
89. Buu-Hoï, N. P., *Nature (London)*, 1958, **182**, 1158.
90. Waradeker, S. and Ranadive, K. J., *J. Natl Cancer Inst.*, 1958, **21**, 1151.
91. Kotin, P. and Falk, H. L., *Cancer*, 1959, **12**, 147.
92. Kennaway, E. L. and Lindsay, A. J., *Br. Med. Bull.*, 1958, **14**, 124.
93. Lyons, M. J., *Br. J. Cancer*, 1959, **13**, 126.
94. Van Duren, B. L., Bilbao, J. A. and Joseph, C. A., *J. Natl Cancer Inst.*, 1960, **25**, 53.
95. Berenblum, I. and Schoenthal, R., *J. Chem. Soc.*, 1946, 1027.
96. Cooper, R. L., *Chem. Ind.*, 1953, 1177, 1260, 1364.
97. Clemo, G. R., *Chem. Ind.*, 1953, 957; 1955, 38.
98. Leiter, J., Shimkin, M. B. and Shear, M. J., *J. Natl Cancer Inst.*, 1942, **3**, 155.
99. Shore, U. C. and Katz, M., *Anal. Chem.*, 1956, **28**, 1399.
100. Kotin, P., *Cancer Res.*, 1956, **16**, 375.
101. Moore, G. E. and Katz, M., *Int. J. Air Poll.*, 1960, **2**, 221.
102. Sawicki, E., Elbert, W., Stanley, T. W., Hauser, T. R. and Fox, F. T., *Int. J. Air Poll.*, 1960, **2**, 273.
103. Waller, R. E., *Br. J. Cancer*, 1952, **6**, 8.
104. Kotin, P., *Acta Unio Int. contra Cancrum*, 1959, **15**, 613.
105. Sawicki, E., Elbert, W. C., Thomas, R., Hauser, M. S., Fox, F. T. and Stanley, T. W., *Am. Ind. Hyg. Assoc. J.*, 1960, **21**, 443.
106. Sawicki, E., Mecker, J. E. and Morgan, M. J., *Int. J. Air Wat. Poll.*, 1965, **9**, 291.
107. Goulden, F. and Tipler, M. M., *Br. J. Cancer*, 1949, **3**, 157.
108. Kuratsune, M., *J. Natl Cancer Inst.*, 1956, **16**, 1485–1496.
109. Kuratsune, M. and Hueper, W. C., *J. Natl Cancer Inst.*, 1958, **20**, 37–51.
110. Fischer, R., *100 Jahre BASF, Aus der Forschung*, 1965, 568.
111. Cardon, S. Z., Alvord, E. T., Rand, H. J. and Hitchcock, R., *Br. J. Cancer*, 1956, **10**, 485.
112. Campbell, J. M. and Lindsey, A. J., *Br. J. Cancer*, 1957, **11**, 192.
113. Kotin, P., Falk, H. L. and Thomas, M., *A.M.A. Arch. Ind. Hyg.*, 1954, **9**, 164.
114. Hoffmann, D. and Wynder, E. L., *Cancer*, 1962, **15**, 93.
115. Kotin, P., Falk, H. L. and Thomas, M., *A.M.A. Arch. Ind. Hyg.*, 1955, **11**, 113.
116. Lickint, F., *Med. Klin.*, 1938, **34**, 1499–1500.
117. Roffo, A. H., *Bol. Inst. de med. exper. para le estud. y trat. d. cancer*, 1938, **15**, 741–762.
118. Chassevant, F. and Heros, M., *Café, Cacao, Thé*, 1963, **7**(4), 349–358.
119. Kuratsune, M. and Hueper, W. C., *J. Natl Cancer Inst.*, 1960, **24**, 463–469.
120. Grimmer, G. and Hildebrandt, A., *Dtsch. Lebensm. Rdsch.*, 1966, **62**, 19–21.
121. Fritz, W., *Umschau Wissensch. Technik*, 1966, **66**, 641.
122. Strobel, R. G. K., *6th Int. Coll. ASIC*, Bogota (Colombia), 1973, 128–134.
123. D'Arrigo, V., *Cuaderni Di Merceologia Bologna Universita*, 1971, **10**, 27.
124. Bracco, U., *La Rivista Italiana Della Sostanze Grasse*, 1973, **L**—Guigno, 166–676.
125. Soos, K. and Fozi, I., *Edesipar*, 1974, **25**(1), 7; (2), 37; (3), 65.
126. Fritz, W., *Ernährungspraxis*, 1975, **20**, 49–51.
127. Lintas, C. and DeMattaeis, M. C., *Food Cosmet. Toxicol.*, 1979, **17**, 325–328.
128. Bories, G. and Gasc, N., *Ann. Falsif. Expert. Chim. Toxicol.*, 1980, No. 789, 487–496.

129. Levi, C. P., *9th Int. Coll. ASIC* (London), 1980, 125–133.
130. Guyot, B., Lahmy, S. and Vincent, J. C., *Café, Cacao, Thé*, 1982, **XXVI**, 199.
131. van der Stegen, G. H. D. and van Overbruggen, G. J. J., *10th Int. Coll. ASIC*, Salvador (Brazil), 1982, 347–354.
132. Ruschenburg, U. and Jahr, D., *Café, Cacao, Thé*, 1986, **XXX**, 3–6.
133. Hischenhuber, C. and Stijve, T., *Dtsch. Lebensm. Rdsch.*, 1987, **83**, 1–4.
134. DeKruijf, N., Schouten, T. and van der Stegen, G. H. D., *J. Agric. Food Chem.* 1987, **35**, 545–549.
135. Eisenbrand, J. and Becker, G., *Dtsch. Lebensm. Rdsch.*, 1967, **63**, 139.
136. Eisenbrand, J. and Baumann, K., *Z. Lebensm. Untersuch.*, 1969, **140**, 158.
137. Gertz, Ch., *Z. Lebensm. Unters. Forsch.*, 1981, **173**, 208.
138. Welling, P. and Kaandorp, B., *Z. Lebensm. Unters. Forsch.*, 1986, **183**, 111–115.
139. Weil-Malherbe, H., *Biochem. J.*, 1946, **40**, 351–361.
140. DeKruijf, N., Schouten, A. and van der Stegen, G. H. D., *Café, Cacao, Thé*, 1987, **XXXI**, 151–154.
141. Hopia, A., Pyysalo, H. and Wickström, K., *J. Am. Oil Chem. Soc.*, 1986, **63**, 889–893.
142. Pertoldi-Marletta, G., *Atti. IV. Convengno Nazionale della Qualita*, 1963, **3**, 15.
143. Calzolari, C. and Pertoldi-Marletta, G., *3rd Int. Coll. ASIC*, Trieste (Italy), 1967, 248–256.
144. Fritz, W., *Die Nahrung*, 1968, **12**, 799–804.
145. Fritz, W., *Dtsch. Lebensm. Rdsch.*, 1969, **65**, 83–85.
146. Sawicki, E., Stanley, T. W., Elbert, W. C. and Pfaff, J. D., *Anal. Chem.*, 1964, **36**, 497–502.
147. Lijinsky, W., Domsky, K., Mason, G., Ramahi, H. Y. and Safari, T., *Anal. Chem.*, 1963, **35**, 953.
148. Sawicki, E., Stanley, T. W. and Johnson, H., *Microchem. J.*, 1964, **8**, 257–284.
149. Eichhoff, H. J. and Koehler, M., *Z. Anal. Chem.*, 1963, **197**, 271.
150. Fritz, W., *Chem. Tech.*, 1966, **18**, 689.
151. Lijinsky, W. and Ross, A. E., *Food Cosmet. Toxicol.*, 1967, **5**, 343–347.
152. Bailey, E. M. and Dungal, N., *Br. J. Cancer*, 1958, **12**, 348.
153. Howard, J. W., White, R. M., Fry, B. E. and Turicchi, E. W., *J. Assoc. Offic. Anal. Chem.*, 1966, **49**, 611–617.
154. Panalacks, T., *J. Environ. Sci. Health*, 1976, **11**(4), 299.
155. Filipovitch, J. and Toth, L., *Fleischwirtschaft*, 1971, **51**, 1323–1325.
156. Berner, G. and Biernoth, H., *Z. Lebensm. Unters. Forsch.*, 1969, **140**, 330–331.
157. Grimmer, G. and Hildebrandt, A., *Z. Krebsforsch.*, 1965, **67**, 272–277.
158. Bolling, H., *Die Mühle*, 1964, **101**, 759–762.
159. Fritz, W. and Engst, R., *Z. Gesamte Hyg. Grenzgebiete*, 1971, **17**, 271–275.
160. Fritz, W., *Naturwissensch.*, 1966, **53**, 132.
161. Meier, H. G. and Stender, W., *Dtsch. Lebensm. Rdsch.*, 1969, **65**, 341–343.
162. Fritz, W., *Nahrung*, 1983, **27**, 965–973.
163. Wickström, K., Pyysalo, H., Plaami-Heikkilä, S. and Touminen, J., *Z. Lebensm. Unters. Forsch.*, 1986, **183**, 182–185.
164. Dennis, M. J., Massey, R. C., McWeeny, D. J., Knowles, M. E. and Watson, D., *Food Chem. Toxicol.*, 1983, **21**, 969–974.

Appendix 1

Glossary of Some Terms and Abbreviations Used

Term (*noun*)	*Meaning*
ADP	Adenosine diphosphate, which occurs in cells and is involved in processes requiring transfer of energy
Adenoma	A benign tumour derived from glandular tissue
Adjuvant	Any substance used in conjunction with another to enhance its activity
Adrenergic system	System of nerve fibres that release noradrenaline as a neurotransmitter
***c*-AMP**	Adenosine monophosphate: occurs in cells and is involved in energy transfer processes
Analgesia	Reduced sensibility to pain
Angina	A sense of suffocation or suffocating pain
Anosmia	Inability to smell, either totally or a particular substance or group of substances
Anoxia	Condition resulting from an inadequate supply of oxygen to the tissues
Aorta	Main artery of the heart, from which all others derive, arising from the left ventricle
Apnoea	Temporary cessation of breath from any cause
Arrhythmia	Any deviation from the normal rhythm of the heart
Arteriosclerosis	Imprecise term for any of several conditions affecting the arteries
Atheroma	Degeneration of the walls of the arteries owing to the formation of fatty plaques and scar tissue, which limit blood circulation and predispose to thrombosis *adj.* **atherogenic**

Term (*noun*)	*Meaning*
Atopic dermatitis	Dermatitis caused by an allergy
c-ATP	Adenosine triphosphate, ATP → ADP or AMP, a reaction reversed by energy from breakdown of carbohydrates or other food substances
Autonomic nervous system	The part of the nervous system responsible for control (by motor nerves) of bodily functions that are not consciously directed
Axon	A nerve fibre; a single process extending from the cell body of a neurone and carrying nerve impulses away from it
Bile	A thick alkaline liquid that is secreted by the liver and stored in the gall bladder, from which it is ejected intermittently into the duodenum via the common bile duct. It has a number of different functions, e.g. emulsification of fats, stimulation of peristalsis *adj.* **biliary**, relating or affecting the bile duct or bile
Blood serum	Fluid that separates from clotted blood or blood plasma that is allowed to stand
Bradycardia	Slowing-down of heart-beat rate (to less than 50 beats per minute)
Cardiovascular system	The heart, together with two networks of blood vessels, which effects circulation of blood around the body, bringing about the transport of nutrients and oxygen, and disposing of waste products
Catecholamines	Group of physiologically active substances acting mainly as neurotransmitters (e.g. dopamine, adrenaline and noradrenaline)
Cell	Basic unit of all living organisms, which can reproduce itself exactly
Cerebral cortex	The elaborately folded outer part of the brain, most directly responsible for consciousness
Cerebrovascular disease	Disorder of the blood vessels of the brain and its covering membranes
CHO cells	Chinese Hamster Ovarian cells
Choleretic	An agent that stimulates the secretion of bile
Cholestaeremia	Excess of cholesterol in the blood
Chromosome	One of the threadlike structures in a cell nucleus that carry the genetic information in the form of genes
Clastogenesis	Giving rise to or inducing disruption, or breakages, in chromosomes

Term (*noun*)	*Meaning*
Convertogenesis	Conversion of normal cells into malignant ones in a test system
Coronary heart disease	Disorder in the arteries supplying blood to the heart, by obstruction as a result of the formation of a blood clot
Cranial nerves	The 12 pairs of nerves that arise directly from the brain and leave the skull through separate apertures; they are conventionally given roman numerals, of which the following are mentioned in this volume: I, V, VII, IX, X
Dorsal, *adj.*	Relating to the posterior part of an organ; or situated at the back of the body
Ecography	Use of ultrasound waves to map out and study the internal structure of the body
EEG	Electroencephalogram, from the apparatus for recording the electrical activity of the different parts of the brain and converting to a tracing
Effector	Any structure or agent that brings about activity in a muscle (e.g. contraction) or gland (secretion)
Endoscopy	Use of any instrument to view the interior of the body
Enterohepatic, *adj.*	(Circulation) within the liver
Epiglottis	A leaf-shaped flap of cartilage situated behind the root of the tongue over the opening from the pharynx into the windpipe
Epithelium (plural: **epithelia**)	Tissue that lines the skin and internal organs of the body
Extrasystole (ectopic beat)	A heart beat from an impulse generated abnormally
Facial nerve	The VIIth cranial nerve (mixed) that supplies, among other regions, the taste buds of the front part of the tongue
Fibrillation	Rapid and chaotic beating of the heart from the ventricular or atrial fibres
Fibrocyst	A benign tumour of fibrous connective tissue containing cystic spaces
Ganglion (plural: **ganglia**)	A nucleus of nerve cells; an enlargement or knot which forms a centre from which nerve fibres radiate
Gavage	Forced feeding

continued

Term (noun)	*Meaning*
Genome	The basic haploid set of chromosomes of an organism. Man has a genome of 23 chromosomes
Germ cell	Any of the embryonic cells that have the potential to develop into spermatozoa or ova
Glossopharyngeal nerve	The IXth cranial nerve (mixed), which supplies sensory fibres to the posterior third of the tongue and the soft palate
Gnotobiotic	Germfree, but inoculated with known micro-organisms
Gonad	A male or female reproductive organ which produces the gametes (ovum of the female, spermatozoon of the male) *adj.* **antigonadotrophic**
HDL-cholesterol	Cholesterol in high-density lipoprotein
Hepatocytes	Large cells in the liver with many metabolic functions
Hyperplasia	Increased production and growth of normal cells in a tissue or organ. The affected part becomes larger, but retains its normal form
Hypertension	High blood pressure
Hypertrophy	Abnormal growth of organs, usually pathological
Hypoglycaemia	A deficiency of glucose in the bloodstream, causing muscular weakness and incoordination, mental confusion and sweating
Hypothalamus	Region of the forebrain, linked with the thalamus (relay station for sensory messages) and the pituitary gland
Hypoxia	Deficiency of oxygen in the tissues
i.d.	intraduodenally administered
Infarction	Death of part or whole of an organ that occurs when the artery carrying its blood supply is obstructed by a blood clot or embolus (e.g. myocardial infarction)
Innervation	The nerve supply to or from an area or organ of the body
Inotropic, *adj.*	Effecting the contraction of heart muscle
Instillation	Application of liquid medication drop by drop
Intubation	Introduction of a tube into part of the body for the purpose of diagnostic treatment
i.p.	intraperitoneally (= through the peritoneum, the membrane lining the abdominal cavity)

Term (*noun*)	*Meaning*
Ischaemia	Inadequate flow of blood to a part of the body caused by constriction or blockage
Isoprenaline	Drug used for heart stimulation
i.v.	intravenously (=into a vein)
LD_{50}	Lethal dose at which 50% of an animal population examined will die, expressed as milligrams of agent per kilogram body weight
LDL-cholesterol	Cholesterol in low-density lipoprotein
Lipaemia	Presence in the blood of an abnormally large amount of fat
Locus coeruleus	Small pigmented area in the floor of the fourth ventricle (fluid-filled cavity) of the brain
Lymphocyte	A variety of white blood cell (leucocyte) present also in lymph nodes, spleen, etc.
Lymphosarcoma (= lymphoma)	Any malignant tumour of the lymph nodes
Medulla oblongata	The extension within the skull of the upper end of the spinal cord, forming the lowest part of the brainstem
Mesenchyme	The undifferentiated tissues of the early embryo
Metastasis	Distant spread of malignant tumour from the site of origin, via mainly (1) bloodstream, e.g. sarcomas; (2) lymphatic system, e.g. carcinomas; or (3) across body cavities
Microsome	A small particle consisting of piece of endoplasmic reticulum (a system of membranes present in the cytoplasm of cells) with ribosomes attached (particles consisting of RNA and protein which are the site of protein synthesis) *adj.* **microsomal**
Mitochondrion (plural: **mitochondria**)	A structure, occurring in varying numbers in the cytoplasm of every cell, that is the site of the energy production of the cell
Mitosis	A type of cell division in which a single cell produces two genetically identical daughter cells. It is the way in which new body (somatic) cells are produced for growth and repair *adj.* **mitotic**
mM	Millimoles weight of substance (i.e. milligram weight divided by the molecular weight of the substance)

continued

Term (*noun*)	*Meaning*
Motility	Spontaneous movement (e.g. intestinal)
Mucous membrane	A moist membrane lining many tubular structures and cavities
Myocardium	The middle of three layers forming the wall of the heart, composed of cardiac muscle; it forms the greater part of the heart wall and is thicker in the ventricles than in the atria
NADPH	Reduction product of nicotinamide dinucleotide phosphate, a co-enzyme in cellular respiration
Nares (plural)	Openings of the nose; external (anterior) are the nostrils; the two internal (posterior) are the openings leading from the nasal cavity into the pharynx
Neonate	An infant at any time during the first four weeks of life
Neoplasm	A new and abnormal growth, or any benign or malignant tumour
Nerve	A bundle of conducting nerve fibres that transmit impulses from the brain or spinal cord to the muscles and glands (motor nerves); or of greater interest in this volume inwards from the sense organs to the brain and spinal cord (sensory nerves). Most large nerves are mixed nerves
Neurone (nerve cell)	One of the basic functional units of the nervous system; a cell consisting of a number of parts, specialised to transmit electrical impulses from one part of the body to another
Nodule	A small swelling or aggregation of cells
Nosology	Naming and classification of diseases
Oedema	Excessive accumulation of fluid in the body tissues (popularly known as dropsy) *adj.* **oedematous**
Oesophagus (gullet)	Muscular tube, about 23 cm long, extending from the pharynx to the stomach
Olfactory nerve	The first cranial nerve; the special sensory nerve of smell. Fibres of the nerve run upwards from receptors in the nasal mucosa high in the roof of the nose, through minute holes in the skull, join to form the olfactory tract, and pass back to reach the brain
Ontogenesis	Development of an individual from the fertilised egg to maturity

Term (noun)	*Meaning*
Osteoporosis	Loss of bony tissue, resulting in bones that are brittle
Pharynx	A muscular cavity lined with mucous membrane that extends from the beginning of the oseophagus (gullet) up to the base of the skull. It acts as passageway for food from the mouth to the oseophagus, and as an air passage from the nasal cavity and mouth to the larynx
Pithed, *adj.*	When the whole of the central nervous system in an animal has been deliberately destroyed
Placenta	An organ within the womb by means of which the embryo is attached to the wall of the womb
p.o.	orally (*per os*)
Portal, *adj.*	(Blood) carried in the portal system of veins, e.g. to the liver
Pressor	An agent that raises blood pressure
Raphé	A line, ridge, crease or seam in a body tissue or organ
Receptor	A cell or group of cells specialised to detect changes in the environment and trigger impulses in the sensory nervous system
Reserpine	A drug extracted from *Rauwolfia* and used to control blood pressure
S-9 mix	Supernatant liquor of tissue homogenate (*ex* ground liver) containing microsomal enzymes
Sarcolemma	Cell membrane enclosing a muscle cell (fibre)
Sarcoma	Any cancer of connective tissue
Sarcoplasm	Cytoplasm of muscle cells
Sleep	State in which the activity of the brain is not apparent (apart from such body functions as breathing). EEG patterns show different stages of sleep. Following increasing depth of sleep, slow-wave or deep sleep is interrupted by paradoxical or REM (rapid eye movement) sleep, in alternating periods, when dreaming occurs; REM sleep constitutes about one-quarter of the total sleeping time
Somatic cells	Non-reproductive cells of the body
Sphincter	Specialised ring of muscle that surrounds an orifice

continued

Term (*noun*)	*Meaning*
Stenosis	Abnormal narrowing of a passage or opening
Sternebra	One of the four parts that fuse during development to form the body of the sternum (breastbone)
Synapse	The minute gap across which nerve impulses pass from one neurone to the next, at the end of a nerve fibre
Tachycardia	Increase in heart rate above normal
Tegmen	A structure that covers an organ or part of an organ
Teratology	Study of developmental abnormalities and their causes, thus **teratogenic**, etc.
Thyroid gland	A large endocrine gland situated at the base of the neck, concerned with the regulation of the metabolic rate by secretion of the thyroid hormone *adj.* **antithyrotrophic**
Trigeminal nerve	The Vth and largest of the cranial nerves which is split into three divisions, ophthalmic, maxillary and mandibular nerves. Its sensory fibres relay information about temperature, pain and touch from the whole part of the head (including the mouth)
Tumour	Any abnormal swelling or growth of a part of the body (usually tissue) which may be benign or malignant
Urticaria	A vascular reaction of the skin
Vagotomy	Surgical cutting of any part of the vagus nerve
Vagus nerve	The Xth cranial nerve (mixed) of which the sensory branch carries the sensation of taste from the mouth
Ventral, *adj.*	Relating to the anterior part of an organ; or situated at the front or nearby
Ventricle	Either of the two lower chambers (left and right) of the heart, which have thick muscular walls *adj.* **ventricular**
Viscera	Organs within the body cavities, especially the abdominal cavity

Appendix 2

Etymology of Some Physiology/Medical Terms

Greek[a]	*Latin*[b]	*Direct English, and typical derivative*[c]
1. Parts of the Body		
nephros; nephritikos, *adj.*	renes, renum; *m. pl.*	kidney [nephritic, Gk]
hēpar, -atos; hēpatikos, *adj.*	jecur, -oris; *n.*	liver [hepatic, Gk]
kardia; kardiakos, *adj.*	cor, cordis; *n.*	heart [cardiac, Gk]
ostoun, osteon	os, ossis; *n.*	bone [osteoporosis, Gk]
trophē	alimentum, -i; *n.* nutrio; *vb*	nourishment [hypertrophy, Gk; alimentary, L.]
pnoē	anima, -ae; *f*,	breath [apnoea, Gk]
kholē	bilis, bilis; *f.*	bile, gall [choleretic, Gk; biliary, L.]
sternon	pectus, pectoris; *n.*	chest [sternum, Gk]
glotta (*also* glossa)	lingua, -ae; *f.*	tongue [epiglottis, Gk; lingual, L.]
pharugx, -ggos	—	pharynx [pharyngeal, Gk]
mus, muos	musculum, -i; *n.*	muscle, sinew [In combination: (myo-), Gk]
sarcx, sarcos; sarcoma, *adj*	caro, carnis; *f*	flesh; becomes fleshy [sarcoma, Gk]
thēlē	papilla, -ae; *f.*	teat, nipple [epithelium, Gk; papilla, L.]
soma, somatos	corpus, corporis; *n.*	body [somatic, Gk]
pneumōn, -monos	pulmo, pulmonis; *m.*	lung

continued

Greek[a]	*Latin*[b]	*Direct English, and typical derivative*[c]
neuron	—	nerve, sinew [in combination: (neuro-), Gk]
stomā, stomatos	os, oris; *n.*	mouth [oral, L.]
kustis; kustē	vesica, -ae; *f.* vesiculus, -ae; *f dim.*	bladder [In combination: (cyst-), Gk]
haima	sanguis, -inis; *m.*	blood [In combination: (haem-); (-aemia), Gk]
enteron	intestinum, -i; *n*	intestine [In combination (entero-), Gk]
gaster, gasteros	venter, ventris; *m.* (*also* alvus)	belly, stomach
—	viscera, -um; *n pl.*	bowels [viscera, L.]
—	bucca, -ae; *f.*	cheek, mouthful [buccal, L.]
—	medulla, -ae; *f*	pith, marrow [medulla, L.]
—	ventriculus, -i; *m. dim.*	cavity [ventricle, L.]
—	cerebrum, -i; *n.*	brain [In combination: (cerebro-), L.]
derma	cutis, is; *f.*	skin [dermatitis, Gk]
cephalē	caput, -itis; *n.*	head
cheir	manus, -ūs; *f.*	hand

2. General

nosos	morbus, -i; *n.*	disease [nosologist, Gk]
apothekē	cella, -ae; *f.*	store-room [cell, L.]
hudōr, hudatos	lympha, -ae; *f.*	water [lymph, L.]
teras, teratos	monstrum, -i; *n.*	monster [In combination: (terato- = abnormal), Gk]
aortē	—	that which is being hung [aorta, Gk]
oidēma, -atos	—	swelling [oedema, Gk]
khumos; *also* khulos	sucus, -i; *m.*	juice [chyme, Gk; in combination: (chylo-), Gk]
aitia	causa, -ae: *f.*	cause [aetiology, Gk]
plakous, -ountos	—	flat cake [placenta, Gk]
osmē	odor, -oris; *m*	smell [anosmia, Gk]
	olfacio, *vb*	(to) smell [olfaction, L.]
klaō; *vb*	frango, *vb*	(to) break [In combination (clasto-), Gk]
—	villus, -i; *m.*	tuft of hair [microvilli, L.]
algeo; *vb*	doleo, *vb*	(to) feel pain [analgesic, Gk]
hapsis, -idos	—	fastening [synapse, Gk]
—	letum, -i; *n.*	death [lethal, L.]
—	fatalis, *adj.*	fatal
—	mutatio, -onis; *f.*	alteration

Greek[a]	*Latin*[b]	*Direct English, and typical derivative*[c]
histemi; *vb*	sedes, sedis; *f.*	(to) place [metastasis, Gk]
khrōma, -atos; khrōmatikos, *adj.*	color, coloris; *m.*	colour [chromatic, Gk]
thureos	scutum, -i; *n.*	shield [thyroid, Gk]
kutos	vas, vasis; *n.*	vessel [In combination: (-cyte), Gk; vascular, L.]
ganglion	nodus, -i; *m.*	knot [ganglion, -ia, Gk]
—	alveus, -i; *m.* alveolus, -i; *m. dim.*	cavity [alveolar, L.]
Prepositions or other adjective		*Used in prefix combination*
a-, or an-	in-	= not [a-, or an-, Gk; in-, or im-, L.]
meta	—	with [in sense of change: (meta-), Gk]
huper	supra	over, above [hyper-, Gk; super-, L.]
hupo	sub	under [hypo-, Gk; sub-, L.]
epi	—	upon, at, in, at the position of, in addition [epi-, Gk]
syn	cum	with [syn-, Gk; con-, L.]
para	ultra	[in senses of] beside, beyond; wrong, irregular [para-, Gk; ultra-, L.]
ectos	extra	outside [ecto-, Gk]
endon	intra	inside [endo-, Gk]
isos	aequus	equal [iso-, Gk; equi-, L.]
allos	allus	other [allo-, Gk]
eu-	—	well, normal [eu-, Gk]
dus-	—	bad, abnormal [dys-, Gk]
melas	niger	black
erythros	ruber	red
xanthos	luteolus (flavus)	yellow
leukos	albus	white
bradus	tardus	slow [brady-, Gk]
takhus	celer	swift [tachy-, tacho-, Gk]
glukus, glukeros	dulcis	sweet

3. Numerals—Cardinal

		Direct English, and Prefix[d]
eis, mia, ēn	unus, una, unum; I	one [uni-, L.]
duo	duo, duae, duo; II	two [duo-, L. and Gk]

continued

Greek[a]	*Latin*[b]	*Direct English, and Prefix*[d]
treis, tria	tres, tria; III	three [tri-, L. and Gk]
tessares, tessara	quattuor; IV	four [quadri-, L.]
pente	quinque; V	five [quinqu(e)(i)-, L.; pent(a)-, Gk]
hex	sex; VI	six [hex(a)-, Gk; sex(i)-, L.]
hepta	septem; VII	seven [hepta-, Gk; sept(em)(i)-, L.]
oktē	octo; VIII	eight [oct(a)(o)-, L. and Gk]
ennea	novem; IX	nine [nan-, Gk]
deka	decem; X	ten [dec(a)-, Gk; deci- (tenth), L.]
pentēkonta	quinquaginta; L	fifty [—]
hekaton	centum; C	hundred [hecto-, Gk; centi- (hundredth), L.]
pentakosioi	quingenti; D	five hundred
kilioi	mille; M	thousand [kilo-, Gk; milli- (thousandth), L.]

[a] Greek nouns in nominative singular form, together with the indefinite form, where this is different, Latinised characters used, with ē (long) = η; e (short) = ε; o (short) = o; ō (long) = ω; kh = χ; ph = ϕ. Greek u = υ (upsilon) is the Latin y.
[b] Latin nouns given in nominative singular and genitive singular forms with their m., f. or n. genders; dim. = diminutive.
[c] Greek prefixes are only usable in combination with Greek-based words, and Latin prefixes with Latin words.
[d] In general, Greek prefixes are used as multiples (e.g. kilo-) and Latin prefixes for fractions (e.g. milli-).

Index

Abbreviations listed, 365–72
2-Acetylaminofluorene, 333
Acidity, 151–2
Acquired immunity, function of lymphocytes in, 248–9
Activity studies, 84–6
Additives
flavour affected by, 164–5
nutritional effects, 126, 133, 137–8
types used, 135–7
usage data, 136, 137
Adenosine
effect of methylxanthines, 95–6, 98–9, 102–3, 104–5
physiological effects, 93, 96
Adenosine receptor blockade, effect of caffeine, 85, 92–3
Adventitious fungi, 228–9
diseases/disorders caused by, 232
Aflatoxin B_1, 21, 22, 23, 235
molecular structure, 298
production on coffee beans, 291, 299, 300
Aflatoxin poisoning, 219, 234–9
acute aflatoxicosis, 236
chronic aflatoxicosis, 236, 238–9
Aflatoxins
biosynthetic pathway for, 298
listed, 235
Aflatoxins—*contd.*
transformation by liver enzymes, 236, 237
Alimentary toxic aleukia (ATA), 218, 241
Aliphatic carbonyl compounds, mutagenic effects of, 206
Aliphatic compounds, metabolism of, 40
Alkaloids, metabolism of, 47–51
Alkylating agents, carcinogenic effects of, 327–8
Allergens, 267–90
green coffee bean, 7–8
Ames mutagenicity test
coffee components tested using, 205, 206, 207, 208
whole coffee tested using, 196–8
Amino acids, metabolism of, 35
Aminophylline, physiological effects of, 95, 96, 101, 102–3
Anaphylactic (Type I) hypersensitivity reaction, 258, 261–2, 259–65
prevention of mediator release, 263–5
reactions in cell wall, 262–3
reactions within cell, 261–2
Anaphylaxis, 245, 246
mediators of, 255, 256

Angina, effects of caffeine, 58, 106
Animal feeding studies
  advantages of, 172
  carcinogenicity studies, 173–84
  teratology studies, 184–91
  toxicity studies, 173–84
Animal models, limitations of, 172
Antibiotics, 221
Antibody specificity, 245
Antihistamines, 264
Arabica coffees
  flavour, 162
  oil content, 12
Aroma
  definition of, 141
  mutagenicity of, 16
  nature of in coffee, 143–4
  physiological effects, 159–61
  types of volatiles listed, 15–16
  see also Olfaction; Smell
Aromatic amines, carcinogenic activity of, 333
Aromatic compounds, metabolism of, 38–9, 40–1
Arthus reaction, 245–6
Ascorbic acid (vitamin C), 127, 128
  carcinogenicity/mutagenicity effects of, 209
Aspergillus spp.
  allergies caused by, 288–9
  coffee beans infected with, 292–7, 302–3
  disorders caused by, 219, 232, 233
  toxins from, 20, 292–301
Asthma, 289
  treatment of, 265
Astringency, 142–3, 161
Atopic individuals, allergy reactions in, 259–65
Atractyligenin, 8, 9, 10, 24, 39
Atractylis gummifera L., substances from, 8, 9
Automotive exhausts, benzo[*a*]pyrene in, 351
Averantin, 298
Averufin, 235, 298

B-lymphocytes, 251
Bags, allergic symptoms caused by, 278, 282
Balkan endemic nephropathy, 219, 243
Behavioural effects, 82–9
Benzacridine compounds, 332, 348, 349, 350
1,2-Benzanthracene. *See* Benzo[*a*]anthracene
2,3-Benzcarbazole, 235
Benzo[*a*]anthracene
  active regions of, 330
  attachment to cells, 330–1
  carcinogenic activity of, 335
  first isolated, 324
  inactivation of, 334
  molecular structure, 325
  tobacco smoke, in, 348
Benzo[*a*]pyrene
  analytical procedures used, 351–2, 353, 355, 357
  automotive exhaust fumes, in, 351
  carcinogenic activity of, 335
  cell binding capacity of, 338–9, 341–2, 344–5
  chimney effluents in, 350
  coffee products, in, 18, 351–7, 359
  content listed for various products, 354, 358
  detection limits for, 353, 357
  discovery of, 324–6
  first isolated, 326
  foodstuffs, in, 358
  formation of, 346–7
  inhibition of, 357, 358–9
  interaction with cytochrome P-450C, 342
  metabolic transformation of, 339–41
  metabolism of, 40–1
  molecular structure, 17, 326
  oils and fats, in, 354
  physiological effects, 17–19
  pyrolysis products, in, 350
  rural air, in, 349, 357
  tobacco smoke, in, 348, 350
  urban air, in, 349, 357
Benzo[*a*]pyrene, 326

3,4-Benzpyrene. *See* Benzo[*a*]pyrene
Beri-beri, 218
see also Cardiac beri-beri
Biochemical effects, 89–93
Bitterness, 151
compounds responsible, 2, 151
Blood pressure, effects of caffeine, 103–5
see also Hypertensives
'Body', meaning of term, 161
Bradykinin, physiological effects of, 262
Brazilian coffee
flavour, 162
mineral content, 131
Breast cancer, effect of coffee consumption, 69–70
see also Fibrocystic breast disease
Brew aroma, 144
Brews
acidity, 152
benzo[*a*]pyrene in, 352, 353–7
flavour, 162–4
Burlap sacks, allergic symptoms caused by, 278, 282

Cafestol, 9, 12, 39
Caffeic acid
metabolism of, 41–3
molecular structure, 5
mutagenic effects, 207
physiological effects, 4, 6, 7
Caffeine
activity affected by, 84–6
alleged diseases caused by, 57, 58, 63
animal growth affected by, 178
behavioural effects, 82–9
biochemical effects, 89–93
bitterness, 2, 83
carcinogenicity studies, 182–3
cardiovascular system affected by, 94–107
central nervous system affected by, 81–94
chlorogenic acids, effects of, 6
content in green coffee, 4
content of various beverages, 2
Caffeine—*contd.*
fungistatic action of, 20
gastric secretion affected by, 111
gastrointestinal tract affected by, 107–16
green coffee bean dust, in, 275–6
individual variations in response, 88–9
mechanisms of action, 85, 92–3
metabolism of, 48–9, 50–1
molecular structure, 3
mutagenic effects, 207
mycotoxin production inhibited by, 293–4
neoplasm incidence affected by, 179, 181
performance affected by, 83–4
positive effects of, 58
procaine-insensitive effects, 97
procaine-sensitive effects, 97
respiration affected by, 93
sleep affected by, 86–8, 89
stimulatory effects, 84, 87–8, 94
teratology effects, 189, 191
tolerance to, 105
Caffeoylquinic acids (CQA)
metabolism of, 41
molecular structure, 5
physiological effects, 4
Calcium mobilisation, effect of caffeine, 92
Campesterol, 36
Cancer
alleged associations with coffee consumption, 58, 62, 70
cell growth of, 326
history of, 322–3
meaning of term, 322
see also Breast...; Liver...; Nasal cancer
Candida spp., disorders caused by, 232, 233, 289–90
Carbohydrates, metabolism of, 34–5
Carboxyatractyloside, 8, 9, 10, 24
Carcinogenesis, two-stage theory of, 334–8
Carcinogenicity
activation reactions, 329, 333–4

Carcinogenicity—*contd.*
animal studies, 172–84
mice used, 179–81, 184
rats used, 173–9, 184
assessment of activity, 336
electronic configuration effects, 330–1
factors affecting, 332
inactivation effects, 334, 335
metabolism effects on, 332–4
mycotoxins, 235–6, 239
structure effects, 329–32
Carcinogens
chemical, 327–9
high-temperature production of, 346
physical, 326–7
polycyclic aromatic hydrocarbons transformed into, 338–43
primary, 327–8
secondary, 328–9
types of, 326–9
viral, 327
Cardiac beri-beri, 242
Cardiovascular system
animal studies, 96–104
contractability affected by caffeine, 96–9, 104
effects on, 94–107
human studies, 104–7
response variability, 98
Castor beans, allergic reactions caused by, 277, 278, 281, 282, 286
Catecholamines
effect of methylxanthines, 96, 100, 101, 105
molecular structure, 264
physiological effects, 263–4
Cell-mediated immunity, 247, 251
Cell-mediated delayed (Type IV) hypersensitivity reaction, 258, 265–6
Cellular immunity, 251
Cellulose, 132
Central nervous system (CNS), effects on, 7, 81–94
Chimney effluents, carcinogens in, 350
Chimney soot, cancer caused by, 322, 323
Chinese hamster lung cells, mutagenicity tests using, 199–200
Chloretic effects, 4–5
Chlorogenic acids (CGA)
allergy caused by, 273, 277–9
caffeine metabolism affected by, 6
carcinogenic effects, 7
central nervous system affected by, 7
clastogenic activity, 6
content in various products, 4
green coffee bean dust, in, 275–6
metabolism of, 5–6, 41–3
mutagenic effects, 7, 207
physiological effects, 3–8
Cholesterol levels, effects of coffee consumption, 71–7, 183
discussion of studies, 76–7
worldwide studies listed, 71–6
Choline, 127, 128
Chrysene, 325, 348
Cigarette smoking
allergic reactions affected by, 282–4
mutagenicity effects of, 203, 204
Cimetidine, 264
Citreoviridin, 242
Claviceps spp., disorders caused by, 218, 240
Clinical studies, 172
Co-carcinogen, meaning of term, 334
Coal fire fumes, carcinogens in, 350
Coal tar, carcinogens in, 348
Coffee oil
benzo[*a*]pyrene in, 352, 354, 355
content in various products, 12
toxicity, 9
vitamins in, 126
Coffee wax, dust containing, 269, 270, 271, 276, 277
Coffee worker's lung, 285
Complement system, 245, 255, 257–8
functions of, 257–8
Constituents
listed, 2–17
metabolism of, 33–51
Consumption data, 134, 135
additives data, 135–7

Coronary flow, effect of methylxanthines, 101–3
Coronary heart disease, effect of coffee consumption, 77, 106–7
Corticosteroids, 263, 264
Cross reacting allergens, 278
Cryptococcosis, 232, 233
Cup of coffee, standard definition of, 108–9
Cyclic adenosine monophosphate (c-AMP), 261
  agents increasing level of, 256, 261
Cycloartenol, 36
Cyclo-oxygenase pathway, 263
Cyclotenes, 14, 15
Cytochrome P-450, polycyclic aromatic hydrocarbons affected by, 339–41
Cytotoxic (Type II) hypersensitivity reactions, 258

Decaffeinated coffee
  animal feeding studies using, 175
  gastric secretion affected by, 111
Decaffeination
  aflatoxigenic potential of coffee affected by, 294
  flavour affected by, 164
  procedures used, 108
Delayed-type hypersensitivity, 245, 246, 266
Deoxynivalenol (DON), 241
Depulping methods, 305–6
Detoxification processes, 343
1,2,5,6-Dibenzanthracene (dibenzo[*a,h*]anthracene), 325, 335
Dibenzo[*a,h*]pyrene, 332
Dietary fibre, 132–3
Digestive system, effects of coffee, 107–16
4-Dimethylamino azobenzene, 335
Dimethylnitrosamine, 328
  carcinogenic activation of, 329
Disaccharides
  metabolism of, 34–5
  sweetness of, 150
Disodium chromoglycate, 263
Diterpene glycosides, metabolism of, 38–9
Diterpenes, 9, 12
DNA repair processes, 343, 346
Dopamine, effects of caffeine, 90, 91
Drosophilia cells, mutagenicity tests using, 202, 203, 207
Dry processing, mould growth potential, 304–5
Drying processes, mould growth potential, 304, 306
Dust
  bean unloading stations, 268–71, 273–4, 277
  green coffee bean
    allergy studies
      (1961–1978), 277–9
      (1979–1982), 279–82
      (1983 onwards), 284–7
      smoking effects, 282–4
    chemical analysis of, 274–6
    contaminants in, 273, 277, 286, 287
    microscopic analysis of, 268–71, 272–4
    symptoms caused by, 278, 281, 287

Ellagic acid, 283, 284
Endemic Balkan nephropathy, 219, 243
Endocrine system, effects of caffeine, 91–2
Epidemiological studies, 57–77, 172
  cholesterol levels, 71–7
  fibrocystic breast disease, 63–71
  hypertensives, 59–63
Epinephrine, 263, 264
Epoxide hydrolase, polycyclic aromatic hydrocarbons affected by, 341–3
Ergot alkaloids, physiological effects, 240
Ergotism, 218, 240, 311
Ethnic differences, 83, 137
Etymology, physiology/medical terms, 373–6

Eugenol
metabolism of, 43–4
molecular structure, 44
Europe, consumption data, 134
Exogenous compounds, physiological effects, 17–24
Extraction yields, 164
Extrinsic allergy alveolitis, 285, 288

Factor XII, 283
Fatty acids, metabolism of, 35–6
Fermentation processes, 305
Ferulic acid, molecular structure, 5
Fibrocystic breast disease, effect of coffee consumption, 63–71
discussion of results, 67–8
literature review, 68–70
methodology of research, 64–5
results of research, 65–7
Flavonoids, mutagenic effects of, 208
Flavour, 161–5
characteristic compounds, 163
definition of, 142
integration effects, 161–2
physiology of, 141–68
Foodstuffs, benzo[*a*]pyrene in, 358
Fungi
diseases/disorders caused by, 228–43
enzyme systems in, 220–1
factors affecting growth, 222–4
importance of, 217–20
nature of, 220–2
primary metabolites necessary, 226
secondary metabolites formed by, 224, 226–8
see also Moulds
Furfurylmercaptan, 14, 15
Fusarium scab, 240–1
Fusarium spp., disorders caused by, 218, 240–2, 311

Gall bladder, effects of coffee, 113–14
Gastric mucosa, effects of caffeine, 112–13
Gastric secretion, effects of coffee, 111–12
Gastrointestinal haematotoxicosis, 241
Gastrointestinal tract, effects of coffee, 107–16
Germany
additives used, 137
consumption data, 134
Glossary, 365–72
Glycosides
molecular structure, 10–11
physiological effects, 8–9
Goodpasture's syndrome, 290
Great Britain
additives used, 136
consumption data, 134
Green coffee beans
acids in, 152–3
allergens in, 7–8
allergic reactions among workers, 276–87
benzo[*a*]pyrene in, 18, 352, 356
caffeine in, 275
chlorogenic acids in, 275
dust. *See main entry*: Dust
mineral content, 129–30
mould infection of, 291–301, 302–3
mycotoxins in, 300
pesticides in, 20
processing of, 304–6
purine alkaloids in, 4
storage of, 307–8
sugars in, 150
transport of, 308–10
vitamins in, 127
Guaiacol, metabolism of, 44–5
Gustatory neural coding, 147

Hageman factor, 283
Harbour facilities, potential for mould growth at, 309, 310
Health effects, assessment of, 171–2
Heart beat rhythm, effect of caffeine, 99, 105–6
Heart rate, effect of methylxanthines, 99, 104

Heartburn, 109, 114, 115
see also Lower oesophageal sphincter
Heterocylic compounds, metabolism of, 45–7
Hippuric acid, 5, 42
Histamine, physiological effects, 262
Hormones, effects of caffeine, 91–2
Hot-water steam treated coffee, 108, 112
Humoral immunity, 247, 251, 255
Hydrocortisone, 263, 264
Hydrogen peroxide, mutagenic effects of, 208, 209
5-Hydroxymaltol, 14, 15, 16
5-Hydroxymethyl-2-furaldehyde
metabolism of, 45–6
molecular structure, 44
m-Hydroxyphenylpropionic acid, 5, 42
5-Hydroxytryptamide, green coffee dust, in, 274, 276
Hypersensitivity pneumonitis, 288, 289
Hypersensitivity reactions, 246
anaphylactic (Type I), 258, 259–65
cell-mediated delayed (Type IV), 258, 265–6
cytotoxic type (Type II), 258
immune complex mediated (Type III), 258, 259, 265
mechanism of, 258–9
Hypertension Detection and Follow-up Program (HDFP), 59–60, 73
Hypertensives mortality study
discussion of results, 62–3
methodology, 59–61
results, 61–2
Hypoxic hearts, effect of methylxanthines, 101

Iced coffee, 162
Immediate-type hypersensitivity, 246, 254
Immune complex (Type III) hypersensitivity reaction, 258, 259, 265
Immunity. *See* Acquired...; Cell-mediated...; Humoral immunity; Hypersensitivity
Immunoglobulin A (IgA), 249, 252, 253, 254
Immunoglobulin D (IgD), 249, 253, 254
Immunoglobulin E (IgE), 249, 253, 254–5
anaphylactic hypersensitivity caused by, 259–61
Immunoglobulin G (IgG), 249, 252, 253
Immunoglobulin M (IgM), 249, 253, 254
Immunoglobulins, 251–5
genesis of, 247, 248–9
properties of, 253
Immunology, 243–66
history of, 244–7
literature about, 243
meaning of term, 244
In vivo mammalian cell test systems, mutagenicity studies using, 199–202
Indian childhood cirrhosis (ICC), 238–9
Industrial effluents, carcinogens in, 350
Inhal, 263, 279, 287
Instant coffee
acidity of, 152
animal feeding studies using, 175, 177, 180
benzo[*a*]pyrene in, 18, 352, 353, 356
caffeine in, 47
chlorogenic acids in, 4
coffee oil added to, 160
dietary fibre in, 132
mineral content, 130–1
sales proportion, 135
trigonelline in, 13
Intestine, effects of coffee, 113
Intolerance studies, 108
Ischaemic hearts, effect of methylxanthines, 106–7
Isoferulic acid, molecular structure, 5
Isoproterenol, 263, 264

Kahweofuran, 14, 15, 39, 47
Kahweol, 9, 12, 39
Kashin–Beck disease, 242
Koch phenomenon, 245

Lactic acid bacteria, 24
Leucotrienes, 262
Liberine, 2, 3, 4
Lignin, 132
d-Limonene, metabolism of, 37, 38
Lipids
  metabolism of, 35–40
  physiological effects, 9, 12
Lipoxygenase pathway, 262
Liver cancer
  aflatoxin-caused, 236, 238
  distribution of, 236, 239
Lower oesophageal sphincter (LES), effects of coffee, 109–10
Lymphocytes
  circulation of, 247, 250
  differentiation and activation of, 250–1
  genesis and function of, 248–9

Magnesium, content in brews, 16
Maillard reaction products
  metabolism of, 46
  mutagenic effects, 205, 209
Maltol, 14, 15, 16
  metabolism of, 45
  molecular structure, 14, 44
Manganese, content in brews, 16
Margarines, benzo[*a*]pyrene in, 354
Maritime traffic zones, 308–9
Mascarocoffea, bitter compounds in, 9
Mast cell activation mechanism, 255, 256, 260
Melanoidins, metabolism of, 46
2-Methylfuran, metabolism of, 45
Methylglyoxal, mutagenic effects, 206, 209
Methylliberine, 2, 3, 4
Methylphenyltriazole, 329
  carcinogenic activation of, 329
Methylsalicylate, 15
Methylxanthines
  allergy-supressing capability of, 264–5, 277
  blood pressure affected by, 103–5
  coronary flow affected by, 101–3
  heart beat rhythm affected by, 99, 105–6
  heart rate affected by, 99, 104
  mutagenic effects, 207
  physiological effects, 81, 264–5
  see also Aminophylline; Caffeine; Theobromine; Theophylline
Mice
  carcinogenicity/toxicity studies using, 179–81, 184
  teratology studies using, 188–9, 190
Milk, nutritional effects, 138
Mineral content, typical values/composition, 34, 129–31
Minerals, 16–17
  metabolism of, 34
  nutritional effects of, 129–31
Miners, cancers in, 322–3
Molecular structure
  mutagenicity related to, 206
  odour related to, 159
  taste sensations related to, 149–53
Monoamines, effects of caffeine, 89–91
Monosaccharides
  metabolism of, 34–5
  sweetness of, 150
Moolgavkar–Knudson carcinogenesis model, 336–8
Mortality studies, hypertensives, 59–63
Moulds
  diseases/disorders caused by, 228–43
  examples of crop extermination by, 217–18
  factors affecting growth, 222–4
  growth on coffee beans, 291–301
  oxygen requirements, 223
  pH effects, 223
  surface infection by, 302–3
  temperature tolerance, 223
  toxins from, 19–24

Moulds—*contd.*
value of crops spoilt by, 217
water requirements, 223, 224, 225
see also Fungi
Mouthfeel characteristics, 161
Mucormycosis, 232, 233
Mutagenicity
bacterial tests used, 196–9
coffee components evaluated, 205–8
in vivo mammalian tests used, 199–202, 203, 204
in vivo tests, 202–4
interaction effects, 208–9, 211
metabolic deactivation effects, 201–2
mycotoxins, 235
risk assessment for, 210
screening tests, 195–6
see also Ames mutagenicity test
whole coffee evaluated, 195–204
Mycetism, meaning of term, 216
Mycoses, 217, 228–33
meaning of term, 216
Mycotoxicoses, 218, 233–43
meaning of term, 216
Mycotoxins, 19–24, 290–310
affected by
caffeine, 293–4
roasting processes, 301–2
secondary diseases caused by, 234–43

Naphthalene, metabolism of, 38
Nasal cancer, 322, 323
Nasal cavity, 154, 155
Neural pathways
olfactory system, 154, 155
taste system, 146
Neurotransmitters, effects of caffeine, 89–91
Niacin (nicotinic acid), 125, 128–9
molecular structure, 13
physiological effects, 13–14
Nicotinamide, molecular structure, 13
Nitrosoamines, formation of, 7
N-Nitrosopyrrolidine, 15–16
Norepinephrine, effects of caffeine, 90, 91
Norsolorinic acid, 235, 298
Nucleic acid repair processes, 343, 346
Nutritional factors, 125–38

Occupational cancers, 322–3
Occupational pulmonary diseases, 287–90
Ochratoxin A, 21, 22, 23
effect of roasting processes on, 301–2
formation of, 243, 244, 293, 294, 296, 297, 300
Ochratoxin B, formation of, 294, 296, 297
Octahydroanthracene, 324
Odour
definition of, 141
perception, 156–8
physiological mechanisms for, 153–61
receptors, 153–6
Oil. *See* Coffee oil
Oils and fats, benzo[*a*]pyrene in, 354
Olfaction
electron donor theory of, 157
interfacial adsorption theory of, 157
physiological processes of, 155
profile functional group theory, 157
stereochemical theory of, 157
theories of, 157–8
criteria to be explained, 156
vibrational theory of, 157, 158
Olfactory effects, 82, 83
see also Aroma
Olfactory receptors, 153–6
Opportunistic fungi, 228–9
diseases/disorders caused by, 232
Osmium tetroxide, benzo[*a*]anthracene affected by, 330
Oxygen, mould growth affected by, 223

Panel testing, 168
Papua New Guinea, allergy study in, 285–6

Paraxanthine, 2, 3, 4
  metabolism of, 49, 50
Partially identified constituents, physiological effects of, 17
Pathogenic fungi, diseases/disorders caused by, 221, 228, 231
Pellagra (niacin deficiency), 125, 129
Penicillium spp.
  coffee beans infected with, 302–3
  disorders caused by, 219, 242, 243
Per capita consumption data, 134, 135
Performance studies, 83–4
Perylene, 326
Pesticides, 19, 20
pH, mould growth affected by, 223
Phagocytosis, 245
Phenanthrene, 325
Phenolic compounds, mutagenic effects of, 207–8
Phosphodiesterase inhibition, effect of caffeine, 92
Physiological effects, 81–116
Physiologically active substances, 1–25
  classification of, 1
Phytate, 17
  metabolism of, 35
  nutritional effects, 133
Polycyclic aromatic hydrocarbons, 321–59
  carcinogenic activity, 346
    structure effects, 329–32
  cell binding capacity, 338–9
  enzymatic transformation of, 338–43
  physiological effects, 17–19
Polysaccharides
  metabolism of, 34
  nutritional effects, 132–3
Potassium, content in various coffee products, 16, 130
Promoting agents, 334, 336
Proteins, metabolism of, 35
Pulping operations, 305–6
  chemical removal methods, 306
  fermentation methods, 305
  mechanical attrition methods, 306
  warm water treatment used, 306
Purine alkaloids
  content in green coffee, 4
Purine alkaloids—*contd.*
  molecular structure, 3
  physiological effects, 2
  see also Aminophylline; Caffeine; Liberine; Methylliberine; Paraxanthine; Theobromine; Theophylline
Pyridines, metabolism of, 46–7
Pyrolysis products, carcinogens in, 346, 347, 348, 350

Quercetin, 283
Quinic acid
  metabolism of, 5–6, 43
  molecular structure, 5
  mutagenic effects, 207
  physiological effects, 4
Quinoline, metabolism of, 39

Rats
  carcinogenicity/toxicity studies using, 173–9, 184
  teratology studies using, 185–8, 190
Reproduction, animal studies on effects of coffee, 185–9, 190
Respiratory system, 267, 272
  effect of caffeine, 93
  green coffee bean dust, effects on, 273, 279–81, 284, 287
Reye's syndrome, 239
Rice, mycotoxins produced on, 294–6
Roasted coffee beans
  acids in, 152
  allergic activity, 284, 285
  benzo[*a*]pyrene in, 18, 351–7
  caffeine in, 47
  chlorogenic acids in, 4
  oil in, 12
  pesticides in, 20
  vitamins in, 127
Roasting processes
  chemical changes during, 33
  flavour affected by, 164
  mycotoxins affected by, 301–2
  polycyclic aromatic hydrocarbons produced, 321, 351–7

Robusta coffees
flavour, 162
oil content, 12
Rural areas, benzo[a]pyrene in air, 349, 357
Rutin, 283, 284

St Anthony's fire, 218
see also Ergotism
Salbutamol, 263, 264
Saliva, action in taste physiology, 145
Salty taste, 152
Saprobic fungi, 221
Scopoletin, 21, 23
Sensory analysis
environment required, 166
factors affecting judgements, 166–8
instrumental methods compared with, 167–8
panels vs. expert judgement, 168
practical procedures, 167
standards on, 165–6
Sensory effects, 82–3
Sensory perceptions, 142–3
Sensory stimuli, 143
Serotonin, effects of caffeine, 89–91
Sister chromatid exchange (SCE) mutagenicity test, 203, 204
Sitosterol, 36, 37
Sleep, effects of caffeine, 86–8, 89
Slurping
effect of, 154, 163, 167
procedure described, 167
Smell, sense of, characteristics of, 156
Smelling
structures associated with, 155
see also Olfaction
Smoking
allergic reactions affected by, 282–4
cancers caused by, 322, 323
carcinogens listed, 348, 350
mutagenicity effects, 203, 204
Sniffing, effect of, 154, 163
Snuff, cancer caused by, 322, 323
Sodium, content in instant coffee, 131
Soot
benzo[a]pyrene in, 349, 350
cancer caused by, 322, 323
Sourness, 151–2
Spontaneous locomotor activity (SLA), effects of caffeine, 84–6
Staling processes, 160, 165
Sterigmatocystin, 21, 22, 23
molecular structure, 298
production of, 235, 298–301
Sterols, metabolism of, 36
Stigmasterol, 36
Stimulatory effects, 84, 87–8, 94
Stinker beans, volatile constituents, 15
Stomach motility, effects of coffee, 110
Storage, green beans
conditions required, 308
harbour facilities, 310
mould growth during, 307, 308
Sucrose
nutritional effects, 137
sweetening power, 149
Sugars, taste of, 149–50
Sweeteners, taste affected by, 149–51
Sweetness sensation, 149–51
Synthetic sweeteners, taste of, 150–1

T-2 toxin, 241
T-cells, types of, 246, 249, 250
T-lymphocytes, 251
Tannins, 143
Tars, composition of, 323–5, 347, 348
Taste, 82–3
buds, 144–7
definition of, 142
innervation of tongue and pharynx, 146–7
interrelationships, 153
perception, 147–8
theories of, 148–9
physiological mechanisms of, 144–53
Tea, cholesterol levels affected by, 73
Temperature, mould growth affected by, 223
Teratology, animal studies, 185–91
mice used, 188–9, 190
rats used, 185–8, 190
Terpenes, metabolism of, 37–8

Tetralin, 324
Theacrine, 2, 3, 4
Theobromine, 2, 3, 4
  metabolism of, 48, 50
Theophylline, 2, 3, 4
  metabolism of, 48, 51
  pharmacological use, 82, 93, 265
  physiological effects, 95, 96, 98–9, 101, 104, 265
Thiamin content
  changes in, 6
  coffee products, 127
Tobacco
  glycoprotein, physiological effects, 283
  pipe smoking, cancer caused by, 322, 323
  smoke, carcinogens in, 348, 350
  snuff, cancer caused by, 322, 323
Tocopherols, coffee oil, 126–7
Tongue
  neural pathways for, 146
  papillae on, 145, 146
  taste perception on, 144–53
Toxicity, animal studies, 172–84
  mice used, 179–81, 184
  rats used, 173–9, 184
Trace elements, 131
Transport
  green beans, 308–10
  temperature changes during, 309, 310
  ventilated container liner systems used, 309–10
Trigeminal nerve response, 142–3, 149, 155–6
Trigonelline
  metabolism of, 47, 50
  molecular structure, 13, 44
  physiological effects, 13
  thermal demethylation of, 128
Tumour promotion/inhibition, effects of coffee, 181–2

Ulcerogenic effects, 112–13
'Umami' sensation, 148
Urban areas, benzo[*a*]pyrene in air, 349, 357
USA
  additives used, 136
  consumption data, 134

Vaccination, 244–5
Vanillic acid, 5, 42
Vegetarians, cholesterol levels in, 72, 73
Versicolorin A, 235, 298
Versiconal nemiacetal-acetate, 235, 298
Viomellein
  molecular structure, 295
  production of, 294, 296, 297
Vioxanthin
  molecular structure, 295
  production of, 294, 296, 297
Viriditoxin, molecular structure, 295
Vitamins, 126–9
  fat soluble, 126–7
  water soluble, 127–9
  see also Ascorbic acid; Niacin; Tocopherols
Volatile constituents
  aroma comprised of, 159–60
  physiological effects, 14–16

Water
  activity, mould growth affected by, 223, 224, 225
  content, mould growth requirement, 291–2, 301
  flavour affected by, 165
Waxes, metabolism of, 36
Wet processing, pulping operations, 305–6

Xanthomegnin
  molecular structure, 295
  production of, 294, 296, 297